Economics and Ecology of Diversification

François Ruf · Götz Schroth
Editors

Economics and Ecology of Diversification

The Case of Tropical Tree Crops

Springer

Editors
François Ruf
TA-C 85-Environnement et société
CIRAD
Montpellier
France

Götz Schroth
Santarém
Brazil

Éditions Quæ, R10,78026 Versailles cedex, France

ISBN 978-94-017-7293-8 ISBN 978-94-017-7294-5 (eBook)
DOI 10.1007/978-94-017-7294-5

Library of Congress Control Number: 2015943802

Springer Dordrecht Heidelberg New York London

Printed on acid-free paper

Springer Science+Business Media B.V. Dordrecht is part of Springer Science+Business Media (www.springer.com)

Contents

Contributors

Axelle Boulay
Inter-American Development Bank
Washington DC, NW, USA

Benito Dzib Castillo
Instituto Tecnológico de Chiná
Campeche, CP, México

Bénédicte Chambon
UPR Systèmes de Pérennes,
Cirad-Persyst, Montpellier, France;
Irad-Ekona, National Rubber
Plantation Programme, Latex Plants
Programme, Buea, Cameroon

Emmanuelle Cheyns
UMR Moisa, TA C-99/15, Cirad,
Montpellier Cedex 5, France

Philippe Courbet
CRD Nestlé Abidjan, ZI Yopougon
Abidjan, Ivory Coast

Richard Democrite
Chambre d'agriculture de
Gaudeloupe Baie Mahault,
Gaudeloupe, France

Pierre Dupraz
INRA, UMR 1302 Smart, Rennes,
France; Agrocampus Ouest, UMR1
302 Smart, Rennes, France

Frank Enjalric
Cirad-Persyst, UPR SIA,
Antananarivo, Madagascar

Laurène Feintrenie
CIRAD-UR B&SEF, Biens et
Services des Ecosystèmes Forestiers
tropicaux, Département
Environnements et Sociétés,
Campus International de Baillarguet,
TA C-105/D, 34398 Montpellier
Cedex 5, France

Kouame Fiko
BNETD, Abidjan, Côté d'Ivoire

Emmanuelle Giry
Institut des Régions Chaudes
(Montpellier Supagro), Montpellier
Cedex 05, France

Simon Gobina Mokoko
National Rubber Plantation
Programme, Latex Plants
Programme, PMB 25, Buea,
Cameroon

Jean-Michel Harmand
Cirad, UMR Eco&Sols—Écologie
Fonctionnelle & Biogéochimie des
Sols & Agroécosystèmes—

(Montpellier SupAgro–Cirad–Inra–IRD), Montpellier Cedex 2, France

Patrick Jagoret
Cirad-Persyst, UMR System, Bâtiment Agro Montpellier SupAgro, Montpellier Cedex 1, France

Claude Jannot
Cirad-Persyst, Montpellier Cedex 05, France

Jean-Marie Kalms
UMR 951 Innovation, Cirad-ES, Montpellier Cedex 05, France

Chaiya Kongmanee
Prince of Songkla University, Hat Yai, Thailand

Éric Malézieux
Cirad-Persyst, UPR Hortsys, UPR Fonctionnement Agroécologique et Performances des Systèmes, de Culture Horticoles, Montpellier Cedex 5, France

Mario Martínez
Carrera Diagonal 16 No. 44-26 Pasto, Nariño, Colombia

Isabelle Michel-Dounias
UMR 951 Innovation, Institut des régions chaudes (Montpellier Supagro), Montpellier Cedex 05, France

Murielle Morisson
18 Avenue Pasteur, Maule, France

Jules-René Minkoua Nzié
Département d'économie publique, FSEG—Université de Yaoundé II—SOA, Yaoundé, Cameroon

Jean Ollivier
Cirad-Persyst, UPR Systèmes de Pérennes, Montpellier Cedex 5, France

Marc Oswald
Istom, Cergy Pontoise, Cedex, France; Apdra Pisciculture Paysanne, Massy, France

Emmanuel Akwasi Owusu
Ghana Rubber Estates Ltd., Takoradi, Ghana

Florie Paul
Fleubaix, France

Fabian Pilet
Cirad-Bios, UMR RPB IRD, Montpellier Cedex 5, France

François Ruf
UMR Innovation, Cirad-ES, TA 85/15 (Bât. 15, Bur. 223), Montpellier Cedex 5, France

Vylie Tientcheu Sayam
Ecole Nationale Supérieure de Statistique et d'Economie Appliquée, Abidjan, Ivory Coast

Götz Schroth Santarém, Pará, Brazil

Laure Steer
Institut des Régions Chaudes (Montpellier Supagro), Montpellier Cedex 05, France

Ludovic Temple
UMR Innovation, Cirad-ES, TA C-85/15 (Bât. 15, Bur. 302), Montpellier Cedex 5, France

Philippe Vaast
Icraf, Gigiri, Nairobi, Kenya; UMR Écologie Fonctionnelle et Biogéochimie des Sols et Agroécosystèmes—(Montpellier SupAgro–Cirad–Inra–IRD), Cirad-ES, Montpellier Cedex 2, France

Armand Yao
Ecole Nationale Supérieure de Statistique et d'économie Appliquée, Abidjan Ivory Coast, Côté d'Ivoire

Yoddang
Cirad, Jakarta, Indonesia

Chapter 1
Introduction—Economic and Ecological Aspects of Diversification of Tropical Tree Crops

François Ruf and Götz Schroth

1.1 What Is Diversification?

Economic diversification reflects an individual's strategy for an improved balance between expected income, risks and various constraints (Barrett et al. 2000). It is thus a process of adjustment to changes in the relative costs of land, labour and capital; in profitability; in market risks as well as in political, climatic and environmental risks; and in uncertainties of the various strategies for increasing household incomes. Economic diversification comes about through the adoption of additional activities such as the inclusion of new crops or livestock into farming systems.

In addition to various forms of agricultural diversification, farmers can also diversify through non-agricultural activities, such as 'vertical' diversification through the processing of agricultural products. However, this book is devoted primarily to 'horizontal' diversification, i.e., crop diversification. It covers major tropical perennial crops: cocoa, coffee, rubber, oil palm and coconut, as well as fruit and timber trees.

Horizontal diversification has only recently become a fashionable concept in the world of tropical agriculture. Indeed, it was not so very long ago (1960–2000) that governments and funding entities encouraged specialization by farmers in what they supposedly did best or was most in demand in the market. Specialization in a cash crop was seen as a factor of economic efficiency having political implications. So, for example, after the independence of Côte d'Ivoire in 1960, President

F. Ruf (✉)
UMR Innovation, Cirad-ES, TA 85/15 (Bât. 15, Bur. 223), 73 rue Jean-François Breton, 34398 Montpellier Cedex 5, France
e-mail: francois.ruf@cirad.fr

G. Schroth
Santarém, Pará, Brazil
e-mail: goetz.schroth@gmail.com

F. Ruf and G. Schroth (eds.), *Economics and Ecology of Diversification*,
DOI 10.1007/978-94-017-7294-5_1

Houphouët-Boigny drew lessons from farmer and migratory dynamics of cocoa cultivation: in the east of the country, growers were specializing in cocoa cultivation on their own and trying to create a 'cocoa belt' around Bongouanou town. The Ivorian president accelerated this trend by implementing migration policies. He encouraged farmers who were migrating to forest areas to give up coffee cultivation and specialize instead in growing cocoa (Ruf 1981; Chauveau and Léonard 1996). In Vietnam, there was a similar process in the 1980s and 1990s, but it promoted coffee plantations, especially those with intensive management and irrigation (Fortunel 2000).

These dynamics of diversification conform to the concepts of classical economics. At any given point in their history, at a given level of available technology and various channels of information, farmers, traders and governments discern comparative advantages and attempt to exploit them.

In interaction with markets and farmer initiatives, these policies of specialization were spectacularly successful, at least in terms of market share. So for some time, tiny Vietnam overtook Brazil as the world's leading producer of coffee. In 1978, Côte d'Ivoire became the world's biggest producer of cocoa at the expense of its neighbour Ghana, the specific politico-economic goal of President Houphouët-Boigny. Between 1960 and 1990, Ivorian cocoa production doubled every 10 years, and then grew a further 60 % in the 2000s.

However, throughout the 1990s, falling prices in the global commodity marketplace destroyed the livelihoods of millions of families and the economy of many tropical countries because of their over-dependence on the production of cocoa, coffee, rubber, pepper, etc. In Côte d'Ivoire, this period marked the beginning of a deep social crisis which would ultimately lead to civil war.

In addition to the uncertainties of prices, specialization can put a country's governance at risk. For example, cocoa monoculture led to chaos in Ghana in the 1970s and 1980s (Mikell 1989). Economic thought then coalesced around the theme of the 'curse' of raw materials or, at least, of economies based on a single commodity, including agricultural ones (Rosser 2006). It reminded us that over-specialization also has its risks, not only for individual farmers but also across the province and the nation. The 1970–2000 period was certainly not the first to show this, given the long history of collapses of economies overly dependent on a single agricultural commodity. For example, Amazonia experienced a disaster when the rubber boom came to an end in the early 20th century. But repeated experiences at the end of the 20th century certainly changed the attitudes of farmers, policymakers and donors towards diversification.

Diversification of crops and farming is, however, an ambivalent process that needs to be analyzed and used with care. On the one hand, reducing the dependency of households and regional economies on a single crop or activity reduces—or can reduce—their vulnerability to ecological and market risks in the sense of 'not putting all eggs in one basket'. On the other hand, economic theory reminds us of what policymakers have long known: specialization has its advantages in terms of technical and economic efficiency. This is the case if the activity generates economies of scale and if producers have access to a secure national or international market.

Moreover, excessive diversification may increase the cost of marketing small quantities of produce, especially in remote locations. Such economic realities set limits to farm diversification.

In view of these economic realities, the farmer may well diversify just as a first step in a gradual process of conversion from one crop to another. With time the new crop would become more and more important and the process may end with the farmer as dependent on a single crop, and thus as vulnerable as he was before. From cocoa farms converted into pastures in Central America (16th century), to cocoa orchards replaced by oil palms and mango trees in Malaysia and the Philippines (late 20th century), there are numerous examples worldwide of diversification being just a stepping stone to conversion. From a social point of view, if the new activities are less labour-intensive than the ones they replace, the conversion not only maintains the inherent economic fragility of the household or the region but also generates unemployment.

In fact, the distinction between diversification and conversion relates to the scale at which we analyze the process. If farmers find that they would be better off by allocating some of their labour or capital to a new crop, they have several options to do so. For example, instead of filling gaps in an old cocoa plantation with new cocoa seedlings, they can introduce an additional crop (banana, fruit, timber or rubber). If they do not slowly phase out the old cocoa trees, the result will be a plot-level diversification, an intercropping system that can last two or three decades. Alternatively, they could decide to cut down the old cocoa trees when rubber trees or fruit trees are nearing the start of production, or even to cut down the old cocoa trees at the outset to make a rubber or teak plantation. The result is then a plot-level monoculture with diversification at the household and farm levels through a mosaic of monospecific plots (Chap. 2). Finally, different farmers in the same village or landscape may specialize in different crops. In this case, households and farms are specialized, but diversification takes place at the landscape level (Chap. 11). In this scenario, the risks of specialization are reduced at the regional level, but they remain high at the household level. The process of diversification is illustrated by the dynamics of an old coffee cultivation region of Côte d'Ivoire (Chap. 2, Fig. 2.1). The regional data on planting rates and plantation ages show that a diversification process is underway, but do not specify its scale. In fact, all the different types of diversification described above are taking place simultaneously. Therefore, understanding diversification requires the analysis of household decisions and their determinants.

In this introduction, we present a conceptual review and analytical framework in order to understand the processes of agricultural diversification, in particular for tropical tree crops. This review often uses cocoa—for which these processes have been particularly well studied—as an example. It starts from the hypothesis that for farmers and policymakers tree-crop diversification can be a tool, a goal or even a necessity at some point in the life history of a tree-crop farm, agricultural region or a country. Nevertheless, diversification of tree crops is not an automatic process and one of the main objectives of this book is to explain why and under what conditions diversification can or cannot take place.

Fig. 1.1 Location of the case study areas in Ghana and Côte d'Ivoire. The *circled* numbers on the map correspond to the chapter numbers. Chapter 2 concerns all the southern regions of Côte d'Ivoire

There is sometimes a confusion between 'diversification', which is essentially an economic concept, and 'diversity' or 'biodiversity', which are ecological concepts. As mentioned above, at the scale of the household or the landscape, diversification may take the form of a mosaic of different monocultures but it does not guarantee an increase in biodiversity. Readers interested in the biodiversity of tropical tree crops and agroforestry systems are referred to the volume edited by Schroth et al. (2004).

1.2 From Forest Rent to Diversification

According to economists, investment decisions—including choices about farming or diversifying crops—are mainly driven by current and expected prices and incomes (Bateman 1965; Berry 1976; Akiyama and Duncan 1982; Trivedi 1992; Gotsch and Burger 2001). However, market forces cannot explain all diversification decisions. For instance, in cocoa-growing areas in Côte d'Ivoire, environmental degradation due to deforestation has greatly reduced the possibility of replanting cocoa trees. Thus, ecological change in these areas has clearly been identified as a factor not only in farm abandonment and migration but also in diversification into alternative crops (Ruf 1987; Colin 1990a, b; Chauveau 1995; Léonard and Oswald 1996; Léonard and Vimard 2005). Here and elsewhere in the world, agricultural frontiers of tropical tree crops have often been driven by access to land and relative advantages for certain crops. Cocoa is a typical example of a 'pioneer crop', one that leverages to good effect the factor of production represented by the forest rent. Countries with forests therefore have a potential relative advantage for the production of cocoa (Ruf 1987, 1995; Clarence-Smith 1996). There have been—and still are–agricultural frontiers of plantations of oil palm, sugarcane, coffee, pasture, soybean, tobacco and other agricultural systems in different places and at different times of history (Dean 1995; Fortunel 2000; Angelsen and Kaimowitz 2001). As this book will show, as agricultural frontiers mature, there is often a process of adjustment to the long-term economic and ecological realities after the disappearance of the forest. These changes often take the form of a process of diversification. At the same time, new agricultural frontiers may move on to follow the receding forest or new frontiers may open somewhere else for growing the initial crop, which thus remains a 'pioneer crop'.

Malézieux and Moustier (2005) identify three main determinants of diversification: public policy, markets and ecological change such as the emergence of a crop disease. Any one of these factors can shift the relative profitability of a crop—or even trigger a collapse in income—and thus force diversification and changes in agricultural activities. Nevertheless, analyses of agricultural frontiers and post-forest dynamics have led to a more structural and interdependent vision of diversification.

Sudden rises in agricultural prices and incentive-based economic policies contribute to the growth of production in the form of the agricultural frontier. This is the basis for migrations and massive clearing of forests. This unbridled expansion leads to ecological changes. The environmental degradation that results can, in turn, drive innovation such as diversification under certain conditions (Ruf 2000). In this scenario, the degradation of natural resources (soil degradation, microclimate change, increased pests and diseases) is attributed to the expansion of tree crop monocultures into forest areas. These degradations trigger, after a certain period, a structural change in the economy previously based on monoculture and it evolves towards a more diversified economy. The changes in the other two determinants of diversification—markets and public policy—interact with these ecological drivers, often moving in the same direction. The rapid growth in the production of a crop can contribute to a further fall in prices of the commodity, and thus also encourage diversification. The next section explores these interactions within the general framework of the forest rent.

Public policies often fall into the trap of over-taxation of the product that created the wealth, thus risking 'killing the goose that lays the golden eggs' (Ruf 1995; Léonard and Vimard 2005). There are many examples that can be used to illustrate the role of public policy in the dynamics of perennial crops, at various stages of the commodity's life cycle. These policies can also facilitate access to forest land for migrant farmers (Chauveau and Léonard 1996; Colin 2004; Ruf 1988; Fortunel 2000) or promote the introduction of new tree crops. Such public policies and their role in diversification are discussed in several chapters of this book.

1.2.1 The Forest Rent: Before and After

Commodity booms, such as of cocoa, coffee and pepper, are almost always based on strong waves of migration. Smallholders usually look to occupy virgin tropical forest, not only for access to land but also to benefit from the forest's many environmental benefits: relatively fertile soils, a favourable microclimate, and low levels of pests and diseases (Box 1.1). As farmers find and adopt techniques to optimize these advantages, they are rewarded by a rapid growth of trees at a low cost, without fertilizer use. But when the forest rent is exhausted, 15–30 years after planting the first crop, the cost of replanting the same crop becomes much higher. The difference between these costs can be considered a 'differential' rent in the Ricardian sense (Box 1.1).

Box 1.1. From the 'differential' forest rent to diversification

The concept of differential rent was introduced by the British economist David Ricardo in 1815. He observed that farmers usually grew wheat first on the most suitable soils. Then, as the population increased and demand for wheat grew, they expanded their crops to less fertile land. This resulted in a cost difference between varying ecological settings. As long as the price of wheat covered production costs in the least suitable areas, farmers cultivating the best lands enjoyed extra profits. This is what Ricardo called differential rent.

Forest rent can be interpreted as a differential rent applied to a commodity. It is defined as the difference between the investment and production costs of producing a quantity of cocoa on a farm established just after a forest is cleared and the costs of producing the same quantity after replanting fallow land or an old plantation. This difference in costs is directly attributable to ecological changes and reduction in the benefits provided by the forest. This is not just a simple problem of soil fertility or erosion. The benefits include low frequency of weeds; good top soil fertility and moisture retention thanks to the abundant presence of organic matter; fewer problems with pests and diseases; protection against drying winds; and the availability of food, timber and other forest products necessary for subsistence during the unproductive phase of a tree crop (Ruf 1987; Ruf and Lançon 2004).

It is a historical reality that many cocoa-growing regions have followed a boom-to bust cycle with limited replanting (Ruf 1995; Clarence-Smith 1996). Replanting difficulties turn investments into a 'lottery', farmers say, especially due to the high risk of mortality of young cocoa trees once the forest has disappeared. This is why farmers often abandon their cocoa plantations or diversify their production systems by planting crops that are more tolerant of this ecological degradation. This then results in a process of diversification (Box 1.2).

Box 1.2. After the forest rent

If the cocoa tree leverages the forest rent—while consuming it at the same time—to provide increased revenues on international markets, what happens in all these regions and countries when producers are confronted with difficulties of replanting? Forced against the wall, producers identify other crops and activities, either recently introduced or which have once again become competitive as a result of ecological and economic changes. The whole rationale for this evolution stems from the increased costs of growing cocoa (at least when measured in labour units) and the fall in cocoa prices.

As far as migration is concerned, producers tend to use the opportunity of a particular crop—whose produce commands a high price—which then becomes ubiquitous in production systems and agricultural landscapes. The goal is to make money as quickly as possible, taking advantage of the combination of favourable prices and forest rent. When the forest rent is exhausted, often in conjunction with a fall in prices, conditions may become ripe for diversification, often coinciding with intensification to overcome the cost of having consumed the forest rent (Ruf 1995).

An additional factor is the change in generation. Replanting, with and without diversification, is not necessarily undertaken by the initial farmer. For instance, in Côte d'Ivoire, the progressive conversion of coffee farms into cocoa farms started in the late 1970s through a market in planted farmland. Older autochthons started to sell their old coffee plantations to younger migrants or ceded them to their sons (Table 1.1).

Table 1.1 Mechanisms of diversification and partial conversion from coffee to cocoa in the centre-west of Côte d'Ivoire in 1980

Starting point	Impact of migration	Result
Old coffee farms	Increasing population and land scarcity	Old coffee farms turned into productive cocoa farms
Old and abandoned coffee farms	Partially sold to migrants	Clearing of coffee plots and complete replanting with cocoa with the technical and financial support of extension services
		Progressive underplanting of cocoa below coffee trees and progressive cutting of old coffee trees
Old coffee farms, still producing but with low yields	Partially sold to migrants	Progressive replanting of cocoa trees –some attempts to rehabilitate coffee trees by cutting down the shade trees (forest trees which had been retained at the clearing stage)
	Partially ceded to sons	

Source Ruf 1981

1.2.2 Bridging the Malthus-Versus-Boserup Debate: The U Curve

In the history of perennial tree crops, such types of processes can be analyzed in the broader perspective provided by the Malthus-Boserup debate (Box 1.3). At which point do farmers diversify and innovate in order to overcome the impact of demographic pressure on their environment, production and incomes?

Box 1.3. Malthus versus Boserup

The Malthusian decline. According to Thomas Malthus, population growth leads to soil degradation. To avoid famine, affected populations move to other lands, which they destroy in turn.

The Boserupian innovation. According to Esther Boserup (1965), population pressure caused by mass migrations induces environmental degradation and productivity declines. Production capacity thus decreases and the farmer population is forced to emigrate to intensify. The path to intensification and innovation usually takes the form of an increase in labour, which is in line with population growth.

These two theories are not as antagonistic as they are sometimes made out to be. According to Boserup (1965), the innovation-intensification phase can follow the beginning of a Malthusian decline. If, after years of declining yields and incomes, farmers cut down their coffee trees to invest in pineapple, they must increase substantially the amount of labour (clearing, site preparation and planting) and capital (especially the high cost of planting material) to be able to innovate and intensify. This is possible in a situation of continued or increasing immigration, and hence of population growth.

More broadly, this idea of a two-step process which 'reconciles' Malthus and Boserup is represented by the 'U curve' (Boissau et al. 1999; Picouet et al. 2004; Box 1.4). This curve shows the shift from a Malthusian extensification strategy—when land is abundant but it cannot be fully exploited due to limited labour availability—to a Boserupian intensification and innovation strategy when the land is scarce but labour relatively abundant. A region dominated by a pioneer crop (for example, cocoa, pepper or coconut) is eventually confronted by environmental degradation, aging plantations and populations, and even land conflicts. When that happens, farmers can migrate to the cities to seek non-agricultural employment or adopt one of three farming strategies:

- abandon their existing region and migrate to a new forest frontier to start a new cycle with the same crop. They thus reproduce the same boom-and-bust cycle. This is the Malthusian strategy;
- replant the same tree crop in the same area with increased investment in labour and capital and using new techniques—such as adding fertilizers—to

compensate for the loss of the forest rent. This is a type of Boserupian innovation, which generates a U curve if successful;
- diversify by replanting other crops in the same region, thereby adapting to economic and ecological changes resulting from the loss of the forest rent. It is a U-shaped curve representing the transition from an agricultural frontier to a diversified agricultural economy.

In the latter case, diversification is accompanied by intensification in labour or capital per unit area such as when a coffee or cocoa farm is converted into a pineapple farm, more demanding in terms of labour and capital. However, there are also Malthusian cases where a perennial tree crop in decline is converted into an activity requiring a lower intensity of labour, such as a pasture. Such a development may exacerbate social differentiation and, consequently, poverty and emigration.

These three strategies or choices can evolve over a timeframe of several decades and can span generations. Replanting and diversification may well be favoured by a change in generation, but they are often constrained by the younger farmers' limited access to land and property rights. Picouet et al. (2004) insist on the articulation between the two phases: 'The transition between phases 1 and 2 does not happen automatically. It depends a great deal on various socio-economic factors. A change of production system cannot be viewed in isolation. Any analysis of it must include social, institutional and political factors, which may be considered as mediators of the relationship between the population and the environment.' The modes of land access are also considered to be key factors in the reversal (or not) of the processes at work (Boissau et al. 1999).

Box 1.4. The U curve
The hypothesis of a U-shaped curve has three characteristics:

- It introduces a relationship between an environment and a population by using an indicator of demographic pressure 'population/cultivated area';
- It expresses the fact that there is no linear or mechanistic relationship between the population and the environment. The relationship depends on many social factors;
- It also attempts to unify the Malthusian and Boserupian theories by providing a common analytical framework (Picouet et al. 2004).

This principle of the U curve applies to the decline and aging of cocoa trees that are replaced with new rubber or oil palm plantations or fish-farming ponds. This replacement helps increase labour productivity, as seen in Côte d'Ivoire (Chaps. 2 and 9).

In the following sections of this book, we will use this framework to analyze the farmers' motivations to diversify. Why do farmers diversify? What types of crops or farming systems do they diversify to? What determines the timing of diversification? What kinds of farmers are most likely to diversify? What prevents farmers

GHANA COCOA BOARD
PRODUCE OF GHANA

◀ **Plate 1.1** Trend towards specialization of family farms 'You Europeans, you will never stop consuming cocoa,' says one Ivorian grower. When thousands of families in a tropical country rely on the global market of a commodity such as coffee, cocoa or rubber, they feel relatively secure. 'The price may fall but there will always be a price,' a farmer in Indonesia told us. This belief in a guaranteed income partly explains the trend towards specialization of family farming. Another important factor that encourages specialization is the combination of a favourable environment such as a fertile rainforest and massive waves of migrants attracted to this market and these favourable conditions. Farmers clear the forests and create quasi-monoculture landscapes over several years. The family farm thus becomes dependent on one main crop (cocoa, coconut, pepper, etc.) for its financial resources. But as decades pass, these farming systems become weakened due to population pressure and deforestation. Farmers too age with their farms. Diversification becomes inevitable and takes place at the first market opportunity

from diversifying? What are the public policies needed to encourage a level of diversification which is deemed desirable by the various stakeholders? The authors of this book attempt to answer these questions on the basis of the experience gained from the 15 case studies presented. Among the countries studied, we concentrate in particular on the two largest producers of cocoa in the world, Côte d'Ivoire and Ghana, devoting Chaps. 2–9 to these two neighbours (Fig. 1.1). But the recurrence of diversification mechanisms observed in Cameroon (Chaps. 10–12), Thailand (Chap. 8), Malaysia (Chap. 15), Indonesia (Chap. 16), Vanuatu (Chap. 14) and Central America (Chap. 13) clearly demonstrates the principle of a general process, of a model of transition from a perennial quasi-monoculture to an agriculture that is more diversified at various scales (Plates 1.1, 1.2, 1.3, 1.4, 1.5, 1.6, 1.7 and 1.8).

1.3 Why Do Farmers Diversify?

1.3.1 To Increase Incomes

Among the most important of any farmer's objectives is that of increasing his income at a limited level of risk. There are innumerable cases of farmers adopting a new crop because of its more favourable price. Thus, the 'cocoa boom' of the 1970s in Côte d'Ivoire can be seen as a diversification at the national level from coffee cultivation to cocoa, not only influenced to a great extent by a decline in world coffee prices compared to those of cocoa, but also by guaranteed procurement prices which encouraged the adoption of cocoa (Fig. 1.2 and Chap. 2).

Similarly, the decline in world coffee prices—partly triggered by the boom in that crop's production in Vietnam—encouraged many farmers in Sumatra (Indonesia) to switch to cocoa cultivation (Chap. 16). In the 1970s and 1980s, the low price of cocoa in Ghana played a role in the emergence of diversified farms cultivating oil palm and orange (Chap. 5). In the 1990s and 2000s, a fall in the price of cocoa led to the same phenomenon of diversification (to oil palm, rubber and fish farming) in Côte d'Ivoire (Chaps. 4, 6 and 9). Another example is the decision of farmers in southern Thailand to convert a portion of their rice paddies to rubber plantations, partly due to the low price of rice and the more attractive price of rubber (Chap. 7).

Plate 1.2 From cocoa to an inevitable diversification into rubber in Côte d'Ivoire In the 1990s, public-private projects resembling forms of contract farming helped launch smallholder rubber cultivation. During 2000–2010, young farmers regained confidence in agriculture and created a spectacular rubber boom due to this remunerative crop and because of the regularity of income from it, so much so that rubber growers called themselves ' civil servants' (*Photos 2* and *4*). Rubber is also easier to replant than cocoa on acidic soils. However, adult rubber trees were sometimes overtapped and over-stimulated due to a lack of skilled workforce (use of Ethrel stimulant, *Photo 3*). *Photo 1* Plantation of young rubber trees in the foreground with fertilizer being spread around them. The rubber trees are gradually replacing old cocoa trees (Gagnoa region 2008). *Photo 2* Young rubber trees in the Soubré region. *Photo 3* Adult rubber trees (Gagnoa region 2009). Tapping with intensive stimulant use. *Photo 4* Adult rubber trees (Gagnoa region 2009)

The increase in the cocoa-rice price ratio (by weight) from 2 to 3 in the early 1980s helped trigger diversification towards cocoa cultivation in Sulawesi (Indonesia). Many farmers either sold their rice paddies or entrusted them to sharecropping contracts to be able to fund their migrations and acquire new lands to cultivate cocoa. Similar to the rice-to-rubber transition in southern Thailand, some

Photo 1

Photo 2

Photo 3

Photo 4

Photo 5

Photo 6

Plate 1.3 From coconut to oil palm, rubber and cocoa. *Photo 1* During the 1990s, coconut trees along the coastal strip of Ghana started aging. *Photo 2* The arrival of lethal yellowing disease was not only a misfortune but also a symptom of the end of a cycle (Chap. 3). *Photo 3* Oil palms emerge in the understory of diseased coconut trees. *Photo 4* Some farmers plant cocoa farther away. *Photo 5* Even if older coconut trees are not attacked by the disease, they are overshadowed by the economic attractiveness of a new perennial crop: rubber. *Photo 6* Diversification into rubber is encouraged by public policies and investments of a private plantation company which provides technical advice and extends credit for growing rubber. This company is itself financially supported by a funding entity (Chap. 7)

Photo 1

Photo 2

Photo 3

Photo 4

Photo 5

Photo 6

Plate 1.4 From aging cocoa trees to oil palm in Côte d'Ivoire and Ghana The adoption of hybrid oil palm in Côte d'Ivoire is once again the result of a combination of the difficulties of replanting cocoa and the opportunity for a new market offered by projects and processing factories. As a result of growing domestic demand, artisanal processing units are also set up to meet the demand of urban consumers, further strengthening the adoption of oil palm cultivation (Chaps. 2 and 4). The process is similar in Ghana: in 40 year-old cocoa farms in the southeast of the country, some farmers try to increase their incomes through the exploitation of wood from residual rainforest trees (*Photo 6*). This illegal trade is not very profitable and is an added incentive to diversify into oil palm if there is an opportunity to do so (Chap. 5). *Photo 1* Aging and degraded cocoa farm. *Photo 2* Private oil palm nursery. *Photo 3* Oil palm plantation. *Photo 4* Artisanal processing unit. *Photo 5* Felling of trees in old cocoa farms. *Photo 6* Adoption of oil palm in Kade

Photo 1

Photo 2

Photo 3

Photo 4

Photo 5

Photo 6

Photo 7

Photo 8

Plate 1.5 From aging cocoa to oil palm: Indonesia. There is an attempt to rehabilitate cocoa through grafting but an inevitable diversification into oil palm takes place on the floodplains. In Sulawesi and Aceh in Indonesia (Chap. 16), a phase of declining cocoa cultivation is the result of aging cocoa trees and insect attacks, here by the pod borer (*Photos 1* and *2*). Techniques to rehabilitate cocoa through grafting were developed in the late 2000s and provided hope for a revival of cocoa production (*Photo 3*), so much so that cocoa trees replaced old orange (*Photo 4*) and were even replanted in wetlands (*Photo 5*). However, at least in the wetlands, in 2012 more and more farmers started diversifying into oil palm (nursery in *Photo 6*, young plantation intercropped with maize in *Photo 7*). The process was also sometimes seen in the hills (*Photo 8*)

Photo 1

Photo 2

Photo 3

Photo 4

Photo 5

Plate 1.6 Rice to perennial crop—and back to rice: Indonesia and Thailand When the cocoa market is favourable and when insects and diseases are not too active, it is very tempting to convert rice paddies to cocoa. This happened frequently in Sulawesi in the 1980s and was still sometimes observed in Aceh in the 2000s (Chap. 16). But this process is a reversible one. In Sulawesi, insect attacks and falling yields—as well as public subsidies—helped in a reversion to rice paddies (Chap. 1). In southern Thailand, despite their seemingly eternal appearance, rice paddies, with little or no irrigation, become degraded and productivity falls. Primarily due to the deterioration of rice cultivation—and high rubber prices—farmers innovate. They find ways to convert part of the rice paddies to rubber. The main innovation is to alternate raised beds and drainage ditches (*Photos 3–5*) (Chap. 7). As in Indonesia (rice to cocoa transition), the individual decision to diversify from rice to rubber is reinforced not only by imitation but also by the impact of each individual farmer's decision on his lowland neighbour. Gradually, as the conversion of rice paddies to cocoa or rubber takes place, the supply of water to the paddies becomes increasingly uncertain. *Photo 1* Conversion of rice paddies to cocoa in Aceh, Indonesia. *Photo 2* Felling of cocoa for reconversion to rice paddies. *Photo 3* Rubber versus rice paddies in southern Thailand. *Photo 4* Making ditches and raised beds for rubber trees. *Photo 5* Motorization and the collective dynamic in the service of diversification

Photo 1

Photo 2

Photo 3

Photo 4

Photo 5

Plate 1.7 Agroforestry-based diversification at the plot scale with coconut/cocoa In Southeast Asia, many cases of coconut-cacao intercropping illustrate a principle of agroforestry-based diversification: the introduction of a new crop on a plot that is already deriving value from existing trees. This has several advantages: the remunerative use of two canopy layers, the possibility of separate ownership of trees, such as coconut harvested by the son (since climbing is necessary) and cocoa harvested by the parents (Chaps. 14 and 15). By adding pepper cultivation, a few Sulawesi farmers have even been able to afford their own vehicles. But despite the security afforded by agroforestry-based diversification, the possibility of a partial return to a monoculture, such as oil palm, always exists. This has been observed in some locations in Peninsular Malaysia and more recently in Indonesia, from Sulawesi to Aceh (Chap. 16). *Photos 1* and *2* Coconut-cocoa intercropping in Papua New Guinea. *Photo 3* Cocoa plants in the understory of coconut trees (South Sulawesi, Indonesia). *Photo 4* Coconut trees, areca palms, cocoa (Aceh, Indonesia). *Photo 5* Vanilla vine, left, as an intercrop

Plate 1.8 Agroforestry-based diversification at the plot scale with coffee/cocoa and trees *Photo 1*. Coffee-based agroforestry system in Chiapas (Mexico). *Photo 2* In Guatemala (*Photo* Henk van Rikxoort). *Photo 3* In Central America. *Photo 4* Coffee cherries. *Photo 5* Cocoa-based agroforestry system in Brazilian Amazonia. The most promising agroforestry systems are probably those involving trees planted for timber production, selected for their high commercial value and complementarity with coffee or cocoa cultivation. It was a Brazilian pioneer, Ze Gaucho, who anticipated the growing demand for wood as he observed rainforests being deforested. But this reasoning requires legislation which recognizes that farmers are entitled to the wood from trees that they themselves plant. Costa Rica is one country where such legislation exists: farmers are allowed to plant shade trees whose wood they can sell (Chap. 13). This type of legislation has been slow to evolve in Africa, e.g., Ghana

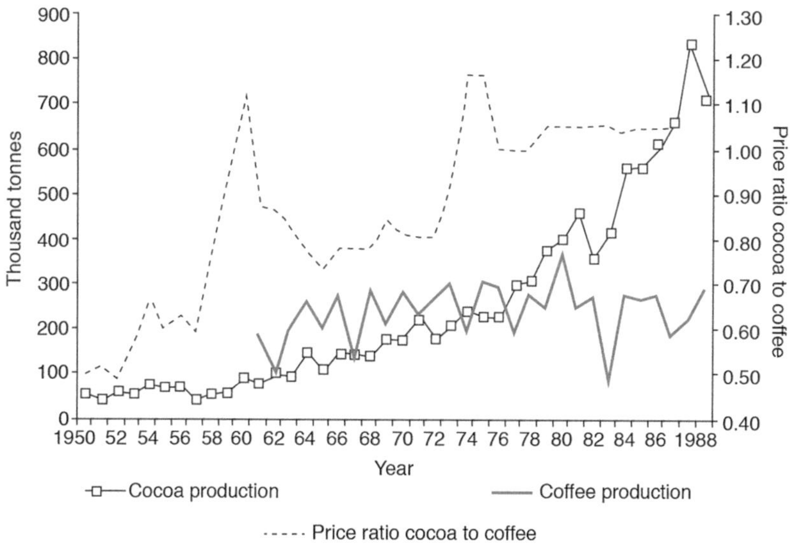

Fig. 1.2 Production of cocoa and coffee, and cocoa/coffee price ratio in Côte d'Ivoire from 1951 to 1990. (*Sources* Cocoa and coffee production: ICCO. Cocoa and coffee prices: Ivorian Ministry of Agriculture, adjusted for 1998–1989 based on data collected by CIRAD)

partially irrigated rice paddies were even converted into drained cocoa fields. However, during periods of more favourable rice prices, as in the early 1990s, the trend could not be reversed: once started, the adoption of cocoa via migrations does not stop or pause because of price fluctuations in some years. This is explained in part by income per unit area and, especially, per unit of labour: due to excellent yields and low pest and disease constraints early in the development of cocoa cultivation, farmers were still able to generate much higher revenues from cocoa than from rice.[1] In 1992, when the price of cocoa was at its lowest level in Sulawesi (equivalent to twice the price of rice by weight), farmers were asked at which point they would possibly give up cocoa cultivation. The response was almost unanimous: 'As long as the price of a kilo of cocoa remains higher than that of a kilo of rice, we will continue to grow cocoa.' But this resolve should be viewed in the context of cocoa cultivation in the pioneering phase, still easy, guaranteeing high returns with few pest or disease problems. With the severe drought in 1997 and the outbreak of cocoa pod borer (*Conopomorpha cramerella*) in the 2000s, production costs began rising as fast as cocoa yields started falling, resulting in a drop in incomes. Despite the rise in cocoa prices, many farmers who had converted their rice paddies to cocoa plots cut down their cocoa trees and reverted back to irrigated rice—and more

[1]Data on changes in incomes are more revealing than those on prices (Chaps. 2, 7, 8 and 9). But they are harder to find or generate because good yield assessments are necessary in order to calculate gross income, incorporating, if possible, production risks. Furthermore, an assessment of the amounts of labour and production costs is necessary to arrive at net incomes.

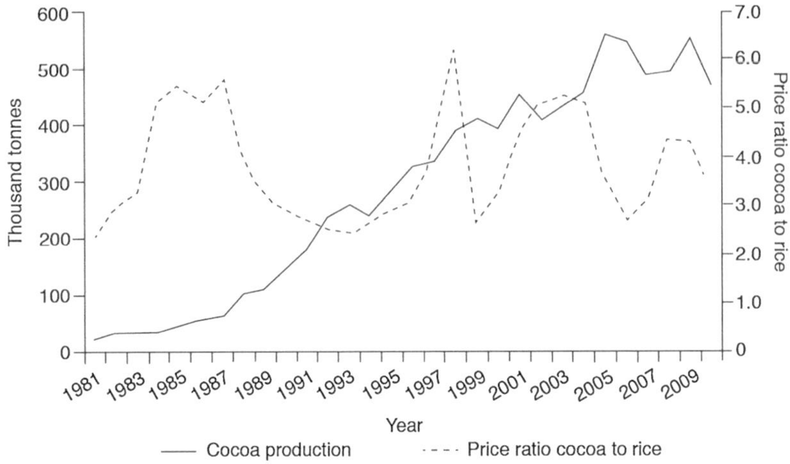

Fig. 1.3 Ratio of cocoa/rice prices in Sulawesi and cocoa production in Indonesia. (*Sources* Production: ICCO; Cocoa and rice prices: Regular data collection by CIRAD from traders in South Sulawesi since 1989)

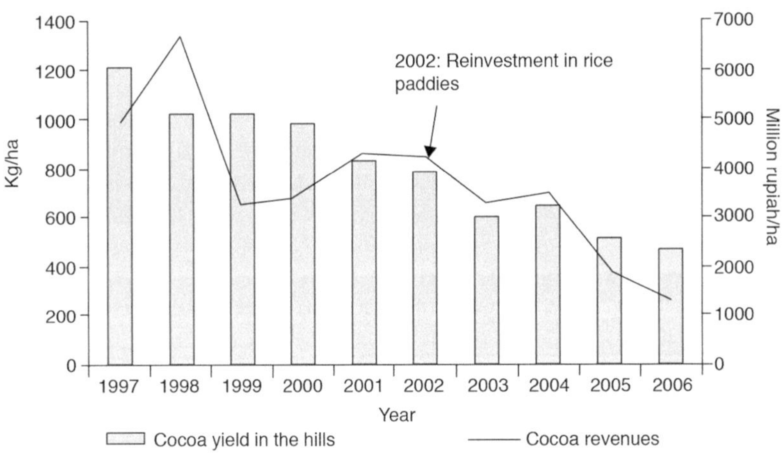

Fig. 1.4 Yields per hectare of cocoa and gross income deflated by the price of rice (Tampumea hills, South Sulawesi 1997–2008). (*Sources* Data collected by CIRAD in South Sulawesi from 1997 to 2006 on a sub-sample of 20 farms in the hills). During the Asian currency crisis of 1997–1998, the Indonesian rupiah collapsed against the US dollar. From 2,500 Rp/$ in early 1997, it went to over 15,000 Rp/$ in the middle of 1998. From 1999, the rate stabilized at between 8,000 and 9,000 Rp/$

recently to oil palm. Some even invested in new paddy fields. These interactions between ecological and economic changes led to the stagnation and decline in cocoa production in Sulawesi in the late 2000s (Figs. 1.3 and 1.4).

All upstream factors that offer opportunities or carry risks bearing on the farmers' earnings expectations influence their diversification decisions. As the 15 case studies in this book illustrate, markets and public policy offer opportunities—and are drivers—for diversification. But the vagaries of the markets and public policies bring with them major risks, too.

1.3.2 To Reduce the Vulnerability to Markets and Public Policies

Closely related to their goal of maintaining or increasing revenue, farmers also seek to reduce risk. Volatility in international markets—partly stemming from boom-and-bust cycles—is one of the risks of producing a commodity like cocoa, coffee, rubber, palm oil, clove or pepper. Fluctuations in farm-gate prices are also linked to national policies, especially taxation policies. From the late 1950s to the 1980s, most coffee and cocoa producers were, in principle, insulated from fluctuations in world prices thanks to interventions of national price stabilization funds. In times of rising prices, as in the 1970s, these structures retained more than 60 % of the cocoa value at the expense of producers. In periods of declining prices, they were sometimes effective in being able to maintain prices for short periods. From 1980 to 1988, for example, the stabilization fund maintained a stable real price of cocoa paid to Ivorian producers even though the world price was in free fall. This protection of farmers encouraged migrations and investments in cocoa—which further contributed to the decline in world prices.

In 1988, the banking system became concerned about the finances of the Ivorian government and lost confidence in Houphouët-Boigny's price-stabilization policy. It blocked new loans to the sector, resulting in a sudden halt to payments to producers. Procurement prices tumbled by 70 %. In most cases, these well-intentioned state structures actually weakened farmers. Another classic example is of Ghana in the 1970s and 1980s, when taxation and the monopoly of the marketing board almost killed off the country's cocoa sector (Bateman 1990).

These market and policy risks are major determinants of diversification of tropical tree crops. Diversified farms are better able to cope with periods of falling prices or unfavourable policies than farms dependent solely on one crop. A typical example is the diversification by clove farmers in Sulawesi. In the 1980s, a decline in clove prices and an unfavourable policy—one of President Suharto's sons had 'hijacked' the clove value chain by establishing a 'clove board' under his direct control—encouraged farmers to diversify into cocoa (Fig. 1.5). A similar diversification took place in Aceh (Chap. 16). Chambon and Mokoko (Chap. 10) show that diversifying into rubber by cocoa growers in Cameroon is partly an anti-risk strategy on prices: whether they are planning for their retirement or thinking of the inheritance they will leave their children, they do not want to be vulnerable to the vagaries of a single market.

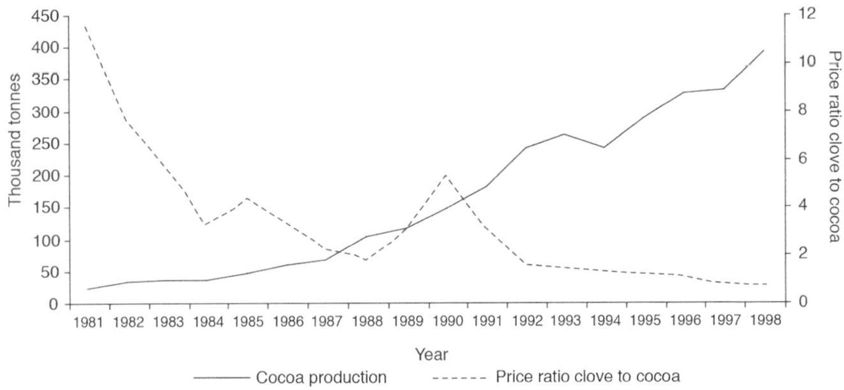

Fig. 1.5 Clove/cocoa price ratio in Sulawesi and cocoa production in Indonesia. Indonesian production can be taken here as an indicator of the production of Sulawesi, the main centre of cocoa production in the country until the 2000s (Ruf and Yoddang 2004)

However, before they can escape the vulnerability inherent in a single market, the farmers have to gain a certain amount of confidence in a new sector and a new market. This is one of the reasons why Ivorian and Ghanaian cocoa farmers continued growing the crop for decades, even after enduring long periods of very low prices. Despite all the vicissitudes, they retained confidence in an established market, and thus in a relatively safe capital and farm heritage. They knew that the price will go up 'one day'. Indeed, it was only after observing this behaviour that economists invented the concept of 'expected price' to explain why producers can maintain their investments when prices are low. This behaviour also reveals that farmers consider a crop they have been cultivating for over a century as a safe haven. To overcome this perception, the new sector has to convince them that there is a long-term, assured market for the new crop before they begin to diversify to it. It is for this reason that rubber took several decades to emerge as a real alternative to cocoa in countries such as Côte d'Ivoire and Ghana. The strategy of a private Ghanaian company to boost the rubber sector is revealing on this point: build or rebuild trust with the farmers (Chap. 8). But it is environmental risks—and therefore production risks—that are often as or more important than market risks as determinants of diversification.

1.3.3 To Reduce the Vulnerability to Environmental and Production Risks

As we have seen, the general model of family plantation economies leads to several consequences: progressive deforestation, loss of forest rent, aging of plantations and of farmers, possibilities of tensions on the labour market, the emergence of institutional problems such as conflicts over land (which will lead to a shifting of centres of production). This model was first described 25 years ago based on observations of cocoa cultivation economies but can be observed in similar form with other tree crops. Farmers respond to environmental and socio-economic pressures by migrating or diversifying. Even though the demographic and social components of this model are significant, we want to emphasize here the key role of the biological component represented by the disappearance of the forest rent. Thus, the diversification of cocoa cultivation towards rubber production observed in West Africa, especially in Côte d'Ivoire, reveals in particular the risks of replanting cocoa in a degraded environment. Diversification is also a response to a production risk which has attained structural levels. Farmers often consider these risks to be more important than those related to markets. For instance, in south-western Côte d'Ivoire, the spread of rubber production across the landscape stems in part from soil degradation and the difficulties of replanting cocoa on land which, once the forest has disappeared, is no longer suitable for this crop (Chap. 2).

A particularly visible aspect of the exhaustion of the forest rent is the almost inevitable increase in weeds, pests and diseases. Thus, the aging tree finds less nourishment and loses some of its ability to defend itself. Furthermore, deprived of their forest habitat, insects or rodents can find a substitute with the perennial crop that replaces the forest. With farmers becoming older and as their incomes decline, their ability to respond to biological pressures decreases. When farmers or their descendants still have the option of investing, diversification is perceived as the best strategy. The partial shift from coconut to oil palm, rubber and cocoa by thousands of farmers in southern Ghana in the 2000s was the result of a diversification process triggered by the combined effects of the spread of lethal yellowing of coconut and the advent of a new generation of farmers (Chap. 3). In Indonesia, problems caused by clove diseases contributed to the diversification into cocoa in the 1980s and 1990s. In the 2000s, the outbreaks of the cocoa pod borer persuaded farmers to diversify into oil palm and rubber (Chap. 16). The witches' broom disease of cocoa, caused by *Moniliophthora perniciosa*, was brought in the 1980s from the Amazon to the cocoa-growing region of southern Bahia. This region was then dependent on a monoculture of cocoa with high susceptibility to the disease, even if these cocoa trees were grown under a highly diversified canopy of forest trees (Schroth et al. 2011). The ensuing cocoa crisis resulted in a diversification into cultivation of Robusta coffee and rubber. However, this diversification did not result in a conversion; the southern Bahia region today remains dependent on cocoa.

A less perceptible environmental change resulting from the progressive replacement of forests by farms at a regional scale is one of a drier microclimate. This can exacerbate the difficulties of replanting crops such as cocoa which have high humidity requirements. Indeed, farmers around the world agree that rainfall patterns have changed after years of deforestation (Léonard and Oswald 1996; Brou 2005). In several of the main cocoa-growing regions of Côte d'Ivoire, Ghana, Cameroon and Indonesia, farmers have moved on to rubber, oil palm, cashew and

1.3.5 To Spread Income and Labour Over Different Times of the Year

Crop diversification usually leads to a better distribution of income and labour over the year. It also helps reduce constraints of peak labour demands such as during harvests. In diversified systems that associate coconut and cocoa in Vanuatu, work schedules are complementary in their distribution during the year, except in September when harvest periods of both crops coincide (Chap. 14). Similarly, in cropping systems in southern Ghana, the combination of different perennial crops (cocoa, oil palm or orange) and annual crops allows farmers to obtain a more regular income while spreading their activities over the entire year (Chap. 6).

When the diversification crop is sufficiently profitable, a farmer can even hire outside labour to meet peak workloads. In Côte d'Ivoire and Ghana, even small cocoa farmers who diversify into rubber cultivation tend to hire workers to tap rubber, while family labour is mainly used to manage cocoa (Chap. 8).

1.3.6 Because of Life Cycles of the Tree Crop and the Family

The physiological cycles of perennial crops often match the biological and generational cycles of farmers and their families, resulting in additional pressures to diversify farming systems. For example, young coconut palms are easily harvested by young farmers who can climb up the trees. But this becomes increasing difficult as the trees grow taller and farmers older. In the early 1980s, for example, farmers planted cocoa in the understory of their coconut trees in Teluk Intan region (Peninsular Malaysia). The goal was to increase family incomes so that the sons could remain on the farm and help harvest coconuts. Without the extra income from cocoa, they would have gone to seek employment in the country's rapidly industrializing economy (Chap. 15; Dupraz and Lifran 1995).

In the 1980s, diversification into sweet banana encouraged sons of older coffee growers to remain on the farms in Madagascar. Here, the coffee trees and their revenues belonged to the fathers, while bananas which were intercropped with coffee were the property of the sons (Blanc-Pamard and Ruf 1992). A generational change often brings with it a change of crops since younger farmers, whether inheritors or buyers, have their own goals.

In southern Côte d'Ivoire, diversification from coffee to cocoa and then to rubber often took place at the change of generations. The farmers of the 1950s and 1960s tended to cultivate coffee rather than cocoa. Starting in the 1970s, they started abandoning their old coffee farms to migrate westward to forested areas where cocoa was the preferred crop. They were followed by their sons and nephews in the 1980s who also became cocoa farmers before turning to rubber in the 2000s. In addition to generational change, replanting and partial diversification of cocoa-growing regions to rubber were also driven by land issues (Box 1.5). To simplify, in many Ivorian villages, the grandfather was a coffee farmer, the father a cocoa grower, while the son became a rubber grower.

Box 1.5. An example of an Ivorian farmer combining strategies of forest encroachment and progressive diversification
In 1953, the father migrated from central Côte d'Ivoire to the forested region of Agboville where he negotiated with the autochthons of the region to obtain 25 ha of forest. Between 1953 and 1970, he cleared the 25 ha and converted them into a coffee plantation. By 1983, a large part of the already aging farm was hit by drought and fires. His replanting attempts failed.

In 1984, father and son migrated to the western region of Soubré and negotiated with autochthons for 30 ha of forest. The father immediately transferred 18 ha to his son.

In 1999, all the 18 ha had been cleared and converted into a cocoa farm. The father had died in the interim. In Agboville, where the father had first migrated to, the autochthons tried to take advantage of the father's death to repossess the partially abandoned 25 ha. The son managed to retain 10 ha and immediately hired a labourer to replant them with coffee, cocoa and rubber. One of his main objectives was to secure the 10 ha after the negotiations with the autochthons. In contrast, in Soubré, the autochthons have not yet attempted to renegotiate the initial sale of 30 ha sold in 1984, probably because there was a formal financial transaction. In Agboville, the sale of forest was more akin to an exchange of gifts: a token gift such as a bottle of alcohol and a case of wine.

1.3.7 To Take Advantage of Site Diversity

There are also more mundane reasons for diversification. Indeed, farms in a landscape may be composed of a variety of plots suitable for different crops, making diversification almost mandatory. Michel-Dounias et al. (Chap. 5) show that cocoa, oil palm and orange occupy different segments and soils on a cultivated slope in eastern Ghana. Cocoa and orange trees are located on the upper and middle slopes, while oil palms are grown on the lower slopes. Ollivier et al. (Chap. 3) describe a similar situation in another region of south-central Ghana, with coconut (before the onset of lethal yellowing disease) and oil palm dominating the valleys and plains while orange trees find place on the hills. Farmers in Ghana's Western region also choose tree crops according to the toposequence, for example, oil palms for low-lying plots (Chap. 8).

1.3.8 Because of the History of the Site and the Cropping Practices

In many cases, the level of diversification is also a legacy of the history of a site and how a tree crop was established on it. If the crop was established by partially clearing the forest and underplanting tree-crop seedlings, as has often been done for cocoa, then the full-grown plantation will have an overstory of diverse forest trees above the cocoa. Simultaneously or alternatively, fruit trees are often planted. For a long time, this was how cocoa plantations, and sometimes coffee plantations, were established. This diversification can be explained not only by a limited ability to clear the forest (for example, before the use of the chainsaw became commonplace) but also by an economic dimension. The fruits are consumed by the family and workers, while some trees can be sold for timber. For example, in Bahia, Brazil, many timber trees in cocoa farms were sold illegally to offset the loss of income during the cocoa crisis of the 1990s (Alger and Caldas 1996).

In most cases, however, remnant forest trees in tree crop farms are rarely managed sustainably for continuous production and revenue. However, in Costa Rica and Guatemala, as in other countries in Central America, farmers have started diversifying their coffee and cocoa farms by planting new timber trees (Chap. 13). And yet, the sale of timber is forbidden by governments in many countries, thus denying any rights to the value of the wood to farmers. This is one reason for the widespread felling of trees at the initial clearing stage in West Africa (Boni 2005; Ruf 2011). Laws are beginning to change but farmers have little knowledge of them or they are rarely applied. In Ghana, some experiments have been undertaken to diversify into forest trees through reintroduction of trees into established cocoa farms (Basterrecha et al. 2011).

The presence of a high density of such 'companion trees' in tree crop farms, which have regenerated spontaneously, may also reflect periods of abandonment or negligence for economic reasons. For example, a high density of jackfruit (*Artocarpus heterophyllus*) and caja (*Spondias mombin*)—both exotic fruit species that regenerate easily—in Bahian cocoa plantations is partly the result of the cocoa crisis of the past two decades (Sambuichi et al. 2012). Similarly, in the 1990s, wild oil palms spontaneously became over-abundant in cocoa farms during periods of neglect but were eliminated a few years later, when the price of cocoa increased and the plot found an heir or buyer. Interestingly, the palms generate significant palm-wine revenue for the new owner when he fells the trees (Ruf 2014).

1.3.9 To Respond to Emerging Market Opportunities, Government Projects and Donors

Farmers therefore tend to diversify into the crop that is the most profitable, has a relatively secure market and does not require a very high capital investment.

The opening up of new markets can encourage farmers to diversify into other crops or agricultural activities. These opportunities can often result from public or public-private projects such as the oil palm and rubber cultivation smallholder schemes in Côte d'Ivoire (Chaps. 2 and 4) and Ghana (Chaps. 5 and 8). Such projects help farmers diversify their cocoa farms by providing them with planting material, capital and information in the form of technical assistance. They also extend long-term credit to farmers which brings the apparent cost of investment down to almost zero. This is where major funding entities play a leading role. It has been so in the case for rubber cultivation in Côte d'Ivoire and Ghana (Chap. 8).

Industrial projects bring with them a guaranteed market and thus constitute a major factor in the farmer's decision to diversify. But the determinants underpinning the farmers' confidence in such a market and sector can prove to be ephemeral. For example, in Cameroon, problems of marketing rubber caused delays in payment to the farmers, who then reverted to cocoa or, in particular, switched to oil palm for which marketing options are more diversified (Chap. 10). In Ghana, differences between rubber estates and family farms have contributed to a return to cocoa cultivation for some smallholders.

Irrespective of such government policies, we have seen that the growth of urban markets in Africa has increased the demand for food, including staples (cereals, tubers, oil, vegetables, fruits, etc.). To take advantage of this opportunity, farmers with sufficient market access are modifying their traditional farming systems through diversification. For example, in the densely populated Southwest region of Cameroon, cocoa farmers have introduced new perennial crops (orange, African plum and oil palm), food crops (cassava and plantain) and vegetable crops (tomato amongst others) (Chap. 11). Similarly, Sonwa et al. (2007) show that cocoa farmers in the proximity of Yaoundé, Cameroon, are increasingly replacing native forest shade trees by fruit trees. It must be noted that this economic diversification has resulted in a biological simplification of these cocoa agroforests.

1.3.10 To Reflect Structural Changes in the Ecological Environment and the Location of Plots

Ecological conditions at a given time limit the diversification options available to a farmer. In the humid tropics, farmers prefer to plant cocoa after the clearing of forest, while less demanding crops such as rubber and oil palm are chosen for plots that have been under other crops or fallow (Chaps. 2 and 10). Thus, crops such as rubber see more interest with growing deforestation and environmental degradation. Soils in the Moyen-Cavally region of Côte d'Ivoire are not conducive to cocoa cultivation so farmers initially planted coffee despite the low prices it fetched and against the trend of decreasing coffee areas in the rest of the country. From the 1990s, a growing number of them adopted rubber, a crop that can be grown on more acidic and less fertile soils (Chap. 2).

Distances also play a role in the choice of diversification. In south-western Cameroon, farmers prefer to plant oil palm on the plots nearest to the villages to minimize costs of transporting the fruit bunches and to reduce the risk of theft. In the more distant plots, they plant cocoa, even if they are not specifically interested in this crop (Chap. 10). Similar reasoning prevails in Ghana for the teak-cocoa combination. To facilitate sales and also to avoid the risk of theft and unauthorized logging at night, the teak plots are on the trail leading to the village. As for the young cocoa trees, they are relegated further away where, as it happens, the forest vegetation is also somewhat less degraded.

Finally, where available land area is small, the choice may be limited to crops that can be grown profitably under such conditions, such as vegetables.

1.3.11 To Obtain Regular Income

One of the advantages of growing rubber is that it generates continuous income throughout the year. In addition, it remains productive for about 30 years (Chap. 10).[2] Its regular income turns the farmer into a 'salary earner' and is one of the reasons for the widespread adoption of rubber cultivation by Ivorian cocoa farmers (Chap. 2). In 2002, 54 % of farmers who adopted rubber in south-western Côte d'Ivoire stated that they did so mainly to benefit from a continuous income over the year, while only 15 % mentioned increased income as a reason. It should be noted, however, that this survey was undertaken when rubber prices were relatively low (Chap. 6). The oil palm is the other preferred crop because of its regular production over the year (Chaps. 4 and 10). On the other hand, cocoa retains the advantage of entering production earlier than rubber (3–4 years instead of 6–7 years).

1.3.12 Because of Diversity of Market Outlets

Crops that offer a range of marketing outlets are more attractive to farmers. One of the attractions of oil palm is that the fruits can either be sold to local factories, to units in the informal sector or processed at home for the sale of oil or for home consumption. In addition, in Africa, the palm trees felled during replanting generate income from the sale of palm wine. This income can cover an unforeseen family emergency or pay for replanting (Chaps. 2 and 4). In 2011, a wild palm tree sold for between 500 and 1000 FCFA in Côte d'Ivoire and a hybrid palm fetched between 2000 and 3000 FCFA. For an oil palm plantation of 150 hybrid palms per hectare,

[2]The life of a plantation is 30 years, subject to a careful tapping of rubber trees. In Côte d'Ivoire, cases of indiscriminate tapping of trees have been observed. This can bring down the tree's economically useful life considerably (Ruf 2011).

the felling generates a capital of 300,000–450,000 FCFA per hectare (450–700 euros per hectare).

The coconut is another crop that generates value through its multiple uses, such as food and building material (Chap. 14).

1.3.13 Imitating Neighbours

Finally, farmers are influenced in their crop choices by the behaviour of neighbours. This principle of imitation is particularly applicable to tree crops, with innovators taking a risk that is proportional to the length of the unproductive stage. The majority of farmers wait for the results and then copy the innovators once they are successful. This is how cocoa cultivation spread in Indonesia (Burger and Pomp 1995). This imitation effect amplifies the impact of government projects when they perform well, with imitators of the direct project beneficiaries also adopting the crop. Such was the case of oil palm in Ghana (Chap. 5), rubber in Côte d'Ivoire (Chap. 7) and Ghana (Chap. 8), as well as rubber, oil palm and orange trees in Cameroon (Chap. 10).

1.4 Factors Affecting the Timing of Diversification

1.4.1 Availability of Investment Capital

If the desire for an increase in income was the sole driver of diversification, farmers would diversify when the prices of their current crops dropped below those of other crops (Trivedi 1992). But the reality is not always so simple. There are often long delays in product diversification to B (a new crop) away from A (the existing crop) even when the B/A price ratio overwhelmingly favours B. As already mentioned above, this can be the result of the interactions between several other factors such as:

- farmers' habits;
- farmers' trust in the market for A;
- the time needed to gain confidence in the market for B;
- beliefs that prices can be very different in the future;
- current earnings and savings.

In fact, Berry (1976) has clearly shown that the timing of the investment period is often dependent on current earnings and savings. The same factors—trust, expected prices and income levels, thus savings—help explain the inverse process: that is to say, why diversification into crop B continues to grow rapidly even while the B/A price ratio is falling.

During the 1990s and early 2000s, the rate of adoption of rubber by cocoa growers tended to be positively correlated to the price of cocoa in the *département* of Gagnoa, Côte d'Ivoire. This suggests that the revenues from cocoa were being directly invested into new rubber plantings (Chap. 8). Unexpectedly, rubber adoption rates tended to go even higher when the cocoa/rubber price ratio increased. The explanation lies partly in the fact that farmers had acquired confidence in the rubber market and had become convinced of the benefits of growing rubber. A subsequent rise in cocoa prices was not enough to slow down this diversification (Ruf 2012; Chap. 8).

1.4.2 Environmental Disasters

As mentioned above, diversification is often spurred by pest or disease outbreaks. These biotic factors reduce the profitability of the original crop or sometimes even destroy it outright. Thus, the infestation by the cocoa pod borer on cocoa estates in Malaysia triggered their conversion to oil palm in the 1990s (Ruf 2000). Parts of Indonesia followed suit for the same reason (Chap. 16). The appearance of the devastating lethal yellowing disease of coconut palms in southern Ghana forced farmers to diversify into other crops, including oil palm and orange (Chap. 3). On the other hand, Michel-Dounias et al. (Chap. 5) show that swollen shoot virus pressure was not an important factor in the diversification into oil palm by cocoa farmers in eastern Ghana. Here, another environmental factor was more important: the massive destruction of cocoa farms across West Africa through wildfires in 1983. These fires triggered a wave of adoption of oil palm, rubber and, to a lesser extent, fish farming in West Africa as it became difficult to replant cocoa in the degraded environment (Chaps. 2, 4, 5, 8 and 9).

1.4.3 Return of Young People to the Villages

Throughout the developing world, young people leave rural areas to try their luck in the cities. But in times of economic crisis, they are often unsuccessful in finding employment and end up back in the village and the farm, usually with a level of education higher than that of the average villager. While some do not fit back into rural life, others try to take advantage of a rent. Nevertheless, at least some of them return with more openness to change and innovation. Chambon and Mokoko (Chap. 10) describe this situation in Cameroon where liberalization of the cocoa sector resulted in a sharp increase in that crop's price. This motivated young people to return to their villages to set up cocoa farms. They brought fresh life and a new dynamism into the old cocoa farms, which they soon diversified by adopting new crops such as oil palm and rubber. We also find the same phenomenon in Côte d'Ivoire with oil palm in the 1990s (Chap. 4).

1.4.4 The Advent of New Actors: "Upper Middle Class Investors"

A new crop with very attractive profits, such as rubber, does not attract just the unemployed youth but a whole cross-section of a country's upper middle class investors (UMCI).[3] Indeed, after a century of rubber cultivation in Thailand, it is rare to find a university professor who is not also a rubber farmer. In Côte d'Ivoire, a decade of politico-military crisis has paradoxically encouraged the emergence of military and civilian UMCIs as new actors in agricultural diversification.

1.4.5 Access to Planting Material

Can conditions of access to improved planting material impact diversification choices? The answer seems to depend on the crop and the quality of planting material received by farmers. In Cameroon, oil palm was the only tree crop for which farmers based their decisions on the availability of improved planting material and, thus, on the money to buy it. For rubber, they recognized the superiority of grafted seedlings but would also plant unimproved seedling if the former were not available. For cocoa, the rule seems to be to plant unimproved local varieties from village plantations (Chap. 10). In Indonesia, the diversification into oil palm can be hampered by limited availability of hybrid material, which farmers recognize as being of superior quality. They thus refuse to plant non-hybrid varieties (Chap. 16).

In Côte d'Ivoire, the hybrid oil palm is by far the selected planting material most sought after by farmers (Table 1.2).

1.5 Factors Determining Which Farmers Diversify?

1.5.1 Age of the Farmer

It is often assumed that young farmers are more innovative and thus are more likely to diversify their crops (or those of their fathers) but empirical evidence is less clear-cut. A study in south-eastern Ghana showed that older farmers did have less

[3]Upper middle class investor (UMCI): We use this term to designate individuals from outside the world of farming who wish to invest in land and farming. They are usually already employed by the government or private entities. Most of them are politically connected to various degrees. Examples include military officers, diplomats, governors, heads of political parties, high-ranking civil servants, lawyers, judges, doctors, professors, private transporters, and, more generally, directors and employees of many companies, including cocoa companies.

Table 1.2 Distribution of farmers by access to planting material in Côte d'Ivoire (Data collected in 2006)

Planting material	Cocoa	Coffee	Oil palm	Rubber
Unselected planting material (%)	75	48	7	21
Planting material which is supposedly selected (informal source, e.g., local nursery) (%)	3	4	16	36
Selected planting material supplied by a recognized institution (%)	21	46	76	40
Planting material whose source is unknown because the plantation was inherited or purchased (%)	1	1	1	0

Sources Surveys conducted on 1045 farms by CIRAD and A&C-Vie in 2006 on behalf of the European Union

diversified farms than younger farmers, who were the first to adopt orange and oil palm as additional crops (Chap. 5). However, the most diversified farms in south-western Cameroon belonged to older farmers. Perhaps because their cocoa was also older, they had encountered problems of replanting earlier on and had to take recourse to diversification (Chap. 11). Sayam and Cheyns (Chap. 4) suggest that retirees in Côte d'Ivoire reinvested their pensions in their farms, often by diversifying into oil palm. Later on, in the late 2000s, it was rubber that took the fancy of the retirees (Chap. 2). Chambon and Mokoko (Chap. 10) show that young and old farmers in Cameroon diversify in different ways. For older farmers, diversification is the result of the aging of their cocoa, whereas the young people returning to the villages do the opposite: they start with an alternative crop such as rubber, oil palm or orange, and then plant cocoa a few years later.

A similar process is at work in Axim in Ghana, one of the coconut-growing regions affected by lethal yellowing disease. If we measure the level of diversification by the number of perennial crops grown on the farm, the degree of diversification of elderly farmers is slightly higher (1.8 against 1.5) than that of the younger ones who did not have coconut as the initial crop. If the level of diversification is measured by the area planted with oil palm, cocoa and rubber, older farmers are, again, a little more diverse because they have easier access to land. However, if we evaluate diversification by the farmers' ability to diversify away from coconut, and thus generate income through other tree crops, we find that the younger farmers have converted or diversified radically into cocoa, rubber and oil palm (Ruf, unpublished observations).

1.5.2 *Size of the Farm and Role of Women*

In south-western Cameroon, the determining factors for diversification are the farm size—the larger ones are more diversified than the smaller ones—and the household's demographic break-up. Households with more working women are more diversified into food and vegetable crops, probably because in Cameroonian society, women are not entrusted with the responsibility for perennial crops (Chap. 11). In Côte d'Ivoire, the larger farms, often associated with polygamous households, are more diversified because they have better access to land, information, financial resources and support organizations (Ruf 1992a, b).

Women have technical mastery over food crops and thus play a key role in the diversification into these crops in perennial-crop farms. They can also play an increasing role in the adoption of perennial crops. In Côte d'Ivoire and Ghana, they are often the most responsive to development projects promoting the cultivation of cocoa, oil palm or rubber (Chaps. 2 and 8).

1.5.3 Farmers with Non-Farm Incomes and Urbanites

Non-farm incomes allow farmers or non-farmers—including bureaucrats and other white-collar managers—to invest in diversified crops whose planting material is expensive. They can do so more easily than aging smallholders whose incomes from their main crops are in decline (Chaps. 2, 4 and 11).

1.5.4 Factors Preventing Farmers from Diversifying or Modifying Their Land Use to Adjust to New Conditions

The lack of capital or difficulty in accessing credit is the major constraint to diversification into rubber for 56 % of farmers surveyed in 2002 in south-western Côte d'Ivoire. Difficulties of access to land accounts for 20 %, the lack—or high cost—of the labour force for a further 14 % (Chap. 6). Shortage of land is due in part to the high proportion of flood-prone areas not conducive to rubber cultivation (Chap. 6). Land issues also hinder the adoption of crops other than cocoa. In some cases, such as in the Tabou region, the autochthons deny the right to immigrants—even to those who have been there for a long time—to plant rubber on the basis that the right to use the land was granted only to grow cocoa (Colin 2008; Ruf 2009).

1.5.5 Role of Non-farm Activities and Incomes

While non-farm diversification as a general strategy is outside the scope of this book, its role in agricultural diversification in at least two countries is addressed: Côte d'Ivoire (Chaps. 2 and 4) and Cameroon (Chap. 11). Studies have been conducted on the adoption of oil palm in Côte d'Ivoire in 2000 (Naï Naï 2000;

Naï Naï et al. 2000) and on Ivorian farmers' access to planting material of the main perennial crops (in 2006). They clearly show that the first to purchase the selected oil palm and rubber planting material are the upper middle class investors. Among the smallholders, if there is no active project support, the early adopters are retirees with pensions. Indeed, these retirees can fund the purchase of planting material themselves. Barrett et al. (2000) ask some relevant questions:

- If non-farm activities are typically correlated to the gross household income, should they be viewed as the key diversification process to escape poverty?
- Or, conversely, does this correlation suggest that the individual who starts off poor in terms of land and savings can never marshal enough investment for non-farm activities?

On the one hand, the history of cocoa and coffee is replete with success stories; it epitomizes the process of crystallization of working capital. The cocoa tree then becomes a source of new income, enabling diversification into rental housing or transport businesses with the purchase of taxis or other vehicles to transport people or agricultural produce (Ruf and Yoddang 2001).

On the other hand, the example of barriers to investment in selected hybrid palm and clonal rubber planting materials also shows the divide between these two processes: people leaving agriculture to try to diversify their activities and income from urban activities that is used to diversify agriculture.

1.6 Conclusion

Diversification of tropical tree crops appears to be an inescapable necessity. Even though diversification may lead to lower economies of scale, it is a valid response to market risks, public policy, and ecological and institutional changes. Research on diversification helps improve our overall understanding of the changes in family farm economies that have taken place over recent years.

The definitions of diversification presented at the beginning of this introductory chapter have been borrowed from the field of economics. They tend to suggest a process of cyclical adjustment to respond to changes in prices and policies or to various drivers of change.

The 15 chapters that make up this book confirm its core hypothesis that as far as tropical tree crops are concerned, diversification is often a structural response to the unfolding of a migration cycle associated with an agricultural commodity and a dominant perennial crop: aging of plantations and of migrants, generational change, pressure on land, environmental degradation caused by deforestation and decades of presence of a crop such as cocoa with little management of the soil capital.

The maintenance—or rather, the rebuilding—of natural capital thus seems to be an essential goal of agricultural policies in the humid tropics. Compared to monoculture, encouraging the diversification of perennial crops is already the first step in this process.

The second step is to reduce the financial constraints through subsidies—made available as part of measures to support diversification—so that farmers can reinvest not only in the planted tree but also in the soil and the environment. Such measures make perfect sense, especially for family farms. This type of policy, necessarily long term, must be formulated at the first signs of adoption of a new perennial crop. Improved planting material, made accessible to as many farmers as possible, who already recognize its merits, is the third step to improved productivity, resistance to pests and diseases, and improved quality of agricultural products.

In many parts of the world, the issue of land conflicts is just as important as that of soil capital. Migrant farmers innovate and often invest in crops and farms without any land-tenure guarantees or security. But they will do so much more effectively if they are convinced that they and their children will be able to stay on their adopted lands. Such an approach also has to take into account the interests of autochthon and migrant smallholders and remain vigilant against the appropriation of land by UMCIs and multinationals.

References

Akiyama T, Duncan RC (1982) Analysis of the world coffee market. World Bank Staff Commodity Working Paper n 7

Alger K, Caldas M (1996) Cacau na Bahía: decadência e ameaça à Mata Atlântica. Ciência hoje, (20):28–35

Angelsen A, Kaimowitz D (2001) Agricultural technologies and tropical deforestation. CABI Publishing, Wallingford 422 pp

Barrett CB, Reardon T, Webb P (2000) Nonfarm income diversification and household livelihood strategies in rural Africa: concepts, dynamics, and policy implications. Food Policy 16(4):315–331

Basterrecha T, Penot E, Ruf F, Le Grusse P (2011) Diversification et durabilité: introduction des arbres à bois dans les cacaoyères. Le cas de Samreboi, au sud-ouest du Ghana. In: Penot E, (ed) Exploitations agricoles, stratégies paysannes et politiques publiques. Les apports du modèle Olympe. Éditions Quae, Versailles, France, pp 77–95

Bateman MJ (1965) Aggregate and regional supply functions for Chanaian Cocoa, 1946-1962. J farm Econ 47:384–401

Bateman MJ (1990) Ghana cocoa pricing policy study. World Bank, working paper WPS, p 429. World Bank, Washington, DC

Berry S (1976) Supply response reconsidered: cocoa in Western Nigeria, 1909–44. J Dev Stud 13 (1):4–17

Blanc-Pamard C, Ruf F (1992) La transition caféière. Cirad-SAR. Collection systèmes agraires n 16:249 pp

Boissau S, Locatelli B, Weber J (1999) Population and environment relationship: a u-shaped curve hypothesis. In: Inra, Cirad, planetary garden. International symposium on sustainable ecosystem management, Chambéry, France, pp 125–128, March 1999

Boni S (2005) Clearing the Ghanaian forest: theories and practices of acquisition, transfer and utilisation of farming titles in the Sefwi-Akan area. Institute of African Studies, University of Ghana, Legon

Boserup E (1965) The conditions of agricultural growth: the economics of agrarian change under population pressure. Allen and Unwin, London, United Kingdom

Brou YT (2005) Climat, mutations socio-économiques et paysages en Côte d'Ivoire. Mémoire de synthèse des activités scientifiques présenté en vue de l'obtention de l'Habilitation à diriger des recherches. Université des Sciences et technologies de Lille, Lille, France, 204 pp

Burger K, Pomp M (1995) Innovation and imitation: adoption of cocoa by indonesian smallholders. World Dev 23(3):423–431

Chaléard JL (1996) Temps des villes, temps des vivres. L'essor du vivrier marchand en Côte d'Ivoire. Karthala, Paris, France. 661 pp

Chauveau JP (1995) Land Pressure, Farm household life cycles and economic crisis in a cocoa farming village (Côte d'Ivoire). In: Ruf F, Siswoputanto PS (eds) Cocoa cycles. The Economics of cocoa supply. Woodhead Publishing, Cambridge, pp 107–123

Chauveau JP, Léonard É (1996) Côte d'Ivoire pioneer fronts: historical and political determinants of the spread of cocoa cultuvation. In: Claence-Smith WG (ed) Cocoa pioneer fronts since 1800: the role of smallholders, planters and merchants. Macmillan, London, pp 176–194

Clarence-Smith WG (ed) (1996) Cocoa pioneer fronts since 1800: the role of smallholders, planters and merchants. MacMillan Press, Houndmills 247 p

Colin J-Ph (1990a) La mutation d'une économie de plantation en basse Côte d'Ivoire. Paris, ORSTOM, 284 pp. plus annexes

Colin J-Ph (1990b) La dynamique des systèmes productifs agricoles en basse Côte d'Ivoire. In: Dynamique des systèmes agraires. La dimension économique. Orstom, CNRS, Paris, pp 165–190

Colin JP (2004) Le marché du faire-valoir indirect dans un contexte africain: éléments d'analyse. Économie rurale 282:19–39

Colin JP (2008) Étude sur la location et les ventes de terres rurales en Côte d'Ivoire. Rapport 1. Diagnostic des pratiques. Abidjan: République de Côte d'Ivoire, ministère de l'Agriculture, Délégation européenne

Dean W (1995) With broadax and firebrand: the destruction of the brazilian Atlantic forest. University of California Press, Berkeley

Dupraz P, Lifran R (1995) The Economic complementarity of cocoa and coconut intercropping: asset strategies of smallholders in Malaysia and implications for cocoa supply. In: Ruf F, Siswoputranto PS (eds) Cocoa cycles, the economics of cocoa supply. Woodhead Publishing Ltd, Cambridge, United Kingdom, pp 281–289

Fortunel F (2000) Le café au Viêt-nam: de la colonisation à l'essor d'un grand producteur mondial. L'Harmattan, Paris, 172 pp

Gotsch N, Burger K (2001) Dynamic supply response and welfare effects of technological change on perennial crops: the case of cocoa in Malaysia. Am J Agric Econ 83(2):272–285

Keli JZ, Assiri A (2001) Cultures associées à base de plantes pérennes en Côte d'Ivoire: essais de bilan et perspectives. In: Yapo A, Ruf D, de Vernou (eds) The future of perennial crops. Investment and sustainability in the humid tropics. Proceedings of the Conference held at Yamoussoukro, Côte d'Ivoire, 5–9 Nov 2001

Léonard E, Oswald M (1996) Une agriculture forestière sans forêt. Changements agro-écologiques et innovations paysannes en Côte d'Ivoire. Natures, sciences et sociétés 4(3):202–216

Léonard E, Vimard P (eds) (2005) Crises et recomposition d'une agriculture pionnière en Côte d'Ivoire. Dynamique démographique et changements économiques dans le bas-sassandra. IRD-Karthala, Paris, 368 pp

Mikell G (1989) Cocoa and chaos in Ghana. Paragon House, 284 pp

Malézieux E, Moustier P (2005) La diversification dans les agricultures du Sud: á la croisée de logiques d'environnement et de marché I. Un contexte nouveau. Cahiers Agricoles 14:277–281

Naï Naï S (2000) Les déterminants de l'adoption du palmier à huile en plantation et en replantation. DAA, Institut national polytechnique, Yamoussoukro, Côte d'Ivoire, 61 pp

Naï Naï S, Cheyns E, Ruf F (2000) Adoption du palmier à huile en Côte d'Ivoire. OCL 7(2):155–165

Picouet M, Boissau S, Brun B, Romagny B, Rossi G, Sghaier M, Weber J (2004) Le renouvellement des théories population-environnement. In: Picouet M et al (ed) Environnement et sociétés rurales en mutation. Approches alternatives. IRD, Paris. Coll Latitudes 23:17–43

Rey de Arce (2007) Diagnostic des systèmes de production cacaoyers dans une vallée productrice de coca; l'Alto Huallagua au Perou. Master Thesis, Ciheam-Iamm n°91, Montpellier, 145 pp
Rosser A (2006) The political economy of the ressource curse: a litterature survey. Working paper No 268. Institute of Development Studies, Brighton, 84 pp
Ruf F (1981) Le déterminisme des prix sur les systèmes de production en économie de plantation ivoirienne. Cahiers du Cires (28–29):35–52
Ruf F (1987) Éléments pour une théorie sur l'agriculture des régions tropicales humides. I - De la forêt, rente différentielle au cacaoyer, capital travail. Agronomie tropicale 42(3):218–232
Ruf F (1988) Stratification sociale en économie de plantation ivoirienne. Doctoral Thesis, université de Paris X, Nanterre, vol 6, 1393 pp
Ruf F (1992a) Après la forêt, quelle stabilisation de l'agriculture de plantation ? Le cas du département d'Abengourou, Côte d'Ivoire. GTZ and Cirad, Eschborn et Montpellier, Cirad-SAR 1992/04, 72 pp. + annexes
Ruf F (1992b) Crises et ajustements structurels spontanés. Le cacao et le département d'Abengourou. In: Crises et ajustements structurels: les dimensions sociales et culturelles. Workshop Proceedings, Orstom, Gidis-Côte d'Ivoire, Abidjan, pp 118–133
Ruf F (1995) Booms et crises du cacao: les vestiges de l'or brun. Paris: Éditions Karthala. Collection Économie et développement, 455 pp
Ruf F (2000) Déterminants sociaux et économiques de la replantation. Dossier: Afrique, plantation et développement. OCL 7(2):189–196
Ruf F (2009) L'appui à l'hévéaculture familiale. Capitalisation sur l'expérience de l'AFD, étude de cas: rapport Côte d'Ivoire. Le processus de l'innovation dans la région de Gagnoa. (Smallholder Rubber Production in Côte d'Ivoire; Innovation in the Gagnoa region). Paris, AFD, Série Évaluation et capitalisation, No 26, 57 pp
Ruf F (2011) The Myth of complex cocoa agroforests: the case of Ghana. Hum Ecol 39:373–388
Ruf F (2012) L'adoption de l'hévéa en Côte d'Ivoire. Prix, mimétisme, changement écologique et social. Économie rurale (330–331):103–123
Ruf F (2014) Cocoa Combined with oil palm: an unexpected resilience. Bois et Forêts des Tropiques 331(3):33–44
Ruf F, Lançon F (2004) From slash and burn to replanting: green revolutions in the indonesian uplands. World Bank, Regional and sectorial studies, Washington 220 pp
Ruf F, Yoddang (2001) From migration to motorization: how market liberalization has benefitted Coffee farmers. In: Gérard F, Ruf F (eds) Agriculture in crisis: people, commodities and natural resources in Indonesia, 1996–2000. Curzon Press, Richmond, pp 197–236
Ruf F, Yoddang (2004) The Sulawesi case. Deforestation, pre-cocoa and cocoa migrations. In: Babin D, (ed) Tropical Deforestation to Forest Cover Dynamics and Forest Development (Chapter 23), UNESCO, pp 277–295
Sambuichi RHR, Vidal DB, Piasentin FB, Jardim JG, Viana TG, Menezes AA, Mello DLN, Ahnert D, Baligar VC (2012) Cabruca agroforests in southern Bahia, Brazil: tree component, management practices and tree species conservation. Biodivers Conserv 21:1055–1077
Schroth G, Fonseca GAB, Harvey CA, Gascon C, Vasconcelos HL, Izac AMN (eds) (2004) Agroforestry and biodiversity conservation in tropical landscapes. Island Press, Washington, D.C. 524 p
Schroth G, Laderach P, Dempewolf J, Philpott S, Haggar J, Eakin H, Castillejos T, Garcia Moreno J, Soto Pinto L, Hernandez R, Eitzinger A, Ramirez-Villegas J (2009) Towards a climate change adaptation strategy for coffee communities and ecosystems in the Sierra Madre de Chiapas, Mexico. Mitig Adapt Strat Glob Change 14:605–625
Schroth G, Faria D, Araujo M, Bede L, Van Bael SA, Cassano CR, Oliveira LC, Delabie JHC (2011) Conservation in tropical landscape mosaics: the case of the cacao landscape of southern Bahía, Brazil. Biodivers Conserv 20:1635–1654

Sonwa DJ, Nkongmeneck BA, Weise SF, Tchatat M, Adesina AA, Janssens MJJ (2007) Diversity of plants in cocoa agroforests in the humid forest zone of Southern Cameroon. Biodivers Conserv 16:2385 2400

Temple L, Fadani A (1997) Cultures d'exportation et cultures vivrières au Cameroun. Économie rurale 239:40–48

Trivedi PK (1992) A case study of cocoa replanting and new planting in Bahia, Brazil 1966–1985. J Dev Econ 39(2):279–299

Chapter 2
Diversification of Cocoa Farms in Côte d'Ivoire: Complementarity of and Competition from Rubber Rent

François Ruf

> *Plantation crops seem to be superimposed on each other and, therefore, seem to be more complementary than competitive. Depending on the region, it is the autochthons or migrants who grow one or other of these perennial crops.*
>
> P. de la Vaissière, 1978, p. 105

In 1978, P. de la Vaissière noted the complementarity between perennial crops, at a time when land and forests were still abundant resources in Côte d'Ivoire. In the early 2000s, multinationals of the cocoa and chocolate sectors still expressed great confidence in the diversification and complementarity between perennial crops. They highlighted this as a necessity and professed a commitment to improving the living conditions of producers (Mehra and Weise 2007). While their objective was laudable, these multinationals did not seem to have yet grasped how village land and producers were being coveted by other sectors. It is only recently that they have started expressing private concerns regarding the supply of cocoa. Even though Côte d'Ivoire seems capable of producing more than one million tonnes of cocoa per year for several years yet, the producing countries as a whole do not seem to be in a position to meet the increasing global demand, estimated to rise by over 25 % (nearly one million additional tonnes annually) between now and 2020.

As far as large chocolate companies are concerned, the willingness to preserve diversification and complementarity at the scale of farms and producer countries has now been replaced by concerns about competition or even a trade war. 'My enemy is not my competitor in the purchase of cocoa, but the rubber industry,' said the representative of a large chocolate company at a conference in Abidjan in 2012. Whereas in the early 2000s, the chocolate giants refrained from any open criticism of public policy, some now refer openly to 'unfair competition'. In fact, for decades now, the tax on cocoa has amounted to almost 30 % of export prices against just 2 % on rubber. In Côte d'Ivoire, ongoing reforms, launched in 2012, should lead to a reduction in this gap but only marginally. Large chocolate groups are therefore

F. Ruf (✉)
UMR Innovation, Cirad-ES, TA C-85/15 (Bât. 15, Bur. 223),
73 rue Jean-François Breton, 34398 Montpellier Cedex 5, France
e-mail: francois.ruf@cirad.fr

F. Ruf and G. Schroth (eds.), *Economics and Ecology of Diversification*,
DOI 10.1007/978-94-017-7294-5_2

compelled to initiate new projects to assist farmers in order to boost cocoa production.

For their part, the rubber companies arrived in Côte d'Ivoire in the 1950s and 1960s, at a time when the country was completely ignorant of rubber. Then, in consultation with public authorities, they set up village projects to cultivate rubber. The first project to introduce and develop rubber cultivation in villages, of the order of 100 ha and covering 33 farms, was set up in 1968 and 1969 in Anguededou, near a large plantation established in 1966 by the government. In Côte d'Ivoire, this early model of private intervention in village agriculture remained specific to rubber; it was linked to this industry's particular history (Losch 1990). These projects were of modest size and seemed to remain marginal for a long time. But 30 years later, their effectiveness has proven to be remarkable (Chaps. 6 and 7). The discourse of rubber companies has become accommodative and sensitive to the interests of producers and competitors. 'Cultivation of rubber helps farmers climb out of poverty but we certainly do not want farmers to abandon cocoa and food crops,' says the director of one of the major rubber companies. But within the rubber sector, privatized and liberalized, they now find themselves lined up against each other, competing to encourage farmers and upper middle class investors (UMCI[1]) to plant rubber or to expand their own rubber estates.

The theme of diversification, when combined with institutional changes in relationships between family farming, industries, governments, the elites or the urban bourgeoisie, therefore represents considerable economic and political challenges. One can hypothesize that training, technological innovation and capital in the form of projects is beneficial to family farming. When unaided, family farming is based primarily on factors of labour, land and forest, with little capital other than that created by the planted and cultivated crop. In other words, the scarcity of land, forest and labour resources introduces a new dimension to diversification: beyond a few cases of land grabbing, farmers—diversifying in a context of limited land and labour resources—encourage agro-industrial groups to increase their support to producers in order to counter competing operators and sectors. This has become a new determinant of diversification.

It is therefore extremely important to evaluate and understand the processes of diversification in a country whose resources of rural populations, labour, land and forests are not inexhaustible. The diversification model proposed in this book's introduction emphasizes ecological change after a few decades of quasi-monoculture. Is it also the primary determinant of diversification for the world's leading producer of cocoa? In addition to factors of price and public policy (for example, the creation of a hybrid oil palm sector in the 1960s), has not the process of diversification also been driven by deforestation and exhaustion of the forest rent, in connection with the procurement strategies of large private groups?

This issue has partially been addressed in the economic literature. In various small regions, the diversification towards oil palm, rubber and annual crops—such

[1]For a definition of UMCI, see Footnote 3 in the Chap. 1.

as cassava and pineapple—is already being interpreted as stemming from the exhaustion of forest rent. It has been facilitated by opportunities that projects and investments offer (Ruf 1987, 1993; Chauveau 1993; Colin 1990a, b; Mollard 1993; Oswald 1997; Léonard 1997; Naï Naï et al. 2000; Léonard and Vimard 2005).

But even more than these observations, which are limited to the local scale, the link between the difficulty of replanting cocoa, public policy and diversification deserves assessment at the national level. What has exactly transpired in the coffee sector? What are the main dynamics of agricultural diversification in the country's cocoa cultivation zone, when considered region by region?

How to understand and evaluate this process of diversification at the national level? The issue of intensification associated with diversification after exhaustion of the forest rent has been little discussed. Can we support a hypothesis of an innovation-diversification-intensification process involving resorting to an increase of labour and capital to replace the land factor? Any innovation-diversification which requires extra labour to deal with the degradation of the environment and to raise production falls under the ambit of a classic and neutral Boserupian process in terms of land consumption, whereas any innovation-diversification which calls for additional inputs and capital can said to be part of a 'modernized' Boserupian process (Couty 1989). In this latter case, the substitution of capital for land and labour will increase productivity as compared to these two factors—a decisive progress since they are become limiting factors in Côte d'Ivoire.

It can also be that diversification is part of an extensification process, by reducing the requirement of labour and increasing the consumption of land. In this case, the process moves away from Boserupian models and increases competition between crops and between sectors, especially at the expense of employment. How to analyze the process of diversification of cocoa towards oil palm and especially towards rubber? Finally, authors such as Jacques Weber show us that these innovation-transitions do not take place as a matter of course. Other than the government and large private groups, who are the family-farming actors driving these transitions?

2.1 Methodological Elements

We first rely on a survey conducted in 2006 over 3 months of a total of almost 1100 farms spread out over 13 administrative regions, 25 *départements* and more than 50 villages. This sample is constructed based on regional averages of the 2001 national agricultural census (NAC) of Côte d'Ivoire, by taking a proportionate number of farms by region on the basis of cultivated areas by crop (Table 2.1).

Of the 14 regions within the forest zone, considered to be part of the cocoa-coffee area (south of the Zanzan region), only the Montagnes region could not be included in the survey. The sampling does not claim absolute statistical representativeness within each *département*. Indeed, a compromise had to be made between statistical rigour and a cost/effectiveness balance, by choosing to limit the number of villages

Table 2.1 Distribution of cultivated areas of cocoa and coffee by *département* in 2000, according to the 2001 national agricultural census

Administrative regions in 2001	Cocoa and coffee crop areas (ha)			
	Cocoa	Coffee	Total	%
Bas-Sassandra (7)	535,267	62,800	598,067	25.13
Haut-Sassandra (9)	214,847	94,740	309,587	13.01
Sud-Bandama (6)	163,635	34,074	197,709	8.31
Moyen-Cavally (13)	104,445	84,946	189,391	7.96
Agneby (2)	133,594	43,485	177,079	7.44
Montagnes (14)	64,072	89,459	153,531	6.45
Moyen-Comoé (3)	125,977	23,573	149,550	6.28
Fromagers (8)	101,658	22,441	124,099	5.22
Sud-Comoé (1)	67,092	54,748	121,840	5.12
Lagunes (5)	87,065	31,551	118,616	4.98
Marahoué (10)	80,529	17,086	97,615	4.10
Lacs (11)	69,908	19,007	88,915	3.74
N'Zi-Comoé (12)	19,117	13,882	32,999	1.39
Zanzan (4) and northern regions	10,344	10,283	20,627	0.87
Total	1,777,550	602,075	2,379,624	100

Source Ministry of Agriculture, NAC 2001
Data 830 ha surveyed, apart from the 230 ha of coconut, and about 80 ha for which farmers did not know the previous crop because the plantation was purchased or inherited

to between three and eight per region. But on the whole, the quality of data and the geographical and ethnic diversity of the surveyed zones allow the main trends in each of the 13 administrative regions to be clearly discerned. Through aggregation, the dynamics at the country level were also assessed in 2006.

Then in 2008, we conducted a fresh survey of a total of 435 farms distributed over 8 of these regions, 12 *départements*, and 30 villages in the coffee and cocoa production area. These data were partially updated in 2010. As in 2006, the sampling criteria only chose family farms. The survey teams were able to contact only those farm managers present in the village. Consequently, these two studies do not include 'absentee' farmers and UMCIs living in towns. But another survey, conducted specifically with UMCIs of the Agneby region sheds light on these actors who have acquired a specific significance in the rubber sector.

Finally, in 2010 and 2011, we began to update the data in a few villages. The regions of Cavally and Montagnes were not sufficiently open to allow quantitative surveys, but we were nevertheless able to make some observations.

The official presentation of the 2001 National Agricultural Census includes a long history of 'coffee-cocoa combination' in which these crops seem inextricably linked at the regional and farm scales. According to this census, coffee was still being cultivated on 600,000 ha. But during the 2000s, this representation will turn into a myth. Indeed, the decline of coffee, already begun in the 1980s, will accelerate and contribute significantly to diversification towards oil palm and, later, towards rubber.

2.2 Plantation Dynamics by Administrative Region in 2006

2.2.1 *Sud-Comoé*

2.2.1.1 Diversification Towards Oil Palm Brought on by Public Policies

In the 1980s, Sud-Comoé, especially its Aboisso *département*, was an area known for its coffee cultivation, then for oil palm development thanks to public policies. Indeed, a second oil palm plan was implemented by the government from 1985 to 1990, particularly in the Aboisso sector (Chap. 4). Oil palm became the dominant crop in the 1980s, and especially so in the 1990s following a second wave of adoption of oil palm cultivation, this time without project support (Fig. 2.1).

2.2.1.2 Diversification Encouraged by the Market, Deforestation and Interests of Labourers

During the 2000s, oil palm continued to be planted despite falling prices of palm bunches. In addition to the ability of family farming to persevere for several years in the hope of an increase in prices, the relative persistence of investments in oil palm in this region was due to:

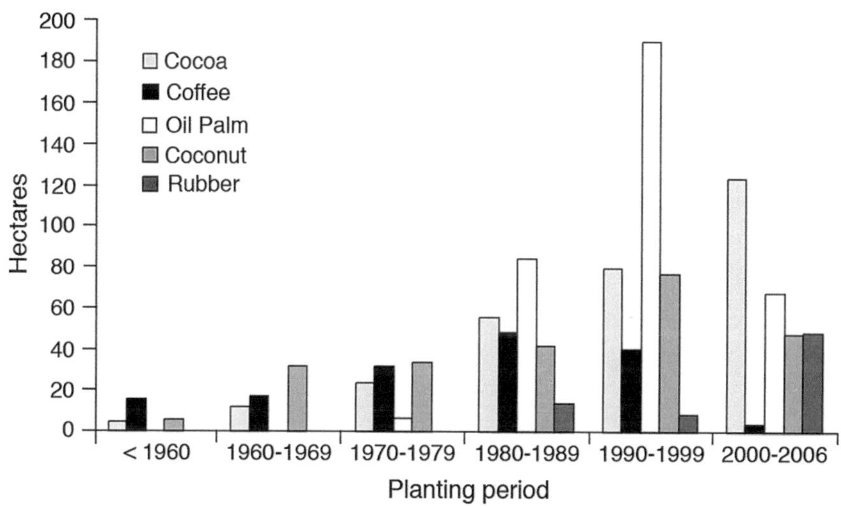

Fig. 2.1 Age structure of perennial crops in the Sud-Comoé region, Côte d'Ivoire, in 2005–2006. (*Sources* Data collected by CIRAD in 2006 on behalf of the European Union, in Ruf 2006a; one-time survey)

- information gained on hybrid oil palm through the project. The second oil palm plan of the 1980s was active in this region and many UMCIs took advantage of it. Village farmers from outside the area also did so. There is quite a bit of knowledge on cultivating hybrid oil palm in this region (Chap. 4);
- the domestic and regional market for red oil (African oil);
- the market for palm wine at the end of the oil palm life cycle. The possibility of obtaining immediate capital by felling the oil palm tree, or rather by negotiating its sale to a team of fellers-extractors-distillers is the great advantage of oil palm in West Africa. It accords flexibility to the farmer since he can count on acquiring capital at the end of the oil palm life cycle. Indeed, the farmer has the option to shorten the life of a part of his oil palm plantation in an emergency;
- ecological change, i.e., the aging of coffee and cocoa trees and lower yields;
- the role of labour and sharecropping employment contracts. Even if farmers were reluctant to diversify and convert old plantations, the labourer situation brought them back to reality. Starting in the 1980s, farmers began finding fewer and fewer candidates to take over their old coffee plantations under the abusa contract, a form of sharecropping with production sharing between the 'taker' (1/3rd of the product) and the owner of the plantation (2/3rd of the product). Also a factor was the abandonment of old cocoa plants by labourers looking instead for oil palms under the abusa contract.

2.2.1.3 Survival of Some Coffee Plantations, Competition from Rubber, Planting-Sharing Contracts

There are few plantations left of coffee, the first of the major crops; they have been partly replaced by oil palm trees. Rubber, introduced in the 1980s, became popular in the 2000s and began to compete with oil palm because of the rise in rubber prices. In conjunction with the increasing adoption of rubber cultivation, there was a simultaneous development of the advent of new actors (retirees and UMCIs, even religious groups) and an institutional innovation: the 'planting-sharing' or 'working-sharing' agreement. This is an arrangement between an owner turning over his land to a 'taker' who clears and cultivates the land. They then share the cocoa, oil palm or rubber plantation when it goes into production (Colin and Ruf 2011).

2.2.1.4 Local Dynamism of the Coconut Sector

Coconut cultivation, concentrated on the coast, extends a few kilometres inland. Even though it is a little over-represented in the sample, there is indeed a dynamism to the coconut sector. Plantations are being established in large numbers even outside coconut programmes launched by Sodepalm in the 1970s. In fact, a market exists due to a mature-coconut sector involving Ivorian, Ghanaian and Nigerian buyers.

2.2.1.5 Cocoa Cultivation in Expansion in the East of the Country

Surprisingly, cocoa cultivation continues its expansion in the east of the country, especially in the east of the *département*, between Mafere and the Ghanaian border, where there were forests yet to be cleared during the 2000s. Several villages and farmers of this region do not even cultivate oil palm; they are specializing once again in cocoa cultivation.

2.2.1.6 Replantation, Diversification and Expansions

This former plantation area is therefore dynamic. In addition to its high level of diversification, the plantations here are relatively young. Taking all perennial crops together, more than half (688 ha) the plantations were established in the 1990s and 2000s of the 1111 ha identified in the 2006 sample.

Even ignoring the role played by public policies in initiating oil palm diversification, this Sud-Comoé region perfectly illustrates a dual process linked to the cocoa cycle (Table 2.2):

- replanting-diversification of the initial—and now aging—cocoa and coffee plantations, conversion to oil palm and rubber;
- expansion of cocoa plantations by forest clearing, on agricultural frontiers, on the outskirts of cultivated areas.

This dual process helps limit the age of cocoa trees to 25 years instead of 35–40 years. The oil palm tree can be felled at 10 or 15 years (instead of 25) if the farmer needs money.

2.2.2 *Agneby*

The situation in Agneby is similar with respect to the age of plantations. Villagers surveyed in 2006 in Agneby had almost no plantations that were established before

Table 2.2 Crops preceding perennial crops in Sud-Comoé

Crop	Preceding crop (ha)					
	Forest	Fallow	Old coffee plants	Old cocoa trees	Oil palm	Total
Coffee	116 (78 %)	3	26	3	0	148
Cocoa	224 (76 %)	7	23	31	8	293
Oil palm	79 (24 %)	49	85	30	82	324
Rubber	12 (19 %)	19	12	9	13	65
Total	431	78	146	73	103	830

Sources CIRAD and A&C-Vie surveys on behalf of the EU in 2006
Data 830 ha surveyed, apart from the 230 ha of coconut, and about 80 ha for which farmers did not know the previous crop because the plantation was purchased or inherited

1970. Coffee plantations of the 1950s and 1960s, the dominant crop of the time, were abandoned or succumbed to the fires of 1983. Others had been converted into cocoa plantations by second generation autochthons, subsequent to the purchase of old coffee plantations by migrants through local land markets. In 2006, the conversion of coffee plantations to those of cocoa which began in the 1970s reached its peak. This took place most notably through the process of sale by the autochthons of old coffee plantations and forests to migrants. Also worth noting is the entry of migrants in protected forests, such as that of Kavi, near Agboville.

As far as family agriculture is concerned, Agneby remained very little diversified in 2006; it relied heavily on cocoa, with coffee surviving in a few outposts. Here, too, a return to cocoa by migrant farmers was observed; they had started with Poyo banana plantations, near the railroad but they finally abandoned this sector because of lack of market access.

2.2.2.1 Investments in Cocoa Plantations

The dominant profile of such a region is therefore the consistency and regularity of investments in cocoa, with new plantations regularly being set up between 2000 and 2005. The total of new areas devoted to cocoa in 6 years is close to the investments made in the 1990–1999 decade. This is amazing for an ‘old’ plantation zone and confirms the revival of cocoa cultivation in the country’s east in the 2000s (Fig. 2.2).

During the 2000s, the revival of cocoa cultivation depended, for the most part, on the clearing of secondary forests. We also observed the recourse to the technique of coppicing—requiring less labour and effort than replanting—favoured by the older farmers. In 2006, rubber was only just emerging, but already this crop was

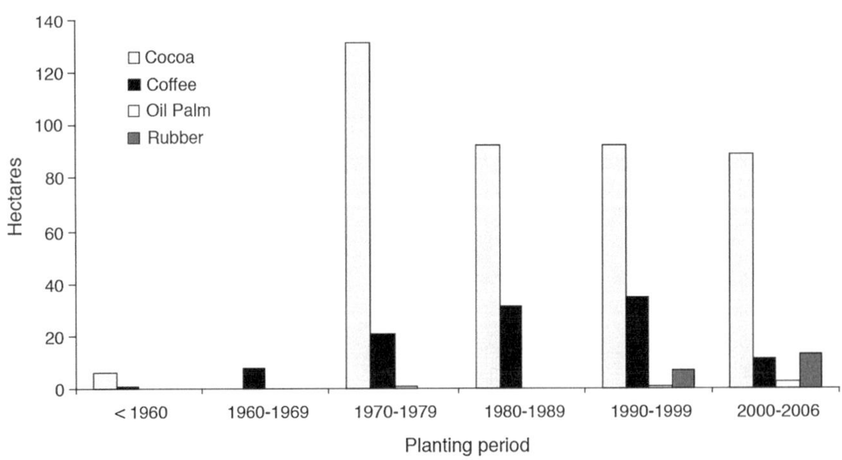

Fig. 2.2 Age structure of perennial crops in the Agneby region, Côte d’Ivoire, in 2005–2006

seen by the farmers as a solution to the problems of replanting cocoa. It is often used as a relay crop on a plot of old cocoa trees: a form of an agroforestry technique, where interplanting between two perennial crops is just a transition between two phases of monospecific plant stands.

2.2.2.2 Rubber Smallholders and UMCIs

Even though family agriculture ventured into rubber cultivation only in 2006, Agneby, near Abidjan, was one of the most coveted regions by UMCIs to set up plantations ranging in size from several tens to several hundreds of hectares. In 2008, the mapping of about 50 plantations owned by UMCIs around the village of Offumpo, 20 km from Agboville, provided insight into the pressure that these new actors are exerting on village lands. This intrusion of UMCIs into village land holdings laid the seeds of future conflicts over land. There were, however, also positive aspects to their arrival, especially the establishment of contacts and exchange of information between farmers and managers of plantations owned by UMCIs, thus facilitating the adoption of rubber cultivation by the villagers.

2.2.3 Moyen-Comoé

In Moyen-Comoé, the situation was observed to be similar to those for the previous two regions. These three regions are part of the large south-eastern region of the country.

- There was a similar rate of cocoa plantations, here with a little more 'cocoa-after-cocoa' replanting. But it was observed that for as much as 63 % of the area, the preceding crop was 'forest' (Table 2.3).
- A relative strength of coffee cultivation well into the 1990s, in part thanks to the peak prices attained in 1994–1995.

Table 2.3 Crops preceding perennial crops in Moyen-Comoé

Crop	Preceding crop (ha)					
	Forest	Fallow	Old coffee plants	Old cocoa trees	Oil palm	Total
Coffee	50 (46 %)	3	34	21	0	108
Cocoa	258 (63 %)	16	25	109	3	411
Oil palm	1 (5 %)	0	0	7	2	10
Rubber	8 (7 %)	41	10	50	0	108
Other	4 (27 %)	0	9	0	0	13
Total	320	60	77	187	5	649

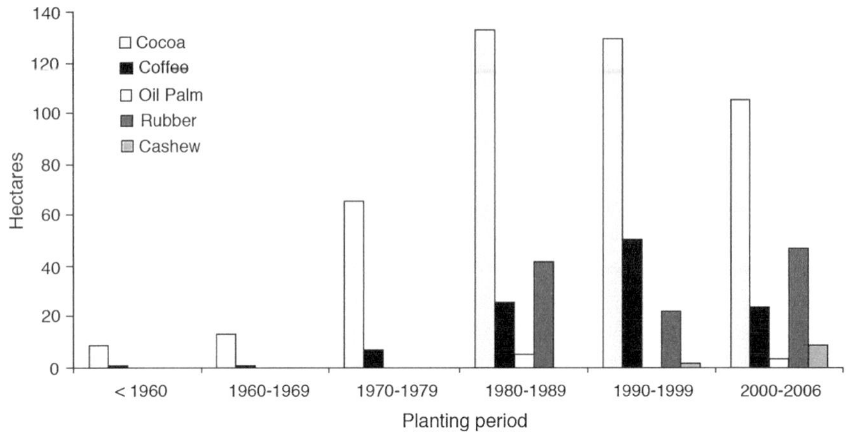

Fig. 2.3 Age structure of perennial crops in the Moyen-Comoé region, Côte d'Ivoire, in 2005–2006

Rubber cultivation began here earlier due to favourable public policy and a project set up in Bettié. After that, it took almost 10 years for independent rubber cultivation—outside projects and without government support—to regain momentum (Fig. 2.3). Almost all rubber plantations were established after clearing old cocoa or on fallow (Table 2.3). The cashew plantations made a hesitant appearance in the 2000s, partly for the good market for the crop, partly for its resistance to drought and fire. Rubber and cashew clearly represent the dimension of ecological change among the major determinants of diversification.

2.2.4 *Zanzan*

With a relatively dry climate, the Zanzan region, also to the east but further north, was more affected than other regions by the drought years, especially by the 1983 fires that destroyed hundreds of hectares of cocoa plantations. Ecological change here clearly determines the process of diversification, with:

- the marked decline of cocoa cultivation;
- the rise of cashew with 95 % of orchards being planted on fallow land (partly created by the burning down of cocoa plantations), after the clearing of old coffee plants and, to a lesser extent, after the felling of old cocoa trees (Table 2.4, Fig. 2.4).

Also observed in 2006 was the survival of old coffee plantations from the 1960s, partially spared by the devastating fires of 1983. This was probably why they have been maintained. After the fires, farmers still needed some coffee to survive.

Table 2.4 Crops preceding perennial crops in Zanzan

Crop	Preceding crop (ha)				
	Forest	Fallow	Old coffee plants	Old cocoa trees	Total
Coffee	16.8 (61 %)	9.8	0	0.5	26.5
Cocoa	56.3 (95 %)	1.5	0.5	1.0	59.3
Oil Palm	0	2.0	0	0	2.0
Rubber	0	0	4.0	0	4.0
Cashew	6.0 (6 %)	54.0	27.5	6.0	93.5
Orange	0	1.3	0	0	1.3
Total	78.5	68.5	32.0	7.5	186.5

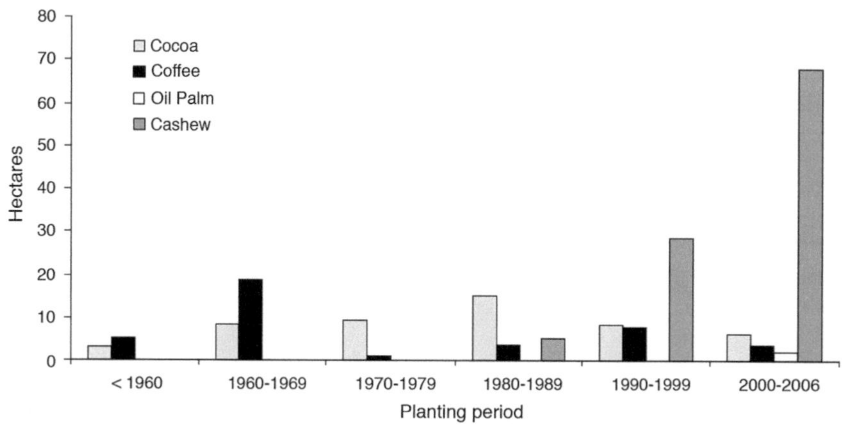

Fig. 2.4 Age structure of perennial crops in the Zanzan region, Côte d'Ivoire, in 2005–2006

2.2.5 *Lagunes*

Located in proximity to the Abidjan markets, Lagunes was recognized as a major cocoa production region. But it was also known for its natural oil palm groves still abundant as recently as the 1950s. Early on, this region benefited from the public policy in the form of the first oil palm plan (1960s) and the first experiments in village rubber plantations (1970s). Thus, the Lagunes region, especially its *département* of Dabou, has every reason to be called one of the most diversified areas of Côte d'Ivoire. In 2006, no trace remained of the oil palm trees of the 1960s and 1970s or even of the 1980s; they were all felled and converted into palm wine (Fig. 2.5).

Each decade saw here a dominant crop:

- 1970s: cocoa;
- 1990s: oil palm, in general, replacing coffee;
- 2000s: rubber.

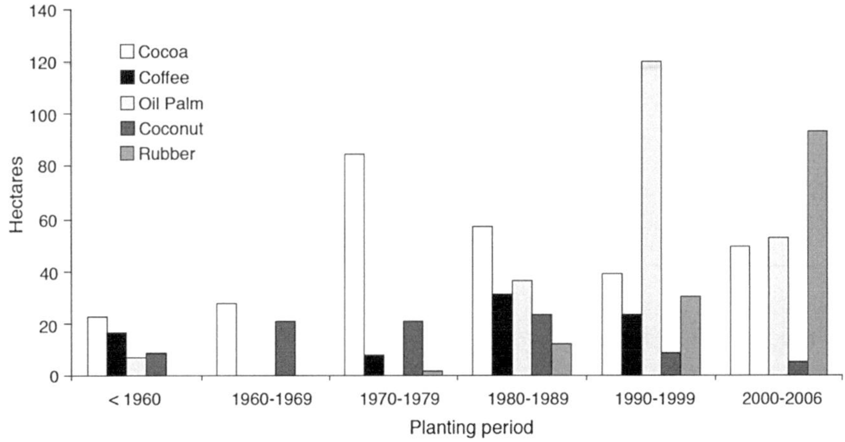

Fig. 2.5 Age structure of perennial crops in the Lagunes region, Côte d'Ivoire, in 2005–2006

A bit like in the Sud-Comoé region, the decline of the old coffee-cocoa combination in favour of the new rubber-oil palm combination is explained by the conjunction between, on the one hand, deforestation and environmental degradation and, on the other, new opportunities provided by markets and favourable public policies. Do the developments in these two regions presage the future of the whole of Côte d'Ivoire?

2.2.6 *Sud-Bandama*

The Sud-Bandama sub-sample is limited to 36 farms, whence a limited representativeness. The importance of oil palm is enhanced by the proximity of the factory of the Palmci group. Nevertheless, the trend of declining cocoa, offset by a rise of oil palm in the 1990s, is very real (Fig. 2.6).

The prominence of the oil palm stems partly from a dual marketing network which reduces market risks:

- the Palmci factory. During the 2000s, as in other regions such as Sud-Comoé (Adiake) and Sassandra, its effectiveness suffered due to transport problems and a shortage of trucks;
- the artisanal sector for red oil and Kabakrou soap (literally soap-stones). It offers reassurance to farmers because of its market security and regularity, and the economic balance that is created within the family. In this sector, women play a leading economic role. Not only do farmers' wives themselves produce and process the oil, the traders from Divo who come to buy it are also mainly women.

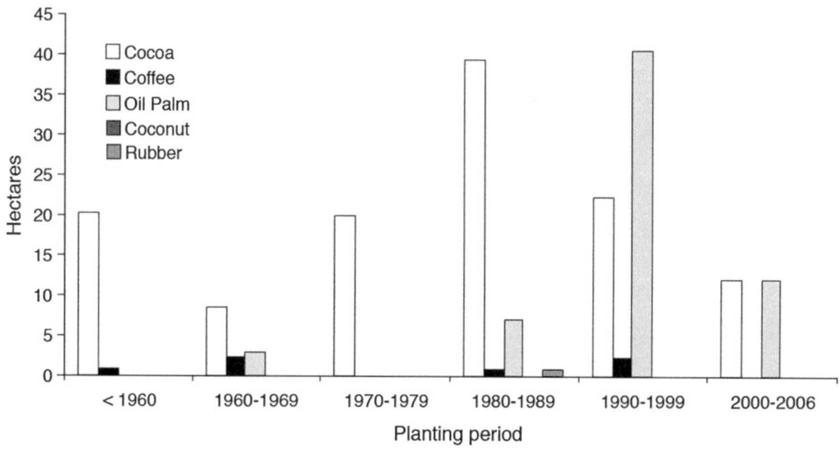

Fig. 2.6 Age structure of perennial crops in the Sud-Bandama region, Côte d'Ivoire, in 2005–2006

However, in addition to price and market factors and public policy, palm diversification was here too accelerated by ecological change and progression of the cocoa life cycle (Table 2.5): the 1983 fires destroyed thousands of hectares of cocoa plants. After a few years, farmers replanted cocoa but encountered the difficulties of doing so in a degraded environment. They thus seized the opportunity to cultivate oil palm instead.

This process of diversification and conversion occurred over a time span which was longer than 50 years and, therefore, over three generations. In a simplified manner, we can say that the grandfather was a coffee grower, the father cultivated cocoa and the son planted oil palm. Very likely the fourth generation will choose rubber.

2.2.7 *Bas-Sassandra*

In this region, which encompasses the three *départements* of Soubré, San Pedro and Sassandra, soils are better suited to coffee cultivation than to cocoa cultivation. As recently as 1970, the area was still covered by rainforest. Forest rent and attractive

Table 2.5 Crops preceding perennial crops in Sud-Bandama

Crop	Preceding crop (ha)					
	Forest	Fallow	Old coffee plants	Old cocoa trees	Oil palm	Total
Coffee	4.0 (80 %)	0.3	0	0.8	0	5.0
Cocoa	34.5 (40 %)	8.0	16.0	28.0	0	86.5
Oil Palm	4.0 (8 %)	22.5	6.5	17.0	1.0	51.0
Rubber	2.0 (50 %)	0	0	2.0	0	4.0
Total	44.5	30.8	22.5	47.8	1.0	146.5

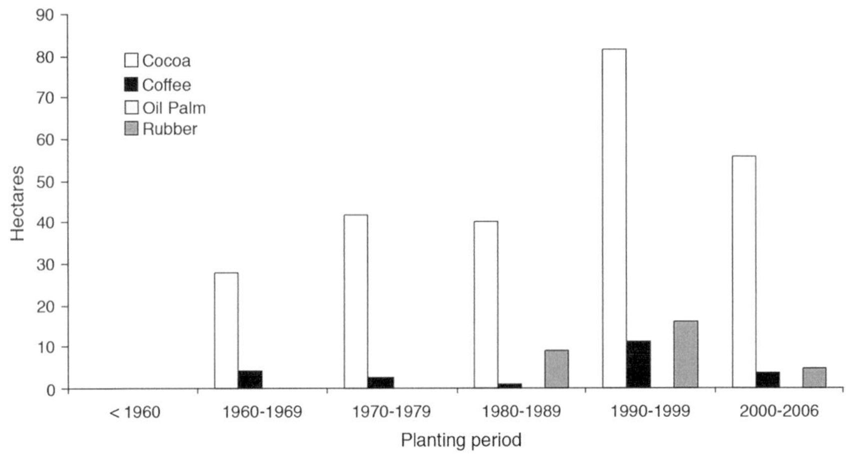

Fig. 2.7 Age structure of perennial crops in the Bas-Sassandra region, Côte d'Ivoire, in 2005–2006

cocoa prices drove cocoa cultivation to prominence from the 1980s to 2000, attracting hundreds of thousands of migrants (more than 20 % annual population growth in the 1970s and 1980s). After cocoa cultivation peaked in the 1980s, a decline in investment in cocoa plantations was observed (Fig. 2.7). The main reasons were land scarcity, loss of forests and its consequences, drop in fertility, increase in pests like termites, all leading to difficulties in replanting.

Farmers began diversifying in the 1990s, first gravitating towards oil palm then to rubber, thanks to public projects or private companies. In addition to assuring a market, projects also helped bring in information on new crops and supply planting material, often the key to investment. In addition to these project opportunities, the first cases of early mortality of cocoa around San Pedro also contributed to the adoption of rubber in the 1990s.

In the *département* of Sassandra, citrus cultivated for essential oil—a crop grown earlier by European planters—saw some development in the 1990s. In 2006, however, the crop became a casualty of the processing factory's closure.

Finally and most importantly, the adoption of rubber cultivation by the villagers was initiated by relatively modest local projects set up by rubber companies in the 1980s. In 2006, rubber was mainly visible along roads, but little inland yet. However, it was obvious that it held a promising future. This is one of the regions that will see an extraordinary acceleration of the adoption of rubber cultivation in the early 2000s, on the outskirts of private estates that provided help to the villagers.

Chocolate companies also helped cocoa farmers effectively through development projects and through certification schemes, which resulted in some technical support and better prices. But in this big region, especially in the San Pedro *département*, diversification into rubber seemed irreversible, sometimes indistinguishable from conversion from one monocrop to another.

2.2.8 *Fromager*

In the Fromager region, some villages benefitted from a rubber project which was set up in the early 1990s. In these villages, rubber cultivation brought economic success to farmers and spread through imitation (Ruf 2012). However, the *département* remained fundamentally cocoa oriented, with renewed investments in the 1990s and 2000s, at least until 2006 (Fig. 2.8).

In the 1960s and 1970s, coffee cultivation declined and then disappeared altogether. Burned or abandoned, it was first replaced by cocoa and then by rubber. As in other regions, there was a revival in the 1990s thanks to the rise in coffee prices in 1994–1995 and efforts undertaken by agricultural extension. But this recovery did not last long.

2.2.9 *Haut-Sassandra*

In 2006, the region of Haut-Sassandra, especially the area around Daloa, exhibited a greater resilience of the coffee sector than elsewhere, but the rapid decline of investments in coffee cultivation was similar to that in other regions. Here too, despite a timid emergence of cashew and teak in 2006, it was cocoa that benefitted from coffee's decline (Fig. 2.9).

2.2.10 *Marahoué*

In 2006, Marahoué was still an 'all-cocoa' region. Here too, there was a revival in investments in the 2000s, after the downward trend experienced in the 1980s due to

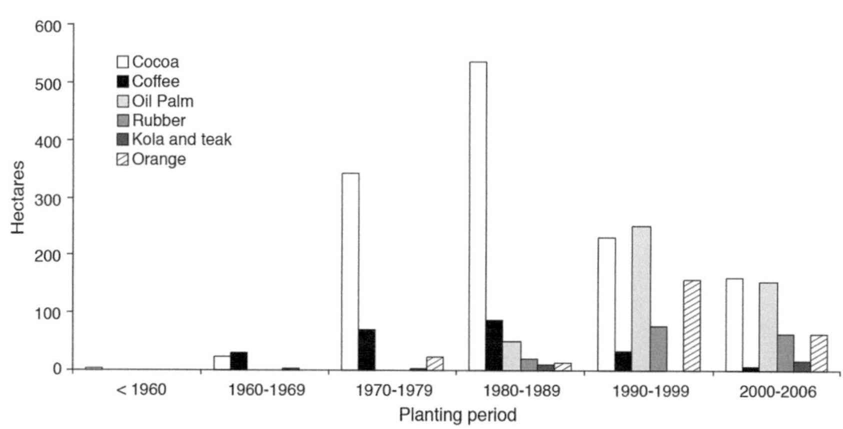

Fig. 2.8 Age structure of perennial crops in the Fromager region, Côte d'Ivoire, in 2005–2006

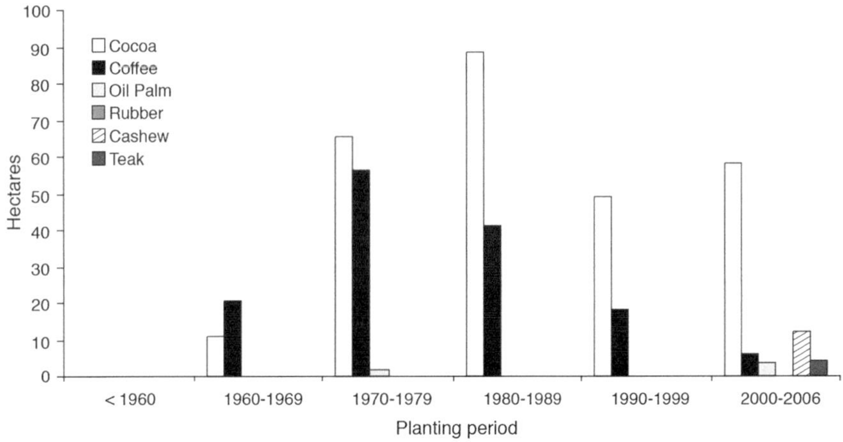

Fig. 2.9 Age structure of perennial crops in the Haut-Sassandra region, Côte d'Ivoire, in 2005–2006

scarcity of land. In addition to the rise in producer prices in 1997–1998, generation change and the subsequent increase of labour and energy played an important role. All the children who came with their parents in the 1980s were in their 30s in the 2000s and found solutions in order to be able to plant and replant.

Nevertheless, Marahoué, once known for its high yields post-forest-clearing, suffered from the decline of production and lower revenues. Cocoa farmers became poorer. The swollen shoot virus was at its peak. Under the impact of environmental change, this region became a candidate for diversification, particularly towards rubber, and saw the emigration of its youth to other regions (Fig. 2.10).

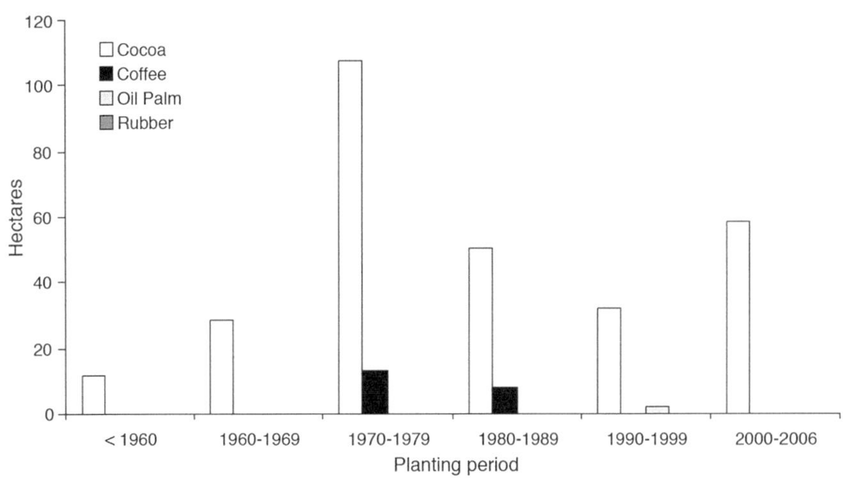

Fig. 2.10 Age structure of perennial crops in the Marahoué region, Côte d'Ivoire, in 2005–2006

2.2.11 *Lacs: The Return to the Village in the 2000s*

Prior to the surveys, we assumed that Lacs was a highly diverse region which was moving away from cocoa. This assumption was contradicted by the facts. In 2006, we observed instead a return to cocoa, in every sense of the term: not only a return of investments, but also return of many migrants to their home villages. They were both disappointed by the decline in production in the west of the country where they had migrated to and concerned about the crisis the country was going through (Fig. 2.11).

2.2.12 *N'Zi-Comoé: The Revival of the Cocoa Belt Until 2006*

The N'Zi-Comoé region was the country's major 'cocoa belt' in the 1950s and 1960s. Its decline began in the mid-1970s, so much so that, in 1979, the then Agriculture Minister, M.D. Bra Kanon, commissioned a study from the research station in Bingerville in charge of cocoa (formerly IFCC-IRCC) to analyze the reasons of this decline. Based on the knowledge of the region, J.M. Gastellu (researcher at Orstom—now IRD) quickly identified the factors responsible. Côte d'Ivoire then began to re-experience the process of migration of farmers from the old planting areas to forests. These new lands were sought by farmers both for the possibility of farming larger areas than in their original farming locations as well as for the good cocoa yields guaranteed by the forest. Production of the old cocoa belt then began to decline because of aging plantations, climate change and decreasing yields (IRCC 1979; Gastellu 1979, 1982; Ruf 1984).

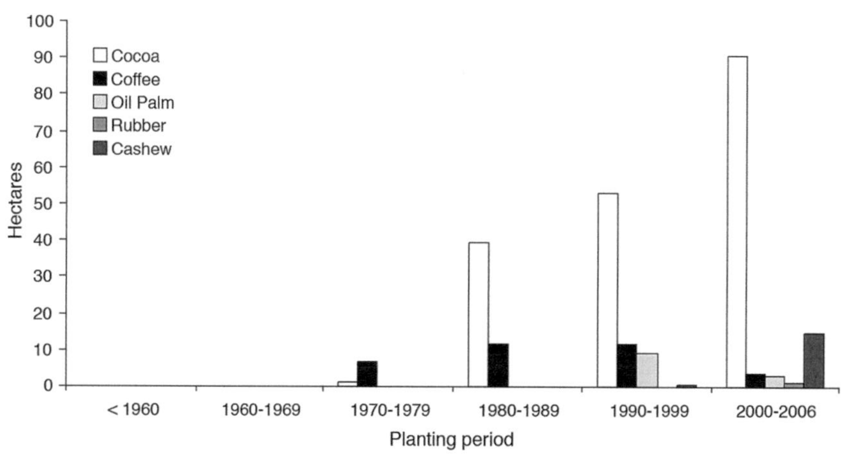

Fig. 2.11 Age structure of perennial crops in the Lacs region, Côte d'Ivoire, in 2005–2006

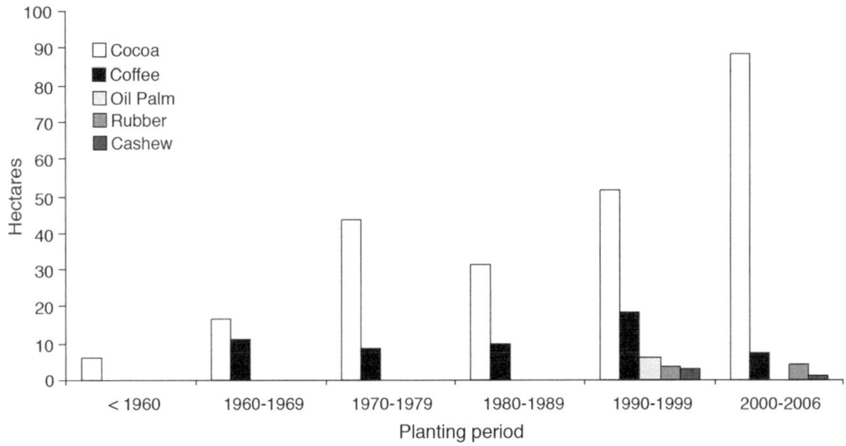

Fig. 2.12 Age structure of perennial crops in the N'Zi-Comoé region, Côte d'Ivoire, in 2005–2006

In the 1970s, migration from N'Zi-Comoé to nearby forests in Moyen-Comoé (near the border with Ghana), in Agneby and particularly in Bas-Sassandra emptied the old cocoa belt of a large part of its workforce, both autochthons and migrants. Workers and Abusan foreigners accompanied these flows, with growers lamenting the lack of labour (Gastellu 1979). As for the Lacs region and its workforce, this exodus of the 1970s and 1980s led to a regrowth of wild vegetation similar to forests on abandoned plantations and fallow land in the 1990s and 2000s. It is these secondary forests which were once again cleared for cocoa: in part by 'emigrants' returning to their villages of origin, but mainly by a second generation, the sons of emigrants or of those who had never left the village. This return to the source of cocoa shows clearly the process of shifting cultivation which applies to a so-called perennial crop at the national level.

Throughout this period, rubber cultivation was still barely visible. It was only from 2005–2006 onwards that rubber cultivation was adopted to replace this revival of investments in cocoa, but this mainly benefitted the UMCIs in Abidjan (Fig. 2.12).

2.2.13 *Moyen-Cavally*

Much like Bas-Sassandra and Haut-Sassandra, the name of the Moyen-Cavally region comes to mind as one of the hotspots of cocoa migration of the 1980s and 1990s. Migrants arrived in their thousands annually. In 2006, there was still a substantial amount of coffee cultivation, here too because of the relative unsuitability of soils for cocoa cultivation. Initially, migrants attempted to grow both

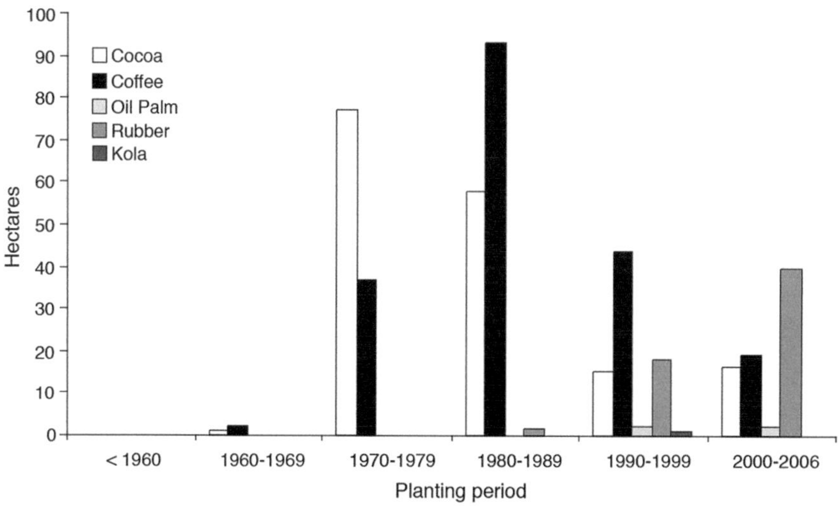

Fig. 2.13 Age structure of perennial crops in the Moyen-Cavally region, Côte d'Ivoire, in 2005–2006

Table 2.6 Crops preceding perennial crops in Cavally

Crop	Preceding crop (ha)				
	Forest	Fallow	Old coffee plants	Old cocoa trees	Total
Coffee	103.5 (58 %)	16.0	6.0	52.1	177.6
Cocoa	136.0 (75 %)	8.0	17.0	20.0	181.0
Rubber	12.5 (23 %)	37.0	2.0	4.0	55.5
Total	252.0	61.0	25.0	76.1	414.1

crops but observed a high rate of cocoa mortality. They then retreated to coffee, to the point that they even replanted coffee in the understory of the dying cocoa trees (Table 2.6). Then, a fall in coffee prices and the continued mortality of cocoa encouraged diversification into rubber cultivation, introduced in the 1980s on rubber estates. However, at least until 2006, coffee remained in second place because it was the only suitable crop to grow during the financially unremunerative period of the rubber cultivation cycle. In January-February, the farmer cannot harvest and sell rubber while coffee can be sold at this time of the year (Fig. 2.13).

2.3 National Aggregates from 2006 to 2011

In 2006, the average farm—assuming such a theoretical entity really exists—at the national scale farmed 6.5 ha of which 3.7 ha were devoted to cocoa cultivation. It still retained 1.0 ha of coffee and had diversified into 1.0 ha of oil palm trees, 0.5 ha

Table 2.7 Cultivated areas of perennial crops for an average farm, based on the 2006 evaluation

Perennial crop	Change in areas (ha) devoted to perennial crops for an average farm						
	<1960	1960–1969	1970–1979	1980–1989	1990–1999	2000–2006	Total
Cocoa	0.1	0.2	0.8	1.0	0.8	0.8	3.7
Coffee		0.1	0.2	0.3	0.3	0.1	1.0
Oil palm				0.2	0.5	0.3	1.0
Rubber				0.1	0.1	0.3	0.5
Others				0.1	0.1	0.1	0.3
Total	0.1	0.3	1.0	1.7	1.8	1.6	6.5

Sources CIRAD and A&V-Vie surveys 2006

of rubber and 0.3 ha of other perennial crops (coconut, cashew, kola tree, teak and citrus) (Table 2.7 and Fig. 2.14).

As we have seen throughout this chapter, these averages are not simple additions of surface areas from one period to the next, but the resultant between the new plantations established on forest or fallow and crop substitution on the plot, to which are added some replantings.

As far as the famous 'coffee-and-cocoa combination' as the historical foundation of the plantation economy of the Ivorian forest area is concerned, it still represented 72 % of the area devoted to perennial crops in 2006. This combination left only 28 % to all the diversification crops consisting of oil palm, rubber and others such as cashew, coconut and teak (Table 2.8). Over the 2000–2006 period, the process of diversification reached 44 %; but the initial, paradoxical, observation remains that of the relative durability of cocoa, at least at a statistical level.

At the beginning of this chapter, we quoted P. La Vaissière—then an economist at the Coffee and Cocoa Institute—on the complementarity of these plantations with respect to the land factor. A statistical analysis of the national agricultural census of 1975 proves him right. In the 1970s, the development of cocoa cultivation in the country was mainly the result of significant migrations. It took off without compromising coffee production which tended to be the domain of autochthons. Similarly, the government's oil palm plan at the end of 1960s was put in place in a context when land and forests were still largely available. At least from the 1980s, a dual process operated at the plot scale and that of the farm. On the one hand, there

Table 2.8 Distribution of perennial crop areas in Côte d'Ivoire

Perennial crop	Area existing in 2006		Area planted during 2000–2006	
Cocoa (%)	57	72	50	56
Coffee (%)	15		6	
Oil palm (%)	15	28	19	44
Rubber (%)	8		19	
Other (%)	5		6	

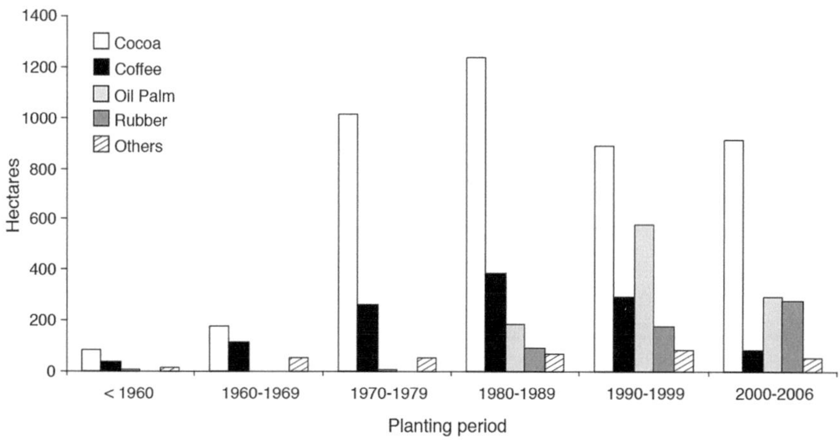

Fig. 2.14 Age structure of perennial crops in Côte d'Ivoire in 2005–2006

was an acceleration of forest clearings almost exclusively for growing cocoa while, on the other, old coffee plants and cocoa trees were felled to be replaced by different perennial crops, mainly oil palm and rubber. Cocoa trees also began to replace old coffee plants, mainly due to disposal and sale of old coffee plantations by autochthons to migrants (see Table 1.1 in the Chap. 1).

An analysis of the preceding crops of plantations demonstrates this dual process. It explains the ability of Côte d'Ivoire to maintain its position as the world's largest cocoa producer, while still proceeding with diversification (Fig. 2.14).

Ultimately, these detailed snapshots of Côte d'Ivoire's cultivated areas, region by region, and national aggregates for the 2005–2006 period have proven invaluable for understanding the dynamics of Côte d'Ivoire's plantation economy in 2012. Contrary to many experts' estimates, cocoa production in Côte d'Ivoire has not fallen. Instead, it even reached new peaks in 2010–2011, with around 1.5 million tonnes. In addition to favourable cyclical climatic factors, there was a significant revival in investment in cocoa cultivation leading up to 2005–2006. This is the main reason that cocoa production in Côte d'Ivoire has been maintained and even increased. According to a universal historical process, these new investments have continued to counteract the phenomena of aging and to increase production.

As for the future, even disregarding potential investments beyond 2006 and given the economically useful life of 25 years of cocoa trees, these analyses of 2005–2006 suggest that the cocoa production should theoretically hold up well for a few years. However, the history of cocoa cycles reminds us of the impact of ecological changes (decreased soil fertility, disease pressure, cocoa mortality). In Côte d'Ivoire, since the late 2000s, this process has been accompanied by an intrusion of rubber cultivation into agricultural landscapes, especially along roads.

2.3.1 Boom in Rubber Cultivation at the Turn of the 2000 Decade

Without going into region-by-region details, the trend confirms an accelerated decline in investments in coffee cultivation, the maintenance of the dynamics of cocoa plantations, especially in certain regions such as Cavally, and exponential growth of rubber plantations from 2006 to 2009 (Fig. 2.15).

Ad hoc surveys conducted in 2010 and 2011 in several small regions highlighted a sharper slowdown in investments in cocoa cultivation, mainly to the benefit of rubber (Figs. 2.16 and 2.17). Figure 2.18 illustrates the particular case of a village where a development project promoted cocoa replanting based on the distribution of hybrid planting material ('18 months cocoa'). It included a demonstration plot and farmer training. The project indeed revived cocoa investment leading to a new peak for cocoa plantations in 2010, especially in Campement Bernard village where plantations were undergoing conversion to rubber. But 2 years later, a very high mortality rate for cocoa was observed. In fact, very few of the plantations set up in 2009 will even reach the production stage. The difficulties experienced in replanting cocoa on degraded soils reinforced the farmers' desire to switch to cultivating rubber.

In 2012, a simple reading of the landscape confirmed these results. In Bas-Sassandra, it was difficult to observe any cocoa along the Soubré-San Pedro axis or even sometimes along tracks perpendicular to it, for example, the track starting from Gabiadji up to 20 km north of San Pedro. And as for rubber plantations, they were largely dominated by young and still-immature plantations. In

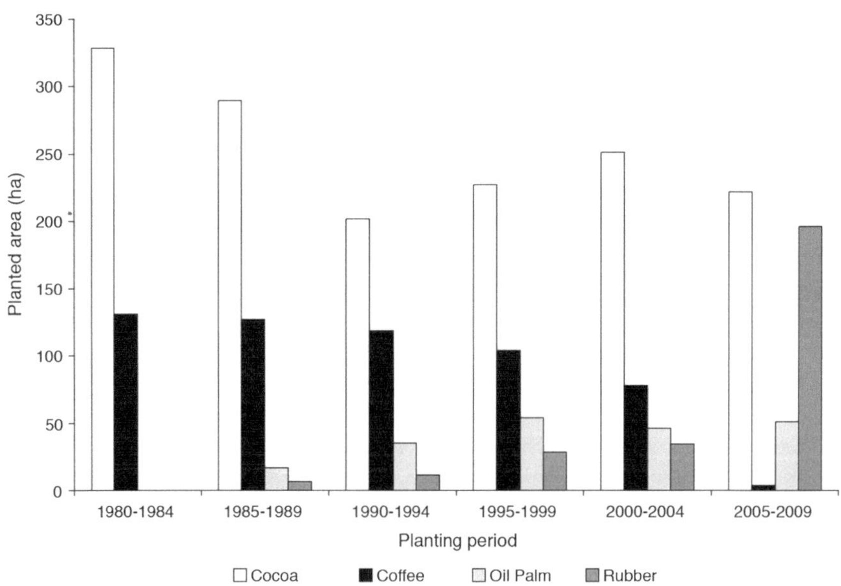

Fig. 2.15 Structure of areas under cultivation for cocoa, coffee, oil palm and rubber in Côte d'Ivoire in 2008–2009 (431 farms). (*Source* Ruf and Akpo 2008; partially updated in 2009–2010)

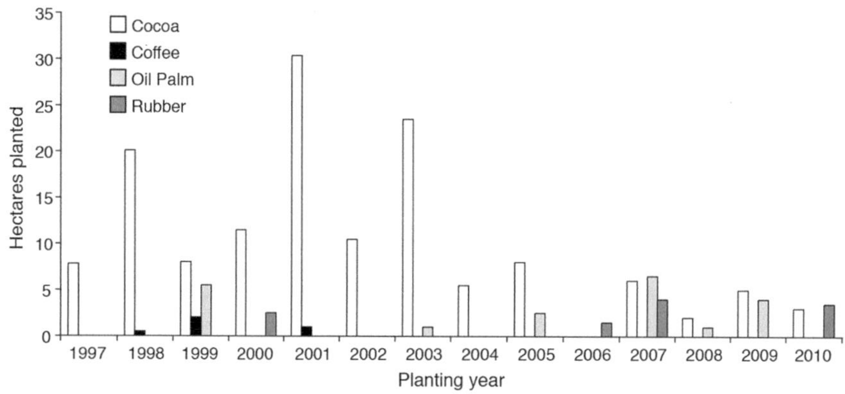

Fig. 2.16 Planting years as reported by farmers in 2011 in the east and centre-west of Côte d'Ivoire in 2011 and from 1997 to 2010. (Data collected from 5 villages and 94 farms in Moyen-Comoé and Haut-Sassandra regions. *Sources* Author's survey, 2011)

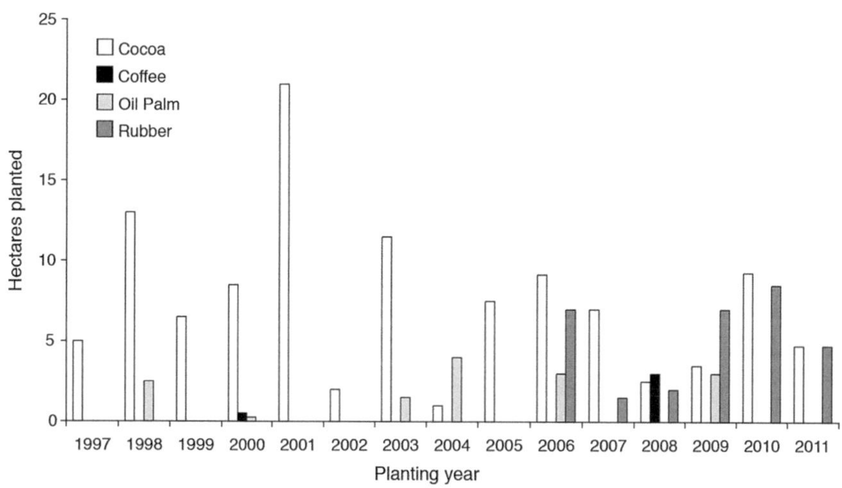

Fig. 2.17 Planting years as reported by farmers in 2011 in the centre-south of Côte d'Ivoire from 1997 to 2011. (Data collected from 4 villages and 70 farms in Tiassalé *département*. *Sources* Author's survey, 2011)

2012, we estimated that two-thirds of rubber areas were immature. This suggests a tripling of production before 2020. While the production of dry rubber was of the order of 235,000 tonnes in 2011, it could well reach or exceed 600,000 tonnes by 2020.

But like in every previous historical period when the world started worrying about its cocoa supplies, cocoa has to be procured from where it is available: far from paved roads and deeper into the bush where the production of adult cocoa

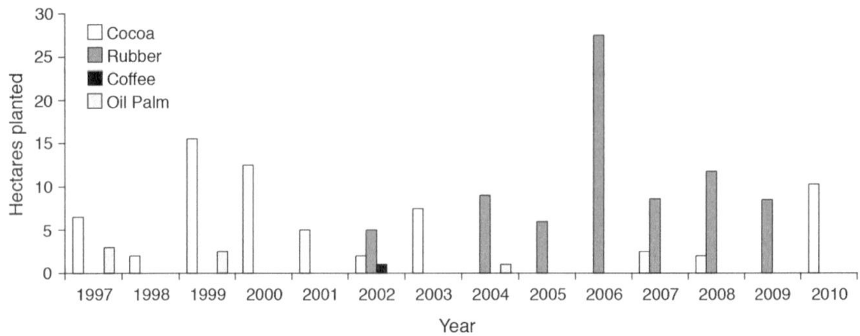

Fig. 2.18 Planting years as reported by farmers in 2011 in the village of Campement Bernard from 1997 to 2010, in south-western Côte d'Ivoire. (Data obtained from 41 farms in the village of Campement Bernard, San Pedro *département*.) In 2010, investments in cocoa cultivation were made through a project. But in 2012, a cocoa mortality rate of more than 60 % was observed. (*Sources* Author's survey, 2011)

trees can still yield around 700 kg/ha with some insecticide use. But these cocoa plantations are mainly found on remote agricultural frontiers, on land obtained by clearing the country's protected forests, far from the existing cocoa belt in Soubré.

2.3.2 *Survival of the Cocoa Sector Through Clearing of Protected Forests*

Because of the political and military crisis, we could not linger in the Cavally and Montagnes regions or in the outlying areas of Haut-Sassandra. But brief forays allowed us to observe a very strong cocoa cultivation dynamic through the resumption of clearing of protected forests, especially from 2008 onwards. Thus, in the protected forest of Goin Débé—which the farmers have humorously nicknamed 'If you want', in the sense of 'if you want, if you are strong, take the risk, buy and clear the land'—open interviews with the autochthons and migrants allowed us to understand the process underway.

To address the issues of diversification, it is important to understand the cycles of cocoa, how this crop can be maintained for a long period in the same country by progressive relocation from one region to another.

In the 1970s and 1980s, having supported the idea of forest protection to protect the environment of the autochthons, the administration of President Houphouët-Boigny allowed Baoulé migrants to enter the forest reserve, over the autochthons' protests. In the 2000s, the frustration of these latter only grew on finding out that the Gbagbo regime did not respond to their complaints to evict the Baoulé or declassify the rest of the forest. This frustration and the low income of autochthons drove them to sell what remained of the forest, mainly to the Burkinabé migrants.

In the same period, the conflicts triggered between the autochthons and the Burkinabé in neighbouring areas created a local demand. Paradoxically, when evicted from a forest where they had started planting cocoa, many Burkinabé migrants were not discouraged and began to look for another forest. These movements promoted contacts between refugees, including in Burkina Faso. They disseminated information about the new availability of forest areas in Goin Débé.

Another amazing paradox is that a forced return to Burkina Faso facilitated contacts between repatriated migrants and labour reserves in the original villages. Once a forest was acquired, it became easier to organize the influx of labour from Burkina Faso by the busloads. The politico-military crisis and the control of the entire northern area by rebel forces facilitated this bus traffic from the north. During the same period, cocoa mortality and falling incomes in Bas-Sassandra triggered internal migrations from the south to Cavally. In this case, migrants who had been settled for a long time had some financial resources. Population pressure, in combination with the cocoa cycle, also played a fundamental role: the sons of immigrants, especially from Bas-Sassandra, wanted to try their luck.

In the old areas under the control of rebel forces or in new areas—which theoretically come under the savannah zone—but where protected gallery forests were still found, clearing by migrants also generated thousands of tonnes of cocoa in the early 2000s.

Finally, international NGOs indirectly provided reassurance to migrants by wanting to settle the conflict; this merely had the effect of attracting new migrants to the area. Thus, several thousands of young migrants converted thousands of hectares of protected forest into cocoa plantations in the late 2000s. At the same time, their parents or brothers in the cocoa-growing areas of Bas-Sassandra and Cavally diversified out to rubber. This process is typical of the history of cocoa cultivation; it combines specific situational factors with structural factors of a cocoa cycle's functioning.

The prospective outlines developed earlier are therefore buttressed. Except in case of an outbreak of a rapidly spreading disease or pest pressure, cocoa production in Côte d'Ivoire is expected to decline only slowly over the next several years, while the production of rubber, now the new solution to youth unemployment, is expected to triple between 2010 and 2020.

2.4 Interpreting Replanting and Diversification Processes

2.4.1 Not Much Replanting, but a Form of Sustainability of the Cocoa Sector

The perception of aging of cocoa trees, even though partially contradicted by the data, persists in the scientific literature. Based on two studies conducted in 1993–1994 and 2001, the researchers concluded as follows:

- 'Even though the average age of harvested trees has not changed much between 1994 and 2001 (it still ranges between 20 and 30 years)—which seems to imply that there is no marked aging (yet?) of the Ivorian cocoa trees,—the share of older orchards, those over 30 years of age, which are seeing a net increase, is a factor to watch. Indeed, once a cocoa tree exceeds 35 years of age, a marked decline in its productivity is observed';
- 'The aging of plantations is nevertheless inexorable. In fact, in 10 years there will be more trees older than 40 years than there will be trees younger than 5 years if the rate of new plantations remains at current levels. Consequently, the national average yield could decrease to about 250 kg/ha. With land reserves decreasing, it is imperative to promote the intensification of technical itineraries on existing cocoa trees and the replanting (or rehabilitation) of old orchards' (Aguilar et al. 2003).

The initial results of surveys presented here shows that the cocoa orchard has not aged as rapidly as expected. The revival of investments in cocoa family farming in the 2000–2006 years has once again pushed back the inexorable deadline of aging. On the other hand, new cocoa plantations were still being set up in the late 2000s at the expense of forests—mainly protected ones—alongside the borders of Côte d'Ivoire with Guinea and Ghana, as well as internal border between the Gbagbo regime and the New Forces.

In 2006, the pace of cocoa investments was maintained primarily through the clearing of forests. But we did see some progress in the process of replanting cocoa in Côte d'Ivoire (Table 2.9).

During the 2000s, these innovations originated primarily from the farmers themselves, without any help from research or the use of selected planting material. They were very dependent on labour. Those growers who succeeded in replanting were generally those who planted on small areas, mainly autochthon youth returning from the city because of the lack of employment there, before they moved on to cultivating rubber. They were also those who still had a sufficient labour force to start replanting several times, forced to do so because of recurrent mortality. Most often, these farmers were migrants from the north, mainly Burkinabé.

Also observed was a cocoa dynamic in the forest-savannah contact zone and in the east. This dynamic resulted from a combination of several phenomena, at the intersection of ecological and social change:

- the recovery of secondary forests. Consistent with the great migrations of the 1970s and 1980s and after the plantation fires of 1983, 20 years of respite for this type of emigration region gave a chance for secondary forests to recover. They represented a new potential for cocoa, a new agricultural frontier and new forms of expansion of cocoa at the expense of a reconstituted forest;
- the progressive learning of how to replant cocoa on fallow and old coffee and cocoa plantations;
- the land saturation in Haut- and Bas-Sassandra. Migrants who left for the west reached the 'end of their forest'; there was nothing left to be cleared; the aging

Table 2.9 Distribution of plantations according to their preceding crop

Type of plantation	Fromager region, soils favourable to the cocoa tree (%)—136 plots	All regions (%)—1869 plots
Plantation created after clearing forest (mainly cocoa)	40	59
Plantation created after clearing fallow without trees, type *Chromolaena odorata* (mainly oil palm and rubber, some cocoa)	31	9
Replanting of the same perennial crop (mainly oil palm followed by oil palm)	18	13
Replanting but changing the perennial crop (mainly diversification from coffee to cocoa, then oil palm to rubber)	12	19
Introduction of a new perennial crop interplanted with an existing crop	<1	<1

Sources Author's survey on behalf of the EU, 2006

and mortality of cocoa was impoverishing them. These migrants were attracted by the forests which had partly recovered in their villages of origin;
- the Ivorian crisis, uncertainty and land-tenure security. Migrants reverted to being autochthons with increased land-tenure security and contributed to the restart of a cocoa cycle. The events of 2002–2011 have increased the role of this 'risk' component;
- rubber as an alternative and with competition still limited in the 2000s. It is mainly cultivated at the expense of coffee.

2.4.2 The Decline of the Coffee Sector

The decline in investments in coffee cultivation was sharper and occurred earlier: abandonment of plantations and felling and substitution by oil palm, cocoa, rubber, cashew, or even kola.

Where the soils were not—or no longer—suitable for cocoa and where rubber had not yet arrived, farmers were forced to keep growing a little coffee, even if they had temporarily abandoned its cultivation. When the price rose, as happened in 2005–2006, they made an effort to harvest a little more. We have also observed cases where the coffee plant is used to provide shade temporarily for replanting cocoa, as in the Abengourou *département*. But on the whole, the country's coffee plantations are gradually becoming placeholders for rubber. As for coffee cultivation, the primary determinant of its decline is the price-income ratio. It is much too low given the high labour requirements, mainly during harvest time.

2.4.3 *Rubber: An 'Employee Farmer' and Easy Replanting on Degraded Soils*

'Growing rubber is just like being a government employee.' One of the primary benefits of cultivating rubber is the quasi-monthly regularity of income.

The difficulties of replanting cocoa after deforestation vary from region to region. It is particularly difficult in Moyen-Cavally and Bas-Sassandra whose soils agronomists have already found to be unsuitable for the cocoa tree. For these soils at the boundaries of the cocoa ecotype, deforestation and loss of forest rents are particularly difficult to overcome. Growers in these regions cite these difficulties more than 50 % of the time against only 25 % in the overall forest area (Table 2.10).

Replanting decisions are also influenced by the alternatives available. If we can draw a parallel with the popular saying, 'Give a dog a bad name and hang him', the attractiveness of the rubber sector can lead farmers to exaggerate the effects of cocoa mortality in a very relevant and precise manner:

- 'the land is not suitable for growing cocoa; it requires frequent pesticide treatment to obtain a good production. I have to treat my cocoa plantation six times a year';
- 'the land in this area (Cavally) is not conducive to cocoa cultivation. Because once it dies, it is difficult to replant';

Table 2.10 Reasons advanced in 2006 by growers (in % for each region) to explain their decision to diversify towards perennial crops

Reasons advanced for choosing to diversify	Zanzan, seriously affected by the 1983 fires (%) (total: 43 plots)	Bas-Sassandra and Cavally (%) (total: 43 plots)	All regions (total: 373 plots)
Price-income factors (mainly for coffee): Price is too low, crop no longer has value or attractive price of the new crop	4	42	51
The labour force no longer wants to work on this crop (mainly coffee, some cocoa)	0	0	10
Mortality of the initial crop, unsuitability of the soil for this crop (mainly cocoa)	84	53	27
Other reasons (decision to diversify, imitation effect, advice, experiment)	12	5	12

Sources Author's survey, 2006

- 'it is cocoa's mortality and the advent of rubber in the area that has led to the clearing of protected forests' (according to growers in San Pedro and Cavally 2011).

One interpretation of these statements is that the arrival of the rubber has ended up by discouraging farmers from replanting cocoa. Outside protected forests, many farmers have benefited from the rubber cultivation projects of private companies. Assistance provided to them in the form of training and planting material was decisive in the early adopters' decision to diversify into rubber cultivation. At the same time and irrespective of any opportunities, the reasons cited by farmers and the way they express themselves are clear: deforestation and soil degradation clearly remain the primary structural factors for diversification into rubber.

A reference book on the agronomy of rubber, published 60 years earlier by Dijkman in Indonesia, is in good resonance with farmer observations in Côte d'Ivoire: 'The good adaptability of *H[evea] brasiliensis* grown in Indonesia has saved many plantations that were financially weak, due to soil conditions which were unfavourable for the production of other competing crops (coffee, cocoa and tobacco). We found, for example, that at the same altitude above sea level in Java, coffee is planted on the best soils and the rubber on the worst. In Sumatra, tobacco plantations in Deli are on the best soils and, again, rubber plantations are on the worst. Because of its adaptability, rubber was more or less considered a weed with relatively low soil requirements. However, it would be wrong to assume that rubber cannot be grown on better soils. Indeed, these examples only indicate that rubber is a crop better adapted to poor soils as compared to coffee, tobacco or rice' (Djkman 1951, 17).

In addition to this structural ecological determinant, the increase in producer prices in recent years have played a role in every sense of the term: the economic success of the first rubber farmers have created ripples, partly at the expense of oil palm.

2.4.4 Difficulties of the Oil Palm Sector

After the success of the two major oil palm plans, liberalization led to an interesting plantation dynamic with a third wave of investments in the sector during the 1990s (Cheyns et al. 2001; Cheyns and Rafflegeau 2005; Naï Naï et al. 2000; Chap. 4 of this volume). The difficulties experienced by the sector were largely confirmed at the national scale by the 2006 survey, mainly because of falling prices. This observation was evidenced by the gradual disappearance of many services formerly provided by Palmindustrie: most notably, transport was less well provided for by the companies that bought up the processing plants. Oil palm cultivation seemed to be in trouble in some regions, encountering stiff competition from rubber whose profitability was clearly visible in the economic success of farmers cultivating it.

However in some cases, as in Sud-Bandama, the challenge of oil palm cultivation was taken on by the 'oil and soap' artisanal sector. The growth in national and regional demand helped offset the decline in exports.

2.4.5 The Dynamics of the Cashew Sector

After the fires of 1983, when the forest could not recover in several areas where forests meet savannah, it became difficult to replant cocoa (and even coffee). The dynamics of cashew then held sway. The cashew tree is known for its resistance to fire. It is often planted without help of support structures, and does not require special planting material. (In any case, there is no specially selected planting material available.) These cashew dynamics constitute another example of ecological change as the primary structural driver of diversification.

2.5 Boserupian Diversifications

Can we interpret the diversification of coffee- and cocoa-cultivating regions towards oil palm, rubber and pineapple as a type of Boserupian innovation? The introductory chapter discussed the Boserup versus Malthus debate and its application to plantation economies. The history of cocoa cultivation, especially in Côte d'Ivoire, teaches us that the most common solution consisted of abandoning the cocoa plantation and establishing another through migration to a new agricultural frontier area, where the process of forest clearing could once again be taken up (Léna 1979; Ruf 1988, 1995; Schwartz 1979, 1993; Chauveau and Léonard 1996). This is, at least in part, the neo-Malthusian process.

If peasant societies—undergoing high population growth as a result of massive migrations—are faced with the deterioration of their environment, their production and their income and thus try innovations on site, a Boserupian process results. Applying the rigorousness of the Boserupian approach of the 1960–1970 period, these innovations involve an increase in labour time, consistent with population growth. In a modernized interpretation of Boserupian theories, this increase in labour is not indispensable; innovation can involve instead an increase in inputs and capital.

The theory of diversification, seen as Boserupian innovation, can a priori be called into question if diversification results in an extensification, in the sense of falling incomes and labour per unit surface area. For example, oil palm cropping systems sometimes have a reputation for not being labour intensive. Furthermore, if there is a change of actors, for example with a growing presence of UMCIs in rubber cultivation and evictions of farmers, the diversification process will be different; it will have a dominating political dimension.

To address these issues, we first examine three attempts—over a period spanning from the 1960s to the late 2000s—to assess labour time and plantation budgets in

Table 2.11 Comparative income per crop, per hectare and per day of labour in 1981

Period	Criteria	Oil palm		Traditional coffee	Traditional cocoa	Rubber
	Yield (kg/ha)	6,000	11,500	350	300	1,500
	Number of days of labour	42	52	70	35	81
1966–1970	Price FCFA/kg	4	4	90	70	–
	Gross income (FCFA/ha)	24,000	46,000	31,500	21,000	–
	Income per day (FCFA/day)	571	885	450	600	–
1976–1980	Price FCFA/kg	10	10	250	250	130
	Gross income (FCFA/ha)	60,000	115,000	87,00	75,000	195,000
	Net income (FCFA/ha)	53,000	88,850	81,000	74,000	186,805
	Income per day of labour (FCFA/day)	1,262	1,709	1,157	2,114	2,306

Sources based on Colin 1990a, b; Hermann 1981; compiled from various sources

Côte d'Ivoire (Tables 2.11, 2.12 and 2.13). These are budgets of plantations which are in the production phase. In all the cases, the decision to fell an old cocoa tree and replant another cocoa tree or replace it with another perennial crop requires an investment of labour and capital—which is already in itself a Boserupian process.

As regards the production phase, the three estimates highlight the process of the increase of labour required when transitioning from cocoa to rubber, or to oil palm

Table 2.12 Gross and net income per hectare and per crop, at Djimini-Koffikro, in 1983–1984

Crops	Yield kg/ha	Price/kg	Per-hectare income		Income per labour day	
			Gross income (FCFA/ha)	Net income (FCFA/ha)[a]	Number of days of labour	Net income/days of labour (FCFA/day)
Oil palm Djimini	9,600	15	144,000	128,000	51	2,509
Coffee Djimini	49	350	17,000	17,000	17	1,000
Cocoa Djimini	63	350	22,000	22,000	13	1,700
Rubber	1,750	240	420,000	343,000	69	4,971
Pineapple (tinned) Djimini	60,000[b]	60,000[b]	780,000[b]	280,000[b]	418[b]	670[b]
Pineapple (export) Djimini	47,000	47,000	1,175,000	750,000	332	2,259

Source Colin 1990a, b

Revenue is presented on an annual basis for perennial crops and on the basis of the duration of the cycle for pineapple (19 months)

[a]Without taking labour costs into account

[b]for a crop grown normally, with proper maintenance and without diversion of inputs

Table 2.13 Estimation of production, labour and labour productivity during the course of the life cycle of a cocoa plantation, replanted with rubber in the 1930–1935 period and from 2008 to 2010

Criteria	Adult cocoa	Aging cocoa	Cocoa affected by mortality	Adult rubber
Age of the plantation (years)	15	25	30–35	40–45
Production of cocoa or rubber (kg/ha)	700	300	150	1,500
Number of labour days	80	45	35	72
Economic conditions in 2008				
Price of cocoa or dry rubber (FCFA/kg)	450	450	450	650
Value realized from one labour day	3,188	2,556	1,643	13,889
Economic conditions in 2009				
Price of cocoa or dry rubber (FCFA/kg)	900	900	900	450
Value realized from one labour day	7,125	5,556	3,571	9,444
Economic conditions in 2010				
Price of cocoa or dry rubber (FCFA/kg)	800	800	800	1,050
Value realized from one labour day	6,250	4,889	3,143	21,319
Economic conditions in 2010 (before tax)				
Price of cocoa or dry rubber (FCFA/kg)	1,150	1,150	1,150	1,040
Value realized from one labour day	9,313	7,222	4,643	21,111

Source Ruf 2012
With the assumption of dry rubber yield of 1500 kg/ha equivalent to 2500 kg of fresh or wet rubber, as cup lump

and, *a fortiori,* to a plot of pineapple. There is sometimes a decrease of gross and net production per hectare when switching to oil palm. But here too, this decline is observed only during a comparison with a traditional plantation in good condition (Table 2.11) or one which is very well maintained (Table 2.12). This is even more so when comparison is made not to a 'traditional' cocoa plantation meeting agreed standards but to an aging cocoa plantation affected by mortality and falling production, in other words, to a cocoa plantation as it appears when the farmer must make the decision to replant or diversify (Table 2.13).

However, the diversification from coffee—which is very labour-intensive—to other crops, including to cocoa, is much less a Boserupian process. It is a matter basically of a market and price mechanism in an environment of low technology use, especially due to the closure of centres producing coffee planting material. It is also a matter of mediocre productivity per hectare and per labour day in Côte d'Ivoire.

This reference to technology and productivity is not limited to coffee; it applies to rubber too. In 2010, the increase in world rubber prices and the subsequent high prices paid to farmers led to labour productivity for clonal rubber that was 3 times higher than for a well-maintained adult cocoa plantation, and 7 times that of an aging cocoa plantation. Even when the effect of taxation was removed, which strongly penalized cocoa farmers, the ratio still ranged between 2 and 5.

These ratios increased further in 2011 thanks to still-higher prices. Of course, these high prices were not sustainable and, of course, at the beginning, rubber cultivation in villages obtained high yields because of very close technical assistance from private rubber companies. But over the medium term, we can probably make the assumption of a shortening of the economic life of a rubber plantation and a possible decline in average yields over the years, especially due to over-tapping, the falling over of rubber trees sensitive to high winds and the inevitable decline in the level of supervision given the growing demand for the crop.

Nevertheless, in 2010–2011, we observed instead an opposite trend with yields often reaching between 3000 and 3500 kg/ha in cup lumps or around 2500 kg/ha for dry rubber (Chaps. 7 and 8). These observations on the current absolute supremacy of the productivity of rubber in Côte d'Ivoire lead us to the notion of rent.

2.6 Diversification Towards Rubber, an Expression of New 'Rents'

Without formalizing them, and despite the collapse of the world price of 2013, we propose some avenues for reflection in this section on the nature of the 'rubber rents' and on their evolution.

2.6.1 New Crop Rent

As a new crop, rubber has not yet attracted major diseases and pests. Just like the cycles of cocoa, we can assume that over the years, insects, fungi or viruses will eventually adapt to this new crop and increase the risks and costs of cultivating it.

2.6.2 Project Rent and Scale

As a new crop, rubber cultivation in villages initially benefitted from financing and a very high quality of supervision and advisory support. Today, the plantations which have been established are losing in quality and therefore in future productivity as a price to pay for massive expansion of the crop.

2.6.3 Cocoa-Taxation Rent

It is said that the heavy taxation on cocoa penalizes it and thus benefits rubber. But, as we have seen, even if the cocoa tax is removed, rubber retains a net advantage in income. Nevertheless, this historical tax on cocoa is helping attract farmers to rubber. A gradual revision of this tax could reduce some of this 'politico-historical rent'. But the fact that influential UMCIs and senior government bureaucrats have invested heavily in rubber will not favour a rebalancing anytime soon.

2.6.4 Rent of the Suitability of the Crop to Transformed Soils

Basically, we have seen that rubber grows and is productive in degraded and acidified soils. It is as if cocoa consumes the forest rent and prepares the ground for rubber, at least in areas with soils that are unsuitable for cocoa. Of course, the cocoa farmers reacted and began to apply fertilizer when their incomes allowed them to, but a 'competing' tree that can produce without fertilizer is in itself a form of rent. And in all the regions studied, we saw that more than prices and projects, it is the problem of soil degradation and of replanting cocoa that is the main driver of diversification, especially towards rubber. This component of the rubber rent, a form of physiological rent, will be very difficult to overcome by the proponents of the cocoa sector, especially in the current cocoa-growing belt. Once deforested, the diagnosis of this region's potential becomes identical to that advanced by agronomists in 1970: the region's soils are not suitable for cocoa cultivation.

2.6.5 Rent of Technology and Technological Progress

To a large extent, the spectacular productivity of rubber is due to technological progress, especially to the creation of high-yield clones. In Asia, the continent of choice for rubber cultivation during the 20th century, the introduction of clones led to an overall increase in yields and labour productivity by factors of 3–4 (Chap. 7). Countries like Côte d'Ivoire have directly benefited from these Asian experiences. Of course, there is also a physiological component. Unlike cocoa, oil palm or coffee, the rubber product is not a fruit. With the current state of technology, when one doubles the yield per hectare of cocoa or coffee, one certainly improves the productivity of the labour day—but we cannot double it. Indeed, a part of the harvest and post-harvest work remains proportional to the number of cocoa pods or coffee beans. But, on the other hand, for rubber, whether we fill a latex cup only to one-third or to the brim does not change the work of the incision on the rubber tree. Moreover, whereas for cocoa cultivation, clonal work is in its infancy, we can consider that the rubber tree-capital has a technological lead, thus generating a form of rent. But for how long? It is evident that without mechanization of the harvest, it is probably not easy to reproduce the impact of technical progress seen on rubber cultivation in terms of labour productivity on trees exploited for their fruit.

The technological advances in the rubber sector and, to some extent, in the oil palm sector illustrate a Schumpeterian view of the economy. The evolution of the Ivorian plantation's economy towards diversification is partly the result of successful innovation in the rubber and oil palm sectors. We can also mention a 'capitalization rent' in the Ricardian sense of the term. We will return to this notion in Chap. 7.

2.6.6 Labour Rent

In Côte d'Ivoire, the gap between the economic performance of rubber plantations and that of cocoa plantations is also explained by the inertia of labour costs generated by lower income from cocoa. In fact, cocoa is still—by far—the dominant crop and thus always plays a role in determining labour costs. Farmers who cultivate rubber benefit directly from this inertia.

In an agricultural economy still dominated by cocoa cultivation, a cocoa plantation owner offering up his plantation for sharecropping still retains two-thirds of the gross income. Taking into account his portion of the cost of inputs and excluding depreciation of the investment in the plantation, his share can be estimated at 55–60 % against 40–45 % for the sharecropper.

In southern Thailand, where the agricultural economy is dominated by rubber cultivation, the ratio is similar: the tapper sharecropper receives between 40 and 50 % of the production.

In Côte d'Ivoire, the proprietor of a rubber plantation made an average of 400 FCFA/kg of wet rubber in 2008, 270 FCFA in 2009, 555 FCFA in 2010 and 800 FCFA in 2011 under the same conditions and without taking depreciation into account. Until 2010, he paid his tapper between 35 and 50 FCFA/kg. During the same period, the cost of production hovered around 70 FCFA/kg taking into account the costs of maintenance, weeding and inputs. A glance at the comparison drawn with Thailand—covered in greater detail in Chap. 7—shows that the ratios of rubber prices paid to farmers and labour costs are revealing (Table 2.14).

After the continued increase in the price of rubber in 2011—it reached 800 FCFA/kg of wet rubber—some tappers earned 100 FCFA/kg, taking the production cost to around 150 FCFA/kg. Even then, the P/C ratio remained greater than 6.

Table 2.14 Ratio of prices and labour cost for rubber cultivation in 2010

	Côte d'Ivoire (FCFA)	South-Thailand (Baht)
Price paid to rubber producers (P)	550	100
Labour cost per kg (C)	70	50
P/C ratio	7.8	2

Without its partial indexation to the price of cocoa, the cost of labour would better follow the progression of rubber prices. Villages where rubber has become the dominant crop have seen an increase in labour costs (to around 100 FCFA/kg). But this increase is still modest. This is one explanation of the origins of rubber rent in Côte d'Ivoire. The regularity of income, which allows the rubber grower to think of himself as a government employee, reinforces this notion of rent.

In the end, even though the notion of 'rubber rent' remains to be explored in depth at the theoretical level, it is already well established as the driver of growth of rubber cultivation. It therefore seems very hard to reject the principle of a rent in explaining the very high incomes of Ivorian rubber grower. The role of 'rubber rent' is more understandable when considered as a factor for attracting UMCIs and retirees to agriculture in recent years. They invest little or nothing in cocoa plantations but do so heavily for rubber. Not surprisingly, more and more farmers are following the same path.

2.7 Dynamics of the Actors Involved by Type of Farm

2.7.1 Rubber Estates

Cultivation of *Hevea brasiliensis* was introduced to Côte d'Ivoire in the form of rubber estates during the 1950–1960 period. For several years, the villagers remained indifferent, convinced that rubber was a preserve of the 'rich'. Moreover, some villagers still remembered that their parents had been subjected to forced labour to collect wild rubber and thus chose to keep away from this crop. And, finally, for several years, rubber estate owners preferred to retain a monopoly over rubber cultivation and the traceability of their rubber, especially to reduce the risk of theft of production. Over the years, the situation changed and these estate owners understood that it was in their interest to encourage not only farmers but their own senior and mid-rank staff and others from the liberal professionals to engage in rubber cultivation.

This presence of the rubber agro-industry in Côte d'Ivoire in the form of a few estates ultimately proved to be an advantage for building a network of monitors. It also helped in the provision of advice, technical support and other services, most notably the supply of high-yield planting material to farmers. The presence of rubber estates is an advantage over the cocoa agro-industry, which has factories at the ports but no estates.

2.7.2 Private Plantations: UMCIs and Villagers

In 2012, there were no statistics to differentiate the production by UMCIs from that by villagers. The statistics provided by Apromac—the Association of Natural Rubber Professionals in Côte d'Ivoire—differentiate only between factory plantations and private plantations. The latter accounted for 42 % of national supply in 2003, a share that increased to 62 % in 2007. From the time of the 2002 politico-military crisis and especially from 2005 onwards, high-ranking civilians and military officers close to the Gbagbo regime invested heavily in rubber (Ruf 2011a). But most of these plantations began production only from 2010–2011. In the late 2000s, 'private' entities producing rubber were still mostly villagers. But who were these villagers?

2.7.3 Autochthons and Migrants

Initially, the rubber agro-industry favoured the autochthons for several reasons. On the one hand, in cases where agro-industrial estates were set up on land on which villagers claimed rights, it was in the interests of the plantations' owners to resolve conflicts by helping villagers plant rubber trees. On the other hand, when agricultural projects which were funded by international donors came up, one of their stated objectives was to reduce the risk of abandonment of rubber plantations due to land disputes. Rightly or wrongly, the autochthons were considered more reliable on this point, as compared to migrants who negotiated access to land with their autochthonous '*tuteurs*'.[2] And, more practically, autochthonous villages are often located closer to roads than are migrant camps. This facilitates contact with their residents and the future cultivation of and trade in rubber. Finally, autochthons grow their food crops in rotation with fallow and thus have recent fallow plots available which are particularly favourable for rubber cultivation.

For their part, the autochthons attempted to become part of village rubber cultivation projects. By claiming to be the holders of land rights and having already sold a lot of land to migrant cocoa farmers, these autochthons saw an opportunity to get back at the latter by adopting rubber cultivation. They tried to slow the adoption of rubber cultivation by migrants; partly to keep or recover the land, to extract a kind of 'land or *tutorat*' rent. According to Colin (2008), the conversion to rubber

[2]Traditionally, for an outsider to the community, land access was part of a broader process of integration into the community, by way of a '*tutorat*' relationship (Chauveau 2006). Through this patronage relationship, an autochthon (the *tuteur*) granted to a migrant rights to land (on uncleared forest taken from the village or lineage land reserve), according to a principle of moral economy, namely, all individuals should get access to the resources necessary for their subsistence. The migrant (or his heirs), in turn, had a 'duty of gratitude' towards his *tuteur* (or the latter's heirs), expressed through everyday civilities and through the offering of gifts after harvests and on important social occasions, such as funerals (Chauveau 2006; Colin 2008).

cultivation is indeed a source of contestation of migrants' land rights, in particular by the heirs of autochthons. In fact, the most intelligent autochthons no longer sell; they instead propose some form of 'planting-sharing' of rubber. This is an arrangement between an owner turning over his land to a lessee who clears it and grows rubber. Both partners then share the plantation when it goes into production (Colin and Ruf 2011).

Nevertheless, an ever increasing number of migrants are diversifying into rubber by clearing coffee plants and cocoa trees, but without making the autochthons really assert their control over the land. In fact, even with rubber, the autochthons remain susceptible to quick cash gains through land sales. For example, in one of the first villages which benefited from the support of an agro-industrial rubber company in the form of a gift of selected planting material, the autochthons sold two-thirds of the plants to migrants. In fact, since the autochthons have sold forest lands in large quantities to migrants, they no longer possessed enough land and labour to plant everything. Since the 2000s, it is often migrants who become proponents of adopting rubber cultivation in the villages of the south-west (Fig. 2.19).

Migrants represent two-thirds of cocoa farmers in the country and over 90 % in Bas- Sassandra. After the autochthons, it is their entry into the rubber sector that is behind the country's rubber boom. Finally, the change of generation among the autochthons and migrants is also a factor in accelerating the felling of old cocoa trees and their replacement by rubber.

We will limit ourselves here to illustrating this through a plot map in the Agboville region, near Abidjan. Figure 2.20 shows a block of more than 20 acres which belonged to an autochthon of the area. This block consisted of cocoa trees, coffee plants and forests. Today the coffee plants have disappeared. The block has diversified into old and new cocoa plantations, recent rubber plantations and fallow land on which food crops rotate. The block is also fragmented in terms of usage and ownership, resulting from the division of the property through inheritance in favour of the original owner's three sons and the sale to a Senoufo migrant. This migrant first purchased part of the land from the father, and subsequently some more from the inheriting sons. Two

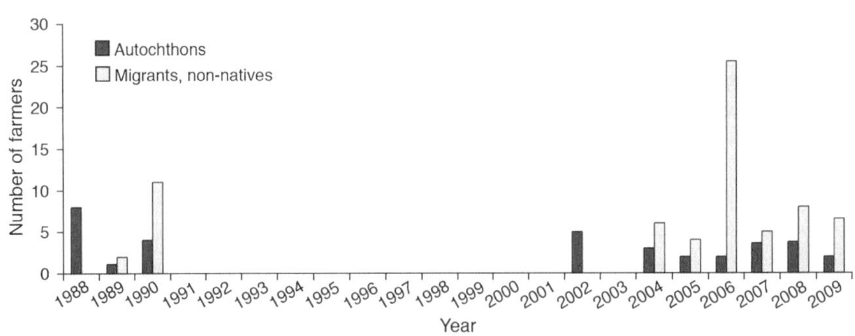

Fig. 2.19 Changes in the place of autochthons and migrants in the process of diversifying into rubber cultivation in 1988–2009. (Data collected in one village and from 41 farmers in the *département* of San Pedro. *Sources* Author's survey, 2011)

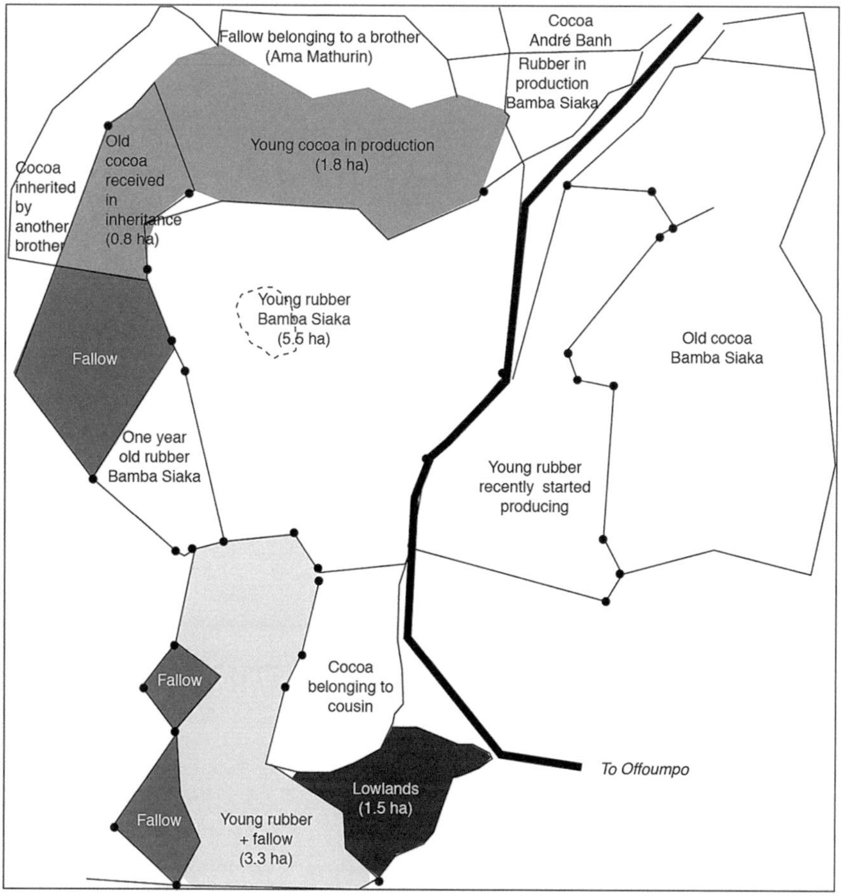

Fig. 2.20 Map of a block of plots dominated by cocoa trees, coffee plants and forests in the Agboville region

of the three sons sold a major part of their inheritances. Only one of the three, Ama, continued with a farm of about 8 ha by jumping on the rubber-cultivation bandwagon. Bamba, the migrant farmer is currently reconstituting a block of more than 20 ha. He is gradually adding to his land by purchasing plots without giving up cocoa cultivation, but he now accords priority to rubber, partially through the clearing of old cocoa trees which he buys and replants with rubber (Fig. 2.21).

This process of felling-replanting-diversifying very often accompanies a change of generation, inheritance or transfer through the land market. Interactions also take place between UMCI farmers and villagers. Even though Bamba, a village farmer and Senoufo immigrant, plays well the role of rubber-adoption pioneer amongst the villagers, it is partly thanks to information he obtains from a manager in charge of 100 ha of rubber plantations (Fig. 2.21). This sharing of information is facilitated by their common Senoufo origin.

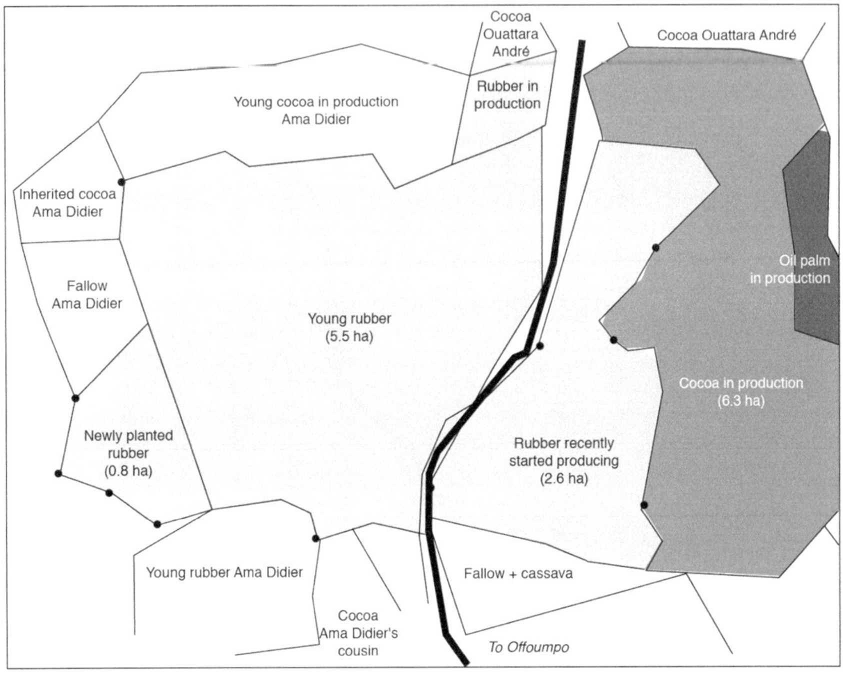

Fig. 2.21 Change in the block of plots after the process of felling-replanting-diversifying

2.7.4 *Retirees, Young People and Changes in Generations*

The villager-UMCI and autochthon-migrant dualism is complicated by changes in generations. Thus, many retirees from relatively modest professions, with available retirement benefits or capital, become villagers specifically to invest in rubber cultivation.

As far as young people are concerned, even if they start off with a small income or capital, often generated by an inherited plot of cocoa or by a first young cocoa plantation, their main goal remains to plant rubber as soon as possible.

2.7.5 *Men and Women*

Slowly but steadily, African women are gaining access to perennial crops. In Côte d'Ivoire, we have observed that projects developed by the private sector draw a much more encouraging reaction from women than from men. Most notably, a project set up by the Mars company and one from the World Cocoa Foundation (WCF) have helped to create or expand their cocoa plantations (Tables 2.15 and 2.16).

Table 2.15 Women growers and membership of the WCF's Cocoa Livelihoods Programme (CLP) project

Village	Project membership	Non-project	Total
Djangobo	16	0	16
Campement Bernard	8	0	8
Total	24	0	24

Table 2.16 Men growers and membership of the WCF's Cocoa Livelihoods Programme (CLP) project

Village	Project membership	Non-project	Total
Djangobo	14	21	35
Campement bernard	19	12	31
Total	33	33	66

2.8 Conclusion

Despite very high taxes on the cocoa sector, the very low prices that reigned between 2004 and 2008, a drastic reduction of extension services available to farmers and in spite of the alarm sounded by the chocolate industry, cocoa production in Côte d'Ivoire still remains very high: over 1.2 million tonnes were being produced annually at the start of the 2010 decade, even exceeding 1.5 million tonnes in some years, including in 2013–2014. This dynamic can be explained to a small extent by the desire of farmers to protect their cocoa heritage by the application of insecticides and a little fertilizer. But it is mainly due to the adoption of the universal model of cocoa economies: new migrations leading to massive clearing of protected forests. This happened in the forests in the west of Côte d'Ivoire, from where more than 300,000 tonnes of cocoa now come. A less significant factor has been the return to secondary forest and fallow to the east, as well as to the forest-savannah boundary in the country's centre.

Thus, in the early 2010s, the diversification/conversion processes undertaken in the old cocoa growing areas did not result in a statistical decline at the national level. The decline in the old regions did not appear in national statistics due to the boom in the new regions. This was, once again, the 'migration/clearing of forests' model at work, unfortunately mainly at the expense of protected forests and national parks.

In the old cocoa growing areas, smallholders were practically left to fend for themselves, without access to advice or technical assistance. Recently, the 'cocoa certification' programme, launched by the chocolate industry, has reintroduced some training and support. This movement has spread rapidly. In 2010, less than 5 % of the farmers had received certification but by 2013–2014, this figure had probably grown to 50 %. But there have been very many irregularities, with many middlemen and traders converting themselves into 'cooperatives' and certifying growers who do not even know they are certified and thus have never received their certification premiums. More fundamentally, the techniques that have been disseminated are of uncertain and varying effectiveness (Ruf et al. 2013; Ruf and Bourgeois 2014).

Until 2005–2006, in all these old cocoa growing regions, investments in new cocoa plantations continued despite the arrival of rubber. Farmers wishing to adopt rubber first replaced their fallow lands and old coffee plants with it—which is what led to the accelerated decline of coffee in the 2000s. Only then did cocoa come into the crosshairs. Since 2009, the lure of rubber cultivation has increasingly gone hand in hand with lower cocoa yields, the fight against insect damage and diseases, and difficulties of replanting. It has led to the clearing of cocoa plantations or the cultivation of rubber in the understory of old cocoa trees waiting to be felled.

The rise of rubber cultivation, the most dramatic diversification of the early 2010s in Côte d'Ivoire, is thus explained by multiple factors which correspond well to the proposed diversification model: (a) ecological change, depletion of soils, pest and disease constraints, leading to a Boserupian mechanism of crop diversification; (b) appeal of the market with prices that had once again become very attractive, at least until 2012, and the regularity of income during the year; (c) economic policies that have largely favoured rubber cultivation, especially through the differential in taxation; and (d) opportunities offered by projects supported by the rubber agro-industry. These four major factors have clearly driven the diversification process at the national level (Chap. 7).

The social dimension of this dynamic is represented by the transition in a big way of 'established migrant farmers' (who had migrated in previous decades for planting cocoa) into rubber cultivation after overcoming the autochthons' resistance. New actors such as UMCIs, their farm managers and labourers have also played a major role in this diversification.

Land policy and the Rural Land Law of 1998 also contributed, indirectly, to the accelerated adoption of rubber since 2006. Migrants, threatened by the autochthons supported by the Gbagbo regime, feared that their fallow land would be seized and resold. Paradoxically, this land risk stimulated investment in tree crops and this is one of the reasons why migrants have shifted to cultivating rubber, despite resistance encountered from the autochthons. It is a classic paradox of the family plantation economy, at least in West Africa: a plantation or a perennial crop marks a certain right to the land and can therefore help remove ambiguity of the property's ownership. This principle was first supported by the policy of Houphouët-Boigny, way back in the 1960s, and the principle of 'the land belongs to those who work it' greatly enhanced the growth of cocoa. Even without this political factor, a perennial crop remains a property marker, more difficult to challenge than an apparently uncultivated land,[3] even if it is being used to grow seasonal food crops in rotation.[4]

Will the fall of the Gbagbo regime in 2011, generally pro-autochthon (at least in its pronouncements), and the return of a generally pro-migrant government change the dynamics of the 2010 decade? Initial observations in 2013 suggest a further acceleration of rubber plantations by migrants, at least in the villages which had 'politico-autochthonous' blocks in place, for example, near Issia, where the former Minister of the Interior owned a large rubber plantation and enforced a ban on planting by non-autochthons. At the same time, due to the collapse of world rubber prices in 2013–2014 and an introduction of a new tax by the new regime, the 'rubber rent' has weakened in Côte d'Ivoire. This could moderate enthusiasm for investment in rubber in the days to come.

Nevertheless, the dramatic increase in rubber production, reaching a level of 300,000 tonnes of dry rubber in 2013, and predicted to reach a level of around 600,000 annual tonnes before 2020, has to be seen with respect to the unfolding of the cocoa cycle and the ecological changes it leads to: the soil degradation and the consumption of forest rent (Chaps. 6–9), which are offset by the rubber rent. Good growth of rubber on degraded and acidified soils, as well as its physiology, contribute to its impressive regularity of production and thus of income. They also contribute towards the potential for improving labour productivity, as well as the technological progress acquired with the creation and selection of improved clones over the last few decades. Thus, diversification into rubber in Côte d'Ivoire partly pertains to the technological advance over the cocoa sector (Chap. 7).

The world of cocoa and chocolate is well aware of all of this. In order to slow down the growth of rubber and revive cocoa productivity, a large section of the chocolate industry is today trying to introduce clonal material and fertilizer appropriate for cocoa cultivation. Will this trend lead cocoa farmers only to diversification or to a more gradual but complete conversion over 20 or 30 years? For many farmers, even more than markets and cocoa prices which need a serious stimulus, the answer rests on the ability of research and the chocolate industry to come up with technological advances applicable at low cost to solve the universal problems of cocoa replantation and rehabilitation. Despite significant resources being deployed by the chocolate industry, it remains a difficult challenge. Because of the new 'competitor', dwindling forests and pest pressure on cocoa (insects, diseases), it seems unrealistic to expect to find a solution in just a few years to a centuries-old problem. Some national institutes may also be reluctant to encourage the adoption of clones for fear of the spread of diseases such as swollen shoot, or more prosaically, for fear of losing their monopoly on planting material.

---

[3]This fundamental role of tree crops as markers of land ownership in Côte d'Ivoire has been widely analyzed by many social scientists (Gastellu, Affou Yapi, Schwartz, Lena, Lesourd, Chauveau, Dozon, Colin Léonard among others). Fallows, on the other hand, can be taken back by autochthons. It has also been identified in Ghana. (Goldstein and Udry 2008).

[4]This development is not, however, without dangers to medium-term food security (Chap. 7).

At the economic and institutional levels, the competition between agro-industrial sectors seems to be favourable to farmers in Côte d'Ivoire. In the Ivorian context, characterized by a phase when land, forest and labour resources have started becoming scarce, diversification is compelling the chocolate and cocoa industry to invest in their sector's upstream (supply of inputs and services to farmers) as well as its downstream (intervention with the government to influence price-tax-subsidy policies). It is imperative that these investments also encompass commercial aspects through contracts with farmers that are financially beneficial to them. For example, if the chocolate industry expects positive results from its programme of mass cocoa certification, it must ensure that the certification premiums are actually and regularly paid. And when they are paid, they have to truly reach the farmers without being misappropriated by intermediaries.

This type of support—in substance and not just in words—extended by agro-industry to producers is piquing the interest of the farmers and, perhaps, even providing them with motivation to diversify without resorting to conversion. From this point of view too, diversification is justified.

References

Aguilar P, Paulin D, Keho Y, N'kamleu G, Raillard A, Deheuvels O, Petithuguenin P, Gockowski J (2003) L'évolution des vergers de cacaoyers en Côte d'Ivoire entre 1995 et 2002. In: Proceedings of the 14th international conference on cocoa research. Accra, Ghana. pp. 1–167, 1–175, 18–23 Oct 2003

Chauveau JP (1993) L'innovation cacaoyère. Histoire comparée de la cacaoculture en Afrique de l'Ouest. In: Byé et Muchnick (eds) Innovations et sociétés. Quelles agricultures? Quelles innovations? Inra/Cirad/Orstom, Montpellier 1:43–58

Chauveau JP (2006) How does an institution evolve? Land, politics, intergenerational relations and the institution of the *tutorat* between autochthons and migrant farmers in the Gban region (Côte d'Ivoire). In: Kuba R, Lentz C (eds) Landrights and the politics of belonging in West Africa, Brill Academic Publishers, African Social Studies Series, Leiden, pp 213–240

Chauveau JP, Léonard E (1996) Côte d'Ivoire pioneer fronts: historical and political determinants of the spread of cocoa cultuvation. In: Claence-Smith WG (ed) Cocoa pioneer fronts since 1800: the role of smallholders, planters and merchants. Macmillan, London, pp 176–194

Cheyns E, Kouamé YS, Verdeaux B, Nai Nai S (2001) Adoption du matériel végétal et itinéraires techniques en plantations villageoises de palmier à huile. In: Assamoi Y, Burger K, Nicolas D, Ruf F, De Vernou (eds) Proceedings of the conference on the future of perennial crops, Yamoussoukro. Cirad, BNETD, ESI-VU, Montpellier, 5–9 Nov 2001

Cheyns E, Rafflegeau S (2005) Family agriculture and the sustainable development issue: possible approaches from the African oil palm sector. The example of Ivory Coast and Cameroon. Oléagineux, Corps Gras, Lipides (OCL), 12 (2):111–120

Colin J-Ph (1990a) La mutation d'une économie de plantation en basse Côte d'Ivoire. Paris, ORSTOM, 284 pp. plus annexes

Colin J-Ph (1990b) La dynamique des systèmes productifs agricoles en basse Côte d'Ivoire. In: Dynamique des systèmes agraires. La dimension économique. Orstom, CNRS, Paris, pp 165–190

Colin JP (2008) Étude sur la location et les ventes de terres rurales en Côte d'Ivoire. Rapport 1. Diagnostic des pratiques. Abidjan: République de Côte d'Ivoire, ministère de l'Agriculture, Délégation européenne

Colin JP, Ruf F (2011) Une économie de plantation en devenir. L'essor des contrats de planter-partager comme innovation institutionnelle dans les rapports entre autochtones et étrangers en Côte d'Ivoire. Revue tiers- monde, (207):169–187

Couty P (1989) Risque agricole, périls économiques. In: Eldin M, Milleville P (eds) Le risque en agriculture. Orstom, Paris, pp 561–568

Djkman MJ (1951) Hevea: thirty years of research in the Far East. Coral Gables. University of Miami Press, Miami, Florida

Gastellu JM (1979) Disparition de la main-d'œuvre étrangère. Cahiers du CIRES (23):17–45

Gastellu JM (1982) Une économie du Trésor: les grands planteurs du Moronou. Abidjan: ORSTOM, 119 pp

Goldstein M, Udry C (2008) The profits of power: land rights and agricultural investment in ghana. J Polit Econ 116(6):981–1022

Hermann E (1981) Analysis of selected agricultural parastatals in the Ivory Coast. Analytical study prepared for USAID

IRCC (1979) Étude de la boucle du cacao. Centre de recherche de Côte d'Ivoire. Bingerville, 42 pp

Léna P (1979) Transformation de l'espace rural dans le sud-ouest ivoirien. Thesis in Geography, Orstom, Abidjan, Côte d'Ivoire, 345 pp

Léonard E (1997) La reproduction de la société agraire dans la région de Sassandra. In: Guillaume A, Ibo J, Koffi NG (eds) Croissance démographique, développement agricole et environnement à Sassandra. Orstom, pp 137–160

Léonard E, Vimard P (eds) (2005) Crises et recomposition d'une agriculture pionnière en Côte d'Ivoire. Dynamique démographique et changements économiques dans le bas-sassandra. IRD-Karthala, Paris, 368 pp

Losch B (1990) La filière caoutchouc naturel en Côte d'Ivoire: originalité et enjeux. In: Griffon M (ed) Économie des filières en régions chaudes: formation des prix et échanges agricoles. Actes du Xe Séminaire d'économie et de sociologie, Montpellier, France, Montpellier: Cirad, pp 785–804, 11–15 Sept 1989

Mehra S, Weise S (2007) West Africa strategy, WCF Partnership Meeting Amsterdam, 28 pp. www.worldcocoafoundation.org/who-we-are/partnership-meetings/pdfs/WestAfrica Strategy.pdf. Accessed 23 May 2007

Mollard E (1993) La différenciation spatiale de l'Économie de plantation. Mappemonde 2:45–47

Naï Naï S, Cheyns E, Ruf F (2000) Adoption du palmier à huile en Côte d'Ivoire. OCL 7 (2):155–165

Oswald M (1997) Recomposition d'une société au travers de plusieurs crises: la société rurale Bété (Côte d'Ivoire). Doctoral Thesis, INA-PG, 290 pp + annexes

Ruf F (1984) What intensification for the ivory coast plantation system? History, farming systems and agricultural policy. L'Agron Trop 39(4):367–382

Ruf F (1987) Éléments pour une théorie sur l'agriculture des régions tropicales humides. I - De la forêt, rente différentielle au cacaoyer, capital travail. *Agronomie tropicale*, 42(3):218–232

Ruf F (1988) Stratification sociale en économie de plantation ivoirienne. Thèse de doctorat, université de Paris X, Nanterre, 6 vol 1393 pp

Ruf F (1993) Adoption et abandon du cacaoyer: cycles d'innovations et ajustements. In: Chauveau JP, Yung JM (eds) Innovations et sociétés. Quelles agricultures, quelles sociétés?, Vol 2 Les diversités de l'innovation CIRAD, INRA, ORSTOM. Montpellier, pp 337–346

Ruf F (1995) Booms et crises du cacao: les vestiges de l'or brun. Karthala (ed) Collection Économie et développement, Paris 455 pp

Ruf F (2006a) Éléments d'une stratégie d'amélioration du peuplement végtal des plantations pérennes paysannes en zone forestière de Côte d'Ivoire, vol 2, La demande en matériel végétal et son utilisation par les planteurs. BDPA and Cirad, 220 pp

Ruf F (2011a) Pareto en Côte d'Ivoire. Cycle du caoutchouc et concentration foncière. In: Jul-Larsen E, Laurent PJ, Le Meur PY, Léonard E (eds) Une anthropologie entre pouvoirs et histoire. Conversations autour de l'œuvre de Jean-Pierre Chauveau. Karthala, Paris pp 59–91

Ruf F (2012) L'adoption de l'hévéa en Côte d'Ivoire. Prix, mimétisme, changement écologique et social. Économie rurale, (330–331):103–123

Ruf F, Akpo JL (2008) Étude sur les revenus et les investissements des producteurs de café et de cacao en Côte d'Ivoire. Report to the European Union.. Cardno Agrisystems, Abidjan, Côte d'Ivoire, 90 pp

Ruf F, Bourgeois M (2014) Pénurie de cacao? Agro-industrie et planteurs villageois. www.inter-reseaux.org/ressources-thematiques/les-notes-d-inter-reseaux/article/penurie-de-cacao-agro-industrie-et

Ruf F, N'DaoY, Lemeilleur S (2013) Certification du cacao: stratégie à hauts risques. Inter-réseaux. http://www.inter-reseaux.org/ressources-thematiques/article/forum-certification-du-cacao

Schwartz A (1979) Colonisation agricole spontanée et émergence de nouveaux milieu sociaux dans le sud-ouest ivoirien: l'exemple du canton bakwé dans la sous-préfecture de Soubré. Cahiers ORSTOM, Ser Sc Hum, XVI(1–2):83–101

Schwartz A (1993) Sous-peuplement et développement dans le sud-ouest de la Côte d'Ivoire. Cinq siècles d'histoire économique et sociale. Orstom, Coll. Études et Thèses, Paris, 490 pp

Chapter 3
Coconut Farmers and Lethal Yellowing Disease: A Case Study in Two Villages in Ghana's Central Region

Jean Ollivier, Philippe Courbet and Richard Democrite

The populations of the coastal zone of Ghana gradually adopted the coconut as a commercial crop in the 20th century. Thanks to its adaptability to the environment and its many social benefits, the 'tree of a hundred uses' has become the main crop of this area. Its main advantages are its long life—which allows the farmer to earn income throughout his life and to leave it as an inheritance to his children—and its adaptation to sandy coastal soils which are unsuitable for other perennial crops. The coconut tree's long life facilitates the transmission of the plantation, and thus the land, from one generation to the next. Nevertheless, during the 1970–2000 period, the dynamics of expansion of coconut cultivation along the Ghanaian coast ran into the problems of scarcity of available land and the lethal yellowing disease. This disease has spread in the coastal Central and Western regions[1] of Ghana which together constitute the main area of coconut production.

In Ghana, the lethal yellowing disease of coconut (LYDC) first appeared in 1932 in Cape St Paul in the Volta region in the country's east. It then spread westwards from Cape Three Points in the 1970s. The disease has not stopped expanding since, progressively decimating most of the coconut groves along its path but in quite a random manner (see colour plate III). The Keea and Aak[2] districts in the Central region are the most affected, and to a lesser extent, Mfantsiman district which is located in an agroecological zone less favourable to coconut cultivation. In the Western region, lethal yellowing is mainly prevalent in the Ahanta and Nzema

[1]Ghana is divided into 10 administrative regions.
[2]Keea: Komenda/Edina/Eguafo/Abirem. Aak: Abura/Asebu/Kwamankese.

J. Ollivier (✉)
Cirad-Persyst, UPR Systèmes de Pérennes, Avenue Agropolis,
TA B-34/02 (Bât. 2, Bur. 110), 34398 Montpellier Cedex 5, France
e-mail: jean.ollivier@cirad.fr

P. Courbet
CRD Nestlé Abidjan, ZI Yopougon Abidjan, Ivory Coast
e-mail: courbetp@yahoo.fr

R. Democrite
Chambre D'agriculture de Gaudeloupe Baie Mahault, Gaudeloupe, France

F. Ruf and G. Schroth (eds.), *Economics and Ecology of Diversification*,
DOI 10.1007/978-94-017-7294-5_3

districts. The advancing front of the disease reached the eastern part of the Jomoro district, where coconut remains the main source of income and wealth. Wassa district, more to the interior, is yet free from the disease. In 1997, it was estimated that the destruction encompassed approximately 3500 ha of coconut trees in the Western region and nearly 2000 ha in the Central region (Dery and Philippe 1997). The disease's impact on production and incomes of thousands of farmers has been disastrous and difficult to quantify at a national level. Indeed, the total surface area currently occupied by coconut groves in Ghana is not precisely known. Furthermore, the disease does not propagate evenly, even within the same area (AFD-Iram 2004).

What are the strategies adopted by farmers to cope with LYDC? Are they aware of its consequences? Do they wish to replant coconut or prefer to move to other disease-free or more commercially remunerative perennial crops? What are the strategies that can most contribute to the reconstruction of farms? The objective of the study of two villages was to identify and clarify the strategies of farmers faced with the possibilities of replanting coconut or of diversifying.

3.1 Typology of Farms and Assessment of Farmer Strategies

From surveys conducted in 2006 and 2007 by the Oil Palm Research Institute (Opri) in collaboration with the Centre for International Cooperation in Agronomic Research for Development (CIRAD), a first typology based on the structure of farms and farmers' strategic choices could be established. The surveys covered about 220 farms spread over four districts of the Western region and three districts of the Central region. They exhibited several regional disparities.

This was followed by a more qualitative study focused on two villages in the Central region to assess the strategies of farmers whose plantations were being decimated by the disease (Fig. 3.1). In addition, this study formed part of a CIRAD programme involving several researchers and students, especially in the Western region. We draw on some results of this programme in this chapter.

3.2 Results and Discussion

3.2.1 Issues Confronting Ghanaian Coconut Cultivation

There does not currently exist any economically viable treatment method or even a technique to stop the spread of LYDC. But tests conducted in Ghana show that early eradication of affected trees through felling, as soon as the first symptoms appear, helps to contain the spread of the disease and prevent devastating outbreaks.

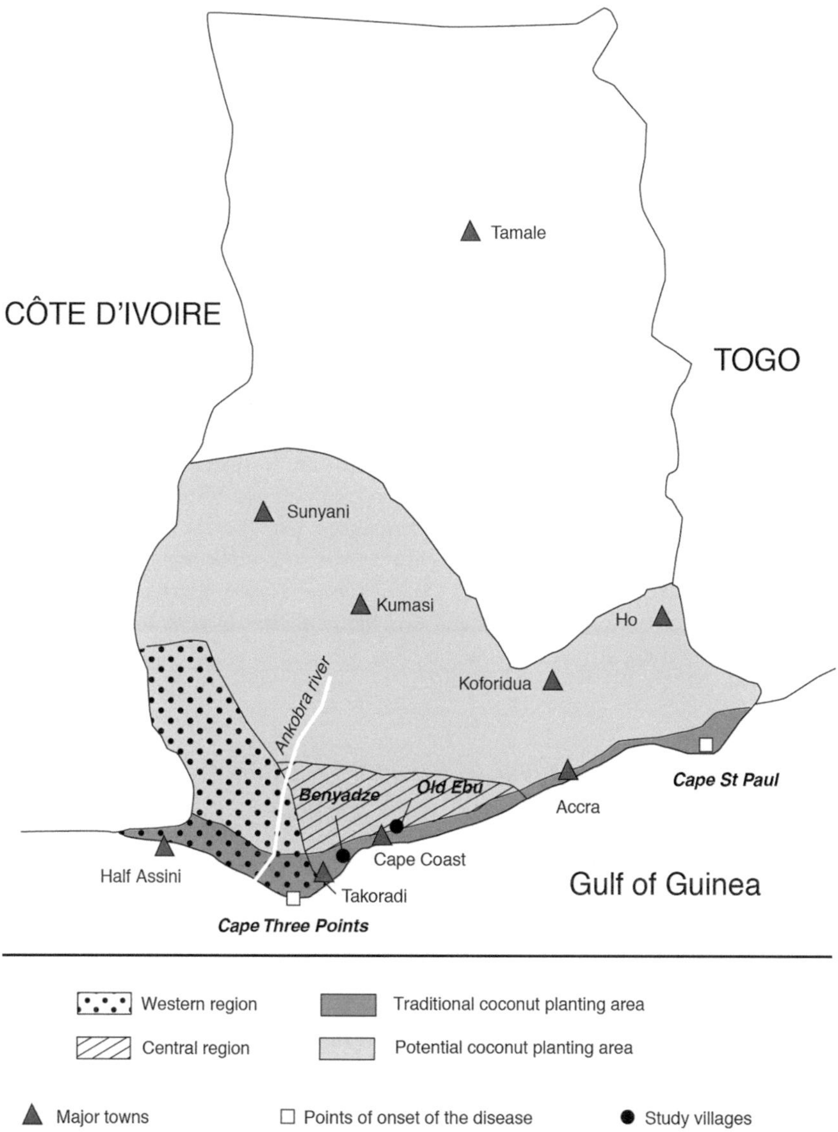

Fig. 3.1 Existing and potential coconut plantation areas and surveyed villages in Ghana

Ultimately, the creation of more or less resistant varieties by country (diversity of phytoplasmas) is the most realistic method of control. No resistant or tolerant variety is currently available in Ghana.

To fight against LYDC, various research efforts have been conducted on varieties and hybrids by the Ministry of Food and Agriculture (MOFA) and Opri. From these experiments, a hybrid MYD × VTT, obtained by crossing Malayan Yellow

Dwarf varieties with Vanuatu Tall, proved interesting. In 1995, tests for adoption of this hybrid were carried out in the field and the findings were encouraging. The MYD × VTT cross represented a new hope for the entire coconut sector. From 1999 to 2004, the Coconut Sector Development Programme (CSDP) funded by the French Development Agency (AFD) aimed to disseminate this hybrid through a technological package (planting material and fertilizer). Nearly 210,000 seedlings were distributed to more than one thousand farmers and 1300 ha were replanted. Unfortunately, this hybrid has belied its promises. Since 2005, some hybrid plantations have been affected by the disease in areas where parasitic pressure is high. Farmers find themselves once again powerless against the disease.

In June 2005, CIRAD and the French Ministry of Foreign Affairs (MAE) set up an institutional support programme financed by the Priority Solidarity Fund (FSP) 'Agronomic research in Ghana in the field of the perennial and food crops'. This project was conducted in scientific partnership with Ghanaian research institutes Opri, the Biotechnology and Nuclear Agriculture Research Institute (BNARI) and the Crops Research Institute (CRI). Its goal was to boost research on the transmission of LYDC and on crop diversification. This project was completed in June 2008. Research continues to test eight new varieties originating from the Marc Delorme Coconut Research Centre in Côte d'Ivoire. A new hybrid, SGD × VTT (Sri Lanka Green Dwarf × Vanuatu Tall), likely to be tolerant to LYDC, was planted in several experimental plots.

3.2.2 Recent Development of Coconut Plantations and Abusa *Institutional Arrangement*

The relatively recent adoption of coconut cultivation is important in order to understand the impact of LYDC on farms. The 1960s saw a rapid development of coconut cultivation in the Western and Central regions, where the humid climate is very conducive to this crop. These regions therefore become the primary centres of Ghanaian production, overtaking the Volta region. Indeed, in this latter region, farmers have replanted coconut trees after their destruction by the disease, but the planted area remains low. On the one hand, confidence in coconut has eroded. On the other, external factors have come into play: non-waterlogged land has become scarce and sea erosion has destroyed many plantations. Above all, however, the dense and growing population has mobilized a large part of the land for housing and vegetable cultivation.

The expansion of coconut cultivation is thus happening in the country's west. On the social level, this expansion often takes the form of institutional arrangements between landowners (often descended from so-called royal families) and migrants or natives who do not have easy access to land. The agreement is a contract resembling sharecropping where the lessee clears and plants. When the plantations start producing, a third of the income accrues to the owner; the lessee keeps the remaining two-thirds. This arrangement continues until the death of the coconut grove. These *abusa* or *abunu* arrangements (the latter is when the sharing is half each) are common in Ghana for perennial crops, including for cocoa (Ruf 2010). It is not strictly speaking sharecropping because the lessee has a right of ownership of the trees planted (Colin and Ruf 2009, 2011). In fact, if the land is sold, the lessee can separately sell the trees as timber.

Due to the long life of the coconut tree and the possibility of replanting before its natural death, the 'pseudo-sharecropper' lessees, initially without access to land, have been able to develop relatively large coconut groves. The coconut tree has therefore led to a certain redistribution of land, and therefore of added value, within the village society. Coconut trees can be transferred by inheritance to the uterine nephew of the *abusa* or *abunu* lessee, but the land can never be sold by the lessee or his heirs. With the increase in population through migration and births, most of the available land has already been planted with coconut trees—with the exception of a few plots which smallholders have reserved for food crops for the family. This is how, in the 1970s, there was no longer any land available for young people wishing to participate in the *abusa* system in the two villages studied. Many young people then left to study or try to earn a living in the cities of the Western region. Some even emigrated to Côte d'Ivoire, mainly to Abidjan which was booming at the time. Others turned to fishing or obtained employment in the local artisanal processing centres (production of oil, alcohol and food).

Nevertheless, the integration of coconut cultivation into the local economic system did create a few jobs and increased the added value produced per hectare of plantation. It thus allowed a growing population to live on a limited area. This integration with coconut cultivation began with the very first plantations. Thus, producers used fibre and palm fronds as fuel for cooking or for smoking fish. They produced oil which they used to fry the fish to be exported; thus increasing the value added to both the fisheries and the coconut sectors. Coconut oil cake was used to feed pigs and, in this way, create additional wealth (Geiger and Lhommet 2005).

This entire economic and social system was shaken by the arrival of LYDC. In addition to the drop in revenues, since a portion of the destroyed plantations were under *abusa* or *abunu* contracts, the death of the coconut trees released landowners of their contractual obligations and they repossessed the plots.

3.2.3 A Rural Economy Undergoing Profound Changes with the Coconut Crisis

Falling prices of copra and coconut oil, lack of credit and *abusa-abunu* arrangements are all obstacles to a rehabilitation of the Ghanaian coconut sector (Dery and Calvez 2002). At present, most of the coconut groves are aging and often located on poor soils. This results in lowered production of copra and fresh nuts, as well as a lack of investment for replanting. Technical itineraries have changed little, with few technical innovations making it from research to production. Dissemination of high-yielding hybrids has been limited, since there is no real guarantee of resistance to the disease. Nut production per tree has fallen sharply.

National agricultural development policies no longer support this sector because of the risk of the disease and competition from more profitable perennial crops (oil palm, cocoa and rubber). In this context of withdrawal of favourable public policies and the ending of AFD's Coconut Sector Development Programme (1999–2004), the village farmers have not picked up the slack. Very little replanting has taken place in the villages on the coast. The tendency is to plant inland on small areas with unimproved or improved planting material (but potentially still susceptible to LYDC).

In this context, there is a gradual substitution of coconut oil with palm oil for food purposes. In addition, the competition from new perennial crops (rubber and oil palm) has led to economic and land-related upheavals in the villages. If alternatives are available, farmers tend to solve the problem of the disease by a complete conversion of the crop, or at least by a process of diversification (Ruf 2006).

Several studies in the country's Western region (Geiger and Lhommet 2005) and its Central region (Democrite 2007) confirm these changes. The dynamic observed in the Western region shows that the actual or anticipated destruction of their coconut plantations leads farmers to look for new perennial crops, not only to ensure their income, but also to secure their land property. Using unselected planting material from existing plantations, oil palm is intercropped with food crops in areas close to the village. Cocoa plantations come up on tree fallows in areas a little farther afield. There is not much replanting with the MYD × VTT hybrid coconut due to the high price of the plants and its susceptibility to the disease. Rubber is cultivated under outgrower agricultural contracts with the Ghana Rubber Estates Ltd (GREL) company (Chap. 7). However, these new plantations require start-up capital. The proximity of an agro-industry in an area affected by the disease allows growers to cultivate a plantation of oil palm or rubber, thanks to credit extended to them.

3.2.4 Strategies Adopted by Producers to Counter LYDC

In the areas affected for a long time by the disease in the Central region, production systems are now diversified. The food sector, very vibrant, meets the demand from cities like Cape Coast, the capital Accra and Kumasi. The Central region is an area of significant production of fruit and cashew. Oil palm cultivation, despite less favourable rainfall conditions than in the Western region, is in expansion. Interest in coconut persists due to demand for nuts for direct consumption with the products finding a good outlet on the domestic markets. In Mfantsiman district, pineapple cultivation predominates in most of the farms surveyed. Even though the agro-climatic conditions for cultivating local coconut are suboptimal, it is still present here. Indeed, LYDC is spreading slowly in this area, perhaps due to the fragmentation of coconut cultivation.

In the Western region, in the districts of Ahanta where the disease is prevalent, there is a strong movement for replacing coconut by oil palm or rubber, the latter partially because of the proximity of GREL. Some farms have quickly converted by intensifying and developing sugarcane cultivation, particularly in the eastern Shama Ahanta district. The dynamic observed in western Ahanta district is similar to that found in eastern Nzema district where oil palm is replacing coconut, as described by Geiger and Lhommet (2005).

Farther inland, as in Wassa district away from the coastal zone and therefore from the disease, the farmers have adopted diversification strategies oriented towards oil palm, cocoa and, for the bigger farms, rubber. Farmers in this area have benefited from the project to develop coconut cultivation and grow this crop, thus confirming the shifting of the coconut production area towards the country's interior (Ruf 2005).

Further west and in the area not yet affected by the disease, especially in Jomoro district, the economy is still closely tied to coconut cultivation and farmers are very dependent on this crop.

Apart from Jomoro district where there really is a continuum of coconut cultivation across the entire coastal strip to the Ivorian border, farms are generally quite diversified and mainly grow perennial crops (hybrid and local coconut, citrus, rubber, oil palm and cocoa). Also found are multiannual crops (sugarcane and pineapple) and annual crops (maize, cassava, peppers, tomatoes and okra). Food and vegetable crops are also very present to meet home consumption needs, with the surplus being sold on local markets or in major consumption areas.

Nine major farm types have been identified on the basis of the dominant crop (Table 3.1).

We are particularly interested in the dynamic observed in Keea district, Central region, to determine the strategies that the farmers have developed. Two villages affected by this disease are taken as examples. The first village is Benyadze, where the first symptoms appeared in 1985. The second is Old Ebu, where the disease broke out in 2000.

Table 3.1 Major farm types on the basis of the dominant crop in the Central region, Ghana

Group	Dominant crop	Districts concerned
1	Rubber	Wassa
2	Cocoa	Wassa, eastern Ahanta, western Ahanta and eastern Nzema
3	Hybrid coconut, food and vegetable crops	All districts except Mfantsiman
4	Oil palm	All districts
5	Sugarcane	Eastern Ahanta
6	Citrus fruit	Aak and Keea
7	Food and vegetable crops	Keea and eastern Ahanta
8	Local coconut	Jomoro, eastern Nzema and Mfantsiman
9	Pineapple	Mfantsiman and Keea

3.3 Case Study in the Central Region: Benyadze and Old Ebu Villages

The historical approach of farming systems helps explain the choices made by farmers to deal with LYDC, irrespective of whether their coconut trees were affected or not. The different strategies adopted by them depended on whether they were able to replant perennial crops and thus on their capital, access to land, location of the plots within the village's territory, their ability to mobilize labour, cost of planting the crop, market opportunities and the presence of any projects proposed by various organizations active in the area.

After the death of coconut trees, replanting coconut trees on the same plot required a transition period to ensure the complete disappearance of the disease and to raise the necessary capital. More generally, the duration of the transition period also depended on the farmers' perception of the resistance or tolerance of new hybrids created by research and on support from processing industries.

As a general rule, farmers did not adopt anticipatory strategies in villages surrounding the outbreaks of the disease to slow its progress. This is partly explained by the lack of communication from research and extension structures on how to control the disease. During the initial phase when the first few cases appeared, the producers had the opportunity to take measures to slow the disease's development. But they did not know how to do so. In addition to this lack of information, the producers were not aware of the seriousness of the situation, given that the damage seemed limited.

On the other hand, they were quick to adapt to business opportunities in the region, especially through specialization in a perennial crop (rubber, oil palm or citrus).

3.3.1 Strategies of Farmers in Benyadze

Before independence in 1957, the majority of farmers in this village farmed small areas mainly devoted to food crops (Fig. 3.2). The food 'gardens' were cultivated for two consecutive years, and then left fallow. In the first year, gardens were used to grow maize and cassava, in the second year only cassava was grown. Faced with the decline in available fallow, some farmers decided to secure their possession of land by venturing into the cultivation of plantain and, later, coconut. From the 1960s, coconut cultivation spread over almost the entire coastal strip for the production of coconut oil.

In 1957, there existed two types of farmers:

- Farmers who belonged to the royal family. They grew maize and cassava in the valleys, vegetable crops, cassava, maize and plantain in the hills, and sugarcane on the floodplains;
- Other families grew maize, cassava, vegetables and plantain in the valleys, and cassava and maize in the hills.

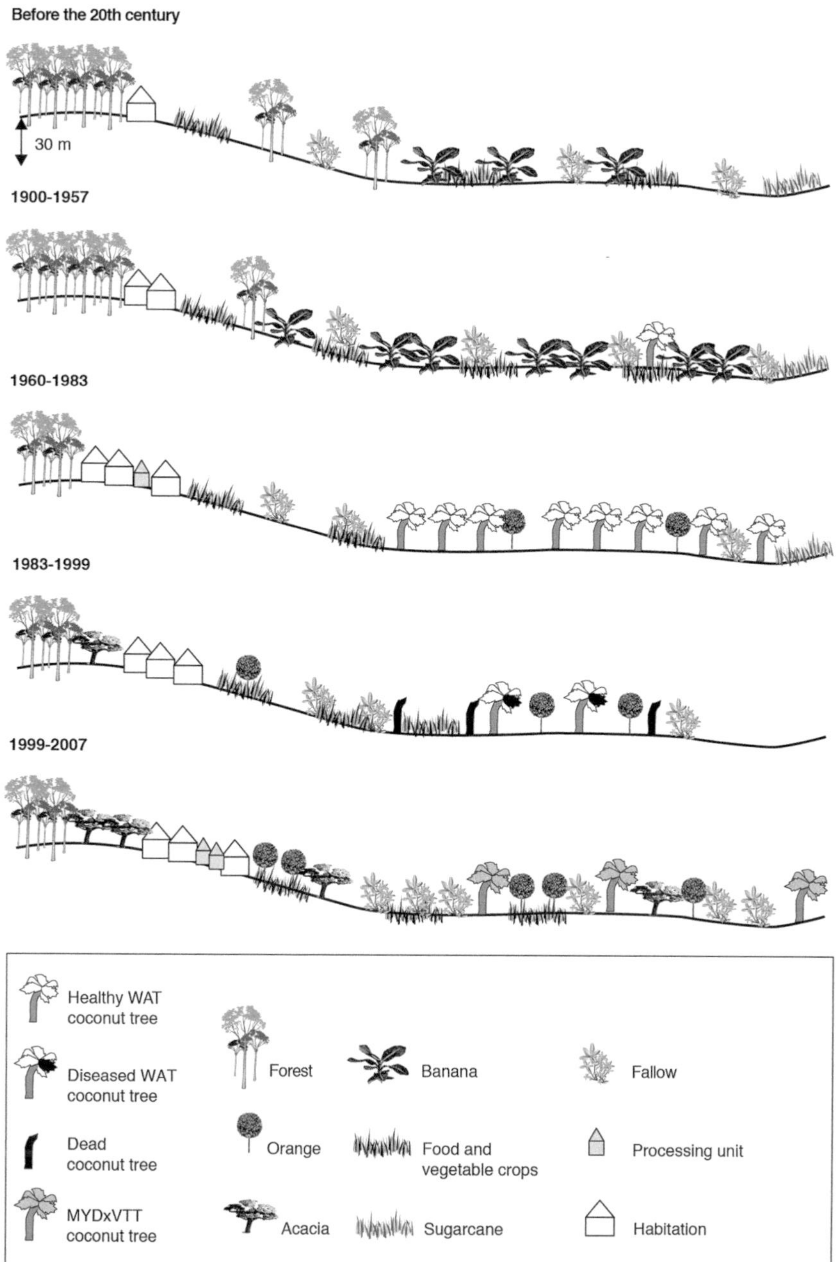

Fig. 3.2 Pattern of evolution of Benyadze's landscape

From 1960 to 1980, as coconut and other perennial crops (oil palm and orange) developed, pressure on land inexorably increased. Production systems became more diversified but significantly increased farmers' workloads. Coconut became the dominant crop of the village and was grown in the valleys (including on lowland floodplains) and on the plains. In the same environment, the Dura oil palm (traditional unproductive variety) was also cultivated but to a lesser degree. Orange groves were planted on the hillsides. Plantain cultivation then disappeared and was replaced by coconut and the Dura palm.

In 1985, LYDC first appeared in Benyadze. However, it was only during the 1993–1998 period that coconut trees were completely destroyed by it. The disease struck farms in a random manner by spreading sporadically closer and closer on the same plot and from one plot to another. When a coconut plot was affected, the disease decimated it within a few months, before eventually spreading to cover the whole village territory.

From that time and depending on social situations—especially access to land—and on farming systems developed before the disease, several strategies were formulated by the producers to deal with the disease.

3.3.1.1 Anticipatory or Short-Term Strategy to Deal with the Disease: Sale of Fresh Nuts

As in the entire region, growers in Benyadze had no anticipatory strategy to deal with the disease. But they did come up with a coping strategy. They realized that once a tree was infected, the mature nuts that would fall from the bunches could not be sold to the coconut oil industry. At the same time, a market for immature fresh nuts emerged and farmers gradually started participating in this new market. They thus opted for a strategy of extracting value from their coconut groves over the short term. Their only objective was to save what was still possible to save. This strategy was adopted not only by the farmers whose plots were being destroyed by the disease but also by those whose trees were unaffected.

3.3.1.2 Strategy of Diversifying into Food and Vegetable Crops on Plots of Ravaged Coconut Trees

Most farmers did not have access to new lands and so they developed a strategy based on the replacement of dead coconut trees by food crops. Their decision on when to fell dead coconut trees depended on their financial and social status. The reason for non-felling was strategic: even a dead but standing coconut tree can mark possession of or right to the land. Moreover, coconut wood is very hard and felling a tree can entail a high cost for a farmer who has just lost his plantation. On the other hand, felling of the dead tree cannot be delayed indefinitely because once the dried crown of the coconut tree falls, its trunks remains standing exposed to rain and sun. It begins to decompose under the action of micro- organisms, fungi and oryctes larvae. A year and a half or 2 years after the tree's death, the action of these secondary detritivore invaders weakens the stipe. It eventually falls, sometimes in large sections which can damage crops in the understory or injure people working underneath.

Farmers whose financial situation was most critical decided to migrate to the cities to find work, as did many younger ones. The goal was to make enough money to tide over this difficult period and replant a perennial crop as soon as possible. The farmer entrusted his plot of dead coconut trees to a family member so it could be used to grow food crops (for 2 years). The plot was then abandoned to fallow until the migrant's return. Generally, these farmers returned to the village 3 or 4 years after their departure. Some young farmers constituted working groups for complete felling of trunks using machetes. Once the coconut trees were felled, most farmers started growing food crops (maize and cassava). The majority of the village's farmers opted for these food crops, but without felling the dead trees.

An intermediate strategy was to fell the trees gradually. Once the coconut trees died, the farmer would begin felling some trunks and planting food crops. Then he would clear a new area to shift the food crops to. He would repeat this process until all the trunks were cleared.

3.3.1.3 Strategies of Spontaneous Diversification to Other Perennial Crops, Oil Palm and Orange, Subject to Access to Land

Farmers who had access to new lands, namely those related to the royal family, abandoned their devastated coconut plantations. Without waiting for the plots to be completely ravaged, they planted orange on hilly areas. Orange thus displaced coconut as their main crop in terms of both cultivated area and income. These farmers interplanted food and vegetable crops during the orange's unproductive phase. They created their own nursery and used the local Rough Lemon seeds to create rootstocks, as well as grafts from Late Valencia and Sweet orange species. 6 months after sowing the local lemon, they could undertake chip-budding. These practices greatly reduced the cost of planting the new crop. At the same time, they proceeded to fell the coconut trunks during the first 2 years after the complete destruction of the trees. For this, depending on their financial resources, they could hire service providers who used chainsaws to quickly remove the trunks.

Those who lacked the necessary capital left fallow the plot of dead coconut trees. 3 years after the destruction of coconut groves, when the trunks had fallen and partly decomposed, farmers returned to their former plots to plant Dura and Tenera oil palms (Tenera is an improved variety that is more productive than Dura).

3.3.1.4 Project-Supported Diversification Strategies

In 1997, the ADRA NGO set up a project to develop perennial crops to promote cultivation of cashew and acacia (*Acacia mangium* used to make charcoal). The project beneficiaries were mainly farmers whose plots had been infected at the beginning of the period (1993–1994) and, to a lesser extent, those who could remove the dead coconut trees quickly enough to benefit from this project.

3.3.1.5 Project-Supported Strategies for Replanting of Hybrid Coconut

In 1999, AFD financed the CSDP project. It provided credit to the farmers in the form of a technological package for planting hybrid coconut. In Benyadze village, 35 farmers participated in the CSDP project and 38 ha were planted. These new plantations were established in valleys and on plains, in former coconut groves and on fallow. Some farmers planted this coconut on lowland floodplains. Today we observe that these latter plots are less developed than those located on plains or in valleys. During the unproductive phase of the hybrid coconut, farmers interplanted food and vegetable crops. Farmers who participated in this project had not joined the ADRA project and, most importantly, they had cut the trunks of coconut trees on their land sufficiently early.

3.3.1.6 Recent Dynamics

During the 1999–2010 period, there was an expansion of perennial crops to the detriment of food and vegetable crops, as well as of perennial crops promoted by projects. The majority of growers replaced completely their destroyed coconut groves and cashew plots by Dura and Tenera oil palms. Orange and acacia groves came up mainly on hilly areas, slopes and upper slopes.

The villagers most affected by the destruction of their coconut heritage decided to convert to non-agricultural activities (transformation). Others migrated to cities which freed up some land. The failure of cashew cultivation to take hold also helped to gradually release land.

In the late 2000s, we could distinguish between three types of farms:

- farms owned by members of the royal family, strongly affected by LYDC and farming from 6 to 10 ha. These farmers replaced the destroyed coconut plantations and their cashew crops by oil palm, mainly of the Tenera variety. They also proceeded to expand their farms by planting acacia and orange. With the influx of migrants, they found willing labour to fell the coconut trees and took advantage of the CSDP project to establish hybrid coconut plantations on fallows or in old coconut groves. In the late 2000s, these farms grew on average 0.4 ha of food crops (with 0.8 ha of fallow); 1.2–2 ha of Dura oil palm; 1.2–2 ha of Tenera oil palm; 0.4–0.8 ha of acacia; 1.2–2 ha of orange and 1 ha of hybrid coconut;
- farms moderately affected by LYDC and farming from 3.5 to 6 ha. These farmers replaced the destroyed coconut by oil palm, mainly of the Dura variety.

They planted hybrid coconut on 1 ha in valleys and on plains. They grew from 0.4 to 0.8 ha of food and vegetable crops (with 0.8–1.6 ha of fallow); 1.2–2 ha of Dura oil palm; 1 ha of hybrid coconut; 0.4–1.2 ha of orange and 0.4 ha of acacia. These farmers also increased their secondary activities by processing agricultural products for sale to intermediaries. They rented out their workforce to farmers of the royal family;

- small farms not directly affected by LYDC and 2.5 ha in size. These growers cultivated approximately 0.8 ha of food and vegetable crops (1.6 ha of fallow), with production destined mainly for processing and sale. This group found support in this approach from medium-sized farms that are oriented towards processing of crops (palm oil, charcoal, gari and kenke). Finally, some growers in this group still rent out their workforce to larger farms.

3.3.2 Farmer Strategies in Old Ebu, a Village with Diversified Crops Before the Arrival of LYDC

The first symptoms of LYDC were observed in this village in 2000. But it was not until 2003 that the disease began spreading actively: it developed in foci scattered on both sides of the territory and at varying degrees of intensity. Farmers sought solutions, but they were clearly much less affected than the farmers of Benyadze because their cropping systems were already highly diversified before the arrival of the disease.

Most farmers abandoned their plots which were devastated by disease to fallow. Others replaced their coconut groves with food crops and specialized in the processing of maize and cassava. Farmers participated in the oil palm and sugarcane projects by planting these crops in the destroyed coconut groves.

Today, three broad groups of farms can be distinguished mainly on the basis of their degree of diversification and infestation of coconut trees by LYDC:

- diversified farms, but which were nonetheless affected. They farm from 4 to 6.5 ha and have 0.4–0.8 ha of coconut; 0.5–0.7 ha of orange; 0.4–0.8 ha of acacia; 0.2 ha of oil palm; 0.4–0.8 ha of food crops (maize and cassava); 0.1 ha of tigernuts and peanuts, and maintain 0.8–1.6 ha of fallow;
- highly diversified farms which were little infected. They consist of plots in which the coconut is not very significant in the production system. They farm from 3 to 5 ha, but with less than 0.4 ha of coconut; 0.4–1.2 ha of sugarcane; 0.4–1.2 ha of orange; 0.4–0.8 ha of acacia; 0.2 ha of Dura oil palm; 0.4 ha of maize and cassava; 0.1 ha of tigernuts and peanuts, and 0.8 ha of fallow;
- relative undiversified farms which were unaffected by the disease. They range in size from about 1–2 ha divided into 0.4–0.8 ha of food crops (with 0.5–1.2 ha of fallow) and less than 0.4 ha of orange. These families depend mainly on non-agricultural activities.

3.4 Conclusion

Reduced incomes from a commercial crop in decline no longer satisfied producers and other actors in the coconut sector. LYDC weakened this sector which already suffered from a structural weakness and low copra prices. Coconut producers would probably have diversified in any case. Unsurprisingly, a village with diversified crops, such as Old Ebu, was much less affected by the coconut disease than a village such as Benyadze, which had opted for a quasi-monoculture.

While emigration provided one solution, diversification was another. It could signify a shift to another crop or, in principle, a shift of focus to other products from the same plant, better suited to global economic conditions, such as virgin coconut oil in the case of coconut. We saw here that the first type of diversification dominated. Proximity to palm oil and cocoa markets was decisive in this choice of diversification. At the same time, we did observe attempts to plant coconut in areas far removed from areas where the disease was concentrated or had broken out.

The history of coconut cultivation in Ghana therefore confirms the theoretical premise of this book. At the end of the first life cycle of a particular perennial crop *A* in a region, the region tends to diversify due to ecological and social necessity. At the same time, crop *A* 'shifts', following its adoption by new regions, to a virgin forest environment.

This process was first verified in the 1930s in south-eastern Ghana, then the home of coconut cultivation. While some farms were able to maintain coconut cultivation with a little bit of replanting, the economic and social dynamic favoured diversification, especially vegetable cultivation.

This process was repeated during the 2000s in the Central and Western regions. In these regions, the overwhelming response from farmers was that of diversification into other perennial crops. Oil palm and orange in the coastal areas, rubber and cocoa in the interior were the main perennial crops the farmers turned to.

Oil palm and rubber are crops supported by industry groups. There is currently not much of replanting of hybrid coconut because of the disease risk, the price of planting material and the plant's lack of hardiness. But a large proportion of farmers would like to cultivate it over the longer term (Geiger and Lhommet 2005). However, there is a transitional phase underway of deriving value from land through a return to food and annual crops around most villages.

Most coconut farmers, either owners or under *abusa* contract, were elderly and impoverished by the partial destruction of their capital or by money borrowed from loan sharks (Ruf et al. 2009). These farmers were therefore not prepared to invest in replanting more productive coconut varieties, including in areas unaffected by the disease. They continued to maintain coconut production with the West African Tall (WAT) variety, highly appreciated by the mature-nuts regional export sector.

For producers under *abusa* contract, the theoretical loss of the right of land use due to death of coconut trees remains to be verified in practice. We refer to the theoretical loss of usage rights, since farmers can benefit from some form of moratorium on land rights from the village or family chiefs. If this loss of land actually takes place, it can have significant consequences. Most notably, if village farmers or new actors such as upper middle class investors (UMCI[3]) seek to acquire land, producers would then have capital to buy land or a usage right. Through the rubber and oil palm projects, they could invest in large cultivated areas, which would tend to increase the inequality of land distribution.

References

AFD-IRAM (2004) Rapport d'évaluation du programme de relance du cocotier

Colin JP, Ruf F (2009) The *ant and Share* contract in Côte d'Ivoire. Incomplete contracting and land conflicts. In: Walter A (ed) 13th annual conference of the international society for new institutional economics. University of California at Berkeley, Haas School of Business, Unites States of America, 18–20 June, 2009

Colin JP, Ruf F (2011) Une économie de plantation en devenir. L'essor des contrats de planter-partager comme innovation institutionnelle dans les rapports entre autochtones et étrangers en Côte d'Ivoire. Revue tiers—monde 207:169–187

Democrite R (2007) Les stratégies des planteurs de cocotiers face à la maladie du jaunissement mortel du cocotier au Ghana. Mémoire IRC. 166 pp

Dery SK, Calvez CH (2002) L'avenir des cultures pérennes, investissement et durabilité en zones tropicales humides, 5–9 November 2001, Yamoussoukro, Côte d'Ivoire. Montpellier, Cirad, Conférence internationale sur l'avenir des cultures pérennes

Dery SK, Philippe R (1997) International workshop on lethal yellowing-like diseases of coconut. Chatham: NRI, 255–260 pp Elmina, Ghana

Geiger R, Lhommet JC (2005) Devenir des exploitations agricoles face à la maladie du jaunissement mortel du cocotier. Mémoire Cnearc, 120 pp

Ruf F (2005) Les déterminants écologiques de la diversification. La région d'Axim au Ghana : de la noix de coco au cacao. Rapport pour le compte du ministère des Affraires étrangères. Montpellier: Cirad, 24 pp

Ruf F (2006) French project contre french project au Ghana ? Développement et antagonismes entre replantation cocotier et diversification hévéa. Communication Cirad, UMR-Innovations, Montpellier, p 15

Ruf F (2010) You weed and we'll share. Côte d'Ivoire and Indonesia. Cirad and GTZ, Montpellier, Land dividing contracts and cocoa booms in Ghana, p 53

Ruf F, Bini S, Ampadu K (2009) Stratégies des planteurs de cocotiers au Ghana face à la maladie du jaunissement mortel. OCL 16(2):76–86

[3]For a definition of UMCI, see footnote no. 3 in the Introduction.

Chapter 4
From the Coffee-Cocoa Combination to Oil Palm Cycles: The Case of Dabou and Aboisso in Côte d'Ivoire

Vylie Tientcheu Sayam and Emmanuelle Cheyns

4.1 Introduction

Even though mineral oil is starting to make an impact on its economy, Côte d'Ivoire is still heavily dependent on agriculture, especially on cocoa. More than 600,000 family farms produce cocoa and provide employment to several million of the country's population. The country is the world's largest cocoa producer, accounting for almost 40 % of the global supply.

A process of diversification into oil palm cultivation by producers of coffee and cocoa has been observed all across the lower part of Côte d'Ivoire (from Bas-Cavally to Aboisso), at least from the 1970s and 1980s. This process was triggered by a proactive policy of the Houphouët-Boigny regime in the late 1960s (Pillet-Schwartz 1972, 1978; Cheyns et al. 2000). As with rubber, the initial diversification into oil palm cultivation was the result of public policy and took place during the great expansion of cocoa production in the country, the glorious era of the 'Ivorian model'.

After a long period of guaranteed procurement prices paid to farmers, an arrangement that was akin to a social pact between them and former president Houphouët-Boigny, the regime was unable to sustain the prices in its final years. The fall in cocoa prices in 1989, the devaluation of the CFA franc in 1994 and liberalization in 1999, quickly derailed by the political crisis, have continued to impoverish cocoa farmers (Ruf 1991; Oswald 1997; Araujeau-Bonjean and

V.T. Sayam (✉)
Ecole Nationale Supérieure de Statistique et d'Economie Appliquée,
Abidjan, Ivory Coast
e-mail: sativy@yahoo.fr

E. Cheyns
UMR Moisa, TA C-99/15, Cirad, 73 rue Jean-François Breton,
34398 Montpellier Cedex 5, France
e-mail: emmanuelle.cheyns@cirad.fr

F. Ruf and G. Schroth (eds.), *Economics and Ecology of Diversification*,
DOI 10.1007/978-94-017-7294-5_4

Chambas 2001). During this difficult time, cocoa production—far from falling as might have been expected—increased again to maintain very high levels during the 2000s. Meanwhile, the oil palm sector also began to encounter some headwinds: the dismantling of structures and privatization, manifested mainly as a withdrawal by the State from the sector.

Given the context of the glorious years, followed by the upheavals of the economy and Ivorian politics, what has been the process of diversification of cocoa farms towards oil palm cultivation? What have been the main motivations for farmers and the major determinants of the adoption of oil palm cultivation?

In comparison with other perennial crops grown in Côte d'Ivoire, the oil palm provides additional diversification through the artisanal extraction of palm oil. How does this activity contribute to household income and, ultimately, to the process of diversification itself?

Finally, can this diversification jeopardize the cocoa economy by becoming a straightforward conversion in the form of the permanent replacement of cocoa by oil palm? Or does income from oil palm cultivation actually promote reinvestment in cocoa cultivation?

This chapter attempts to address these issues based on two surveys of village farmers in 2001 and 2002. One was conducted in the Lagunes region, the other in the Aboisso area. The impact of the social and political crisis that resulted from the coup d'état in September 2002 will not be addressed in this chapter.

4.2 Hypotheses and a Review of the Literature

Consistent with the analytical framework of this book, the main factors that were studied were the ecological difficulties of replanting cocoa and its aging farms, as well as the volatility of world prices and the uncertainties resulting from the privatization of the sector. Land constraints and opportunities offered by other crops were also considered.

4.2.1 Constraints of Replanting Cocoa Orchards

The rise of Ivorian cocoa was largely dependent on the forest rent. Indeed, the establishment of a cocoa plantation after clearing of fallow requires much more labour and inputs than its establishment after clearing of forest (Ruf 1987, 1995; Oswald 1997; Léonard 1997). In West Africa in general and in Côte d'Ivoire in particular, farmers—a significant number of whom are immigrants who have come to cultivate cocoa—have maximized the short-term effect of the forest rent on their net incomes (Ruf 2000). Their forest-clearing techniques have led to a gradual disappearance of the rainforest. The forest offered them a guarantee for cocoa's rapid growth at minimal cost: availability of organic matter, protection against wind and weather hazards, high rainfall, etc. All these factors help reduce investment costs and temporarily increase yields and have been grouped under the term 'differential forest rent' (ibid.). Replanting cocoa, on the other hand, requires greater quantities of inputs and labour than does planting after forest clearing and, more generally, has significantly higher capital requirements.

Starting from the time of the country's independence, the 'diversification' sectors such as oil palm and rubber benefitted from the enormous advantage of being able to offer credit to farmers for the purchase of planting material, herbicides and fertilizers (Ruf 1987; Naï Naï et al. 2000). The need for land tenure security in a context of scarcity and conflicts over resources also often influenced the choice of crops to grow. Farmers could decide, to begin with, to turn to major projects which provided a certain land guarantee (Chap. 8 of this book).

In a more general way, prices and expected incomes could influence the decision between replanting and diversifying. This is what happened in the 1980s when old coffee and cocoa plots were converted to oil palm (Naï Naï et al. 2000). Price expectations can influence decisions in one of two ways:

- relatively low prices may encourage a decision oriented towards replanting when income loss is minimized and when growers expect higher prices in times to come. In this case, it makes sense to plant when prices are low;
- a drop in prices, resulting in reduced income, very often leads farmers to reduce the maintenance of their plantations. This lack of maintenance then leads to lower yields, which may lead to the decision to replant with a crop other than cocoa.

4.2.2 Adoption of Oil Palm as a Diversification Crop

4.2.2.1 Public Policies

In Côte d'Ivoire, the oil palm sector developed as the result of a sustained political will. Large-scale subsidized projects were set up in the context of oil palm cultivation plans of 1963 and 1985. Côte d'Ivoire thus acquired an agro-industrial palm oil production sector, making the country Africa's leading exporter of palm oil. The policy environment encouraged the organization of an oil palm sector which could supply growers with inputs on credit and technical information. It has largely contributed to the adoption of this crop in the areas of influence of the state-owned Sodepalm corporation (and later Palmindustrie). Information on the quality of planting material, capital and availability of inputs were key factors in the decision to grow oil palm in the village environment (Naï Naï et al. 2000).

4.2.2.2 Aging of Cocoa Trees and Ecological Changes

The hybrid oil palm was often used as a first-cycle crop, planted after the clearing of the forest, just like cocoa. Nevertheless, the adoption of oil palm also coincided with the end of the first cocoa and coffee cycles. Diversification, especially towards oil palm, also appeared as a solution to the problems of aging and replanting of cocoa (Ruf 1987).

In Sassandra, most of the area under oil palm has been planted after 1985. Léonard (1997) emphasizes that, to a large extent, the popularity of this crop corresponds to a goal of conversion or diversification when confronted with aging coffee and cocoa orchards and the difficulties of replanting. The same impetus was observed at Aboisso (Nai Nai 2000; Naï Naï et al. 2000).

4.2.2.3 Relative Prices

Changes in the relative prices of agricultural raw materials are one of the key determinants in the decision to adopt a crop. In this regard, the changes in prices of the cocoa-coffee combination in Côte d'Ivoire are exemplary, with a general transition from coffee to cocoa starting in the 1970s when the price of cocoa increased (Ruf 1995). The price of palm bunches is a determinant for the adoption of oil palm not only because it is high, but also because of its consistency. During the 30 years preceding the privatization of Palmindustrie (1997), government management of the sector guaranteed farmers a certain consistency in the procurement prices paid to them. This state of affairs persisted until the end of 1999, with a procurement price fixed every 6 months that increased steadily between 1992 and 1999 (Naï Naï et al. 2000).

4.2.2.4 Regularity of Income

The harvesting of palm bunches is spread out over the year, with peak production during the first 6 months of the year. Producers thus enjoy a monthly income. This regularity also provides them with a guarantee of solvency and dependability when accessing credit. Regularity of income therefore appears as one of the determinants in the choice of this crop (Colin 1990; Naï Naï et al. 2000).

4.2.2.5 Diversification of Markets

From all the perennial crops cultivated in Côte d'Ivoire, oil palm is among those which present the most market opportunities to farmers (Naï Naï et al. 2000). For a long time, factories run by state-owned enterprises provided growers with a guaranteed market. The purchase of bunches from villagers by women artisans who extract oil, as well as oil self-produced for food consumption and production of soap form alternative, though partial, outlets for their crop. Finally, the felling of the oil palm tree is an opportunity for the farmer to obtain one-time capital through the extraction of palm wine. Exploiting a hectare of oil palm trees for extracting wine brings growers an amount between 350,000 and 420,000 FCFA in areas close to urban centres. This money can contribute to social spending and can also finance replanting (Naï Naï et al. 2000).

In summary, the studies cited above suggest that the adoption of oil palm by the farmers was influenced by several determinants following a historical process:

- initial policy decisions to promote the development of oil palm cultivation in the form of projects. They played an essential role in bringing information and capital to this new sector;
- a specific historical development. In Ivorian post-independence civil society, high-level state-employees benefitted by being accorded priority in receiving information and project capital, as well as access to forests for oil palm cultivation;
- the ecological and historical determinism of the unfolding of the cocoa-coffee cycle. Paradoxically, the adoption by farmers of oil palm corresponds instead to an opposite process, that of the initial difficulties of replanting cocoa and, to a lesser extent, coffee in the context of deforestation;
- prices and other economic determinants. Relative prices play a complementary role since cultivation of the oil palm leads simultaneously to an increase, diversification and consistency (monthly revenue) of income. To this is added the uniqueness of the oil palm in Africa: the ability to build up capital through harvesting and selling of palm wine;
- social determinants. The return of young people to the villages in the 1990s seems to have contributed to the revival of investment in oil palm cultivation;
- a return to political factors in 1995. Despite constraints of capital, the privatization of the oil palm sector and access to improved seeds seem to have favoured the adoption of oil palm. Projects played a significant role in the dissemination of information about this new crop.

Given this wealth of knowledge already acquired on the process of adoption of oil palm, it remains for us to test the hypotheses mentioned initially and extend the observation period to the early 2000s in our case studies.

4.3 Study Methodology

4.3.1 Survey in the Lagunes Region

The first survey was conducted in January 2001 in the Lagunes region.[1] It covered 150 oil palm growers in the Dabou and Anguededou areas, of which 75 growers were part of the agro-industrial scheme in Dabou and 75 belonged to the

[1]Survey conducted as part of the research programme 'Organization and development of the oil palm sector', CIRAD, University of Bouaké—Laboratory of Rural Economics and Sociology, 1999–2003 (see Cheyns et al. 2001).

corresponding scheme in Anguededou. During this survey, 12 villages were selected on the basis of criteria of remoteness, size, etc. Then farmers from these villages were selected at random from the list of farmers (official and unofficial) in proportion to the total number of oil palm growers in each village.

4.3.2 *Survey in the Aboisso Area*

The second survey was conducted in the Aboisso area (extreme south-eastern Côte d'Ivoire in the Comoé region) in July 2002 and covered 117 farmers from three villages (Koffikro, Baffia and Assouba). The Aboisso area was chosen because of the dynamics of oil palm adoption by former cocoa producers. These three villages were chosen to be able to survey three different types of farmers:

- cocoa farmers who had adopted oil palm cultivation;
- cocoa farmers who had not adopted oil palm cultivation;
- farmers who no longer grew cocoa and cultivated oil palm only.

Since we had no reference survey data based on this farmer typology, we conducted a preliminary survey of farmers who had already planted cocoa at least once. They were asked to enrol after the presentation of the study. Of the 239 farmers short-listed, a random sample of 117 farmers was selected, with a quota of two-thirds of farmers who had already diversified into oil palm and one-third of those who had not.[2] Together these growers farmed a total area of 1963 ha, across all crops (in addition to forest reserves). The village of Baffia represented approximately 50 % of the total area cultivated by the surveyed farmers.

4.4 Results and Discussion

4.4.1 *The Lagunes Region*

In the area of Dabou and Anguededou, the diversification process was both structural and contextual. The decision to diversify into oil palm depended on factors internal to the farm, institutional factors (development projects) and fluctuations in commodity prices. In these areas, this process pertained to a transition from cocoa towards oil palm or rubber.

[2]The questionnaire was based on (and adapted from) the work done during the workshop to launch the 'Cocoa Net' network at Takoradi (June 2002) with the objectives of comparability of different studies conducted in several different countries. This questionnaire consisted of three parts: (1) socio-economic and demographic characteristics of the farmer and general information on the farm, (2) characterization and dynamics of the farm's plots, and (3) characterization of the diversification (planting dates, areas under cultivation, inputs, preceding crops, etc.).

In the Lagunes region, the curves representing the cumulative cultivated areas of cocoa and coffee that were planted are linear and continuous, and show moderate growth (Fig. 4.1). This indicates that farmers were no longer creating large coffee or cocoa plantations. By contrast, the Lagunes region has known three periods of high oil palm adoption. The 1963–1980 period was the time of the first oil palm plan and the 1985–1990 period that of the second. The latter period seemed to have coincided with the appearance of the first difficulties in replanting cocoa. In any case, the pace of establishment of oil palm plantations accelerated in 1986 and relegated cocoa and coffee to a status of secondary crops. The third period, starting in 1995, was characterized by the liberalization of access to seeds and favourable procurement prices for palm bunches. This state of affairs continued until the end of 1999. Finally, rubber emerged in the mid-1980s and it too overtook the historical crops of cocoa and coffee. A few cocoa plots were indeed planted during this period, but at a much lower rate than for oil palm. The great majority (79 %) of cocoa and coffee was planted following deforestation. In contrast, oil palm was planted after the destruction of old cocoa trees or coffee plants or after fallow (Table 4.1).

When a cocoa or coffee orchard aged and was replaced, the replacement crop tended to be oil palm or rubber. This confirms the observations found in the literature and is in line with the theory of differential forest rent. Most crops benefited from advantages such as organic fertilizer, plant cover and ease of weeding

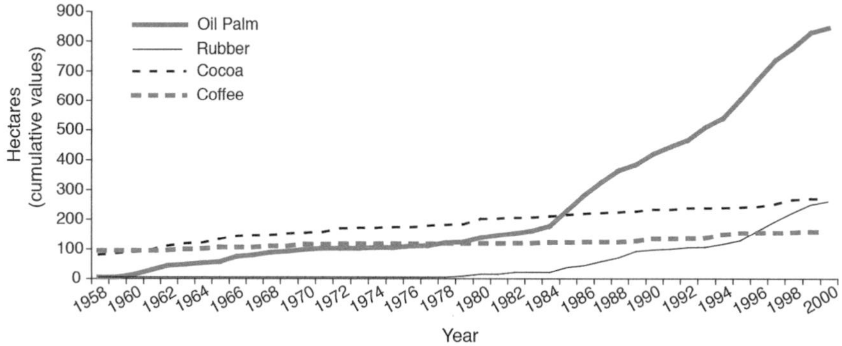

Fig. 4.1 Cumulative change in areas planted with oil palm, rubber, cocoa and coffee in the Lagunes region. (*Source* survey by CIRAD and University of Bouaké, 2001)

Table 4.1 Distribution of plots (in %) by crop and preceding crop in the Lagunes region

Preceding crop on the plot	Forest	Fallow	Old cocoa or coffee	Other[a]	Total
Cocoa, coffee	79	7	4	10	100 (n = 174)
Oil palm	12	11	28	49	100 (n = 396)
Rubber	21	9	23	47	100 (n = 117)

Source survey by CIRAD and University of Bouaké, 2001

[a]Lowlands, burned out plots, fallow unidentified as such, oil palm, food crop, banana, etc

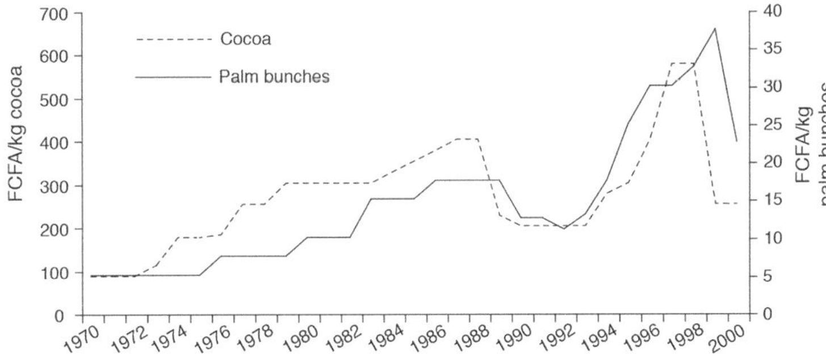

Fig. 4.2 Changes in procurement prices of cocoa and palm bunches (current prices paid to producers). (*Sources* Ministry of Agriculture; Ruf and Akpo 2008)

during the early years. But cocoa is the one crop that benefited the most. As a corollary, it is cocoa which suffered the most from the exhaustion of the forest rent. These results confirm the hypothesis that diversification often occurs late in the cocoa life cycle, thus becoming a solution to the problems of replanting it. This solution is supported by the change of crop, capital and information provided by the projects (Ruf 1987, 2000).

While the first oil palm period (1963–1978) focused mainly on the development of industrial farms, the second period (1985–1990) marked a rapid expansion of smallholder areas (Figs. 4.2 and 4.3). This increase was closely linked to the supply of credit from plantations and the supply of inputs to smallholders.

Economists define investment as a decision taken by an entrepreneur based on estimated costs and expected financial returns. But are farmers like other entrepreneurs? If yes, their investments should depend primarily on existing and expected market prices, but also on support that a development project may be able to offer them.

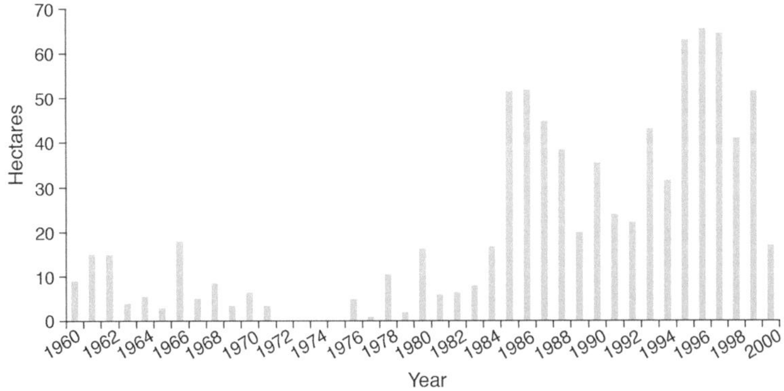

Fig. 4.3 Cultivated areas of oil palm planted annually in the Lagunes region, 1960–2000. (*Sources* surveys by CIRAD and University of Bouaké, 2001)

Fluctuating prices of cocoa and palm bunches did not seem to have played a very important role in the decision to diversify (Fig. 4.3). Before 1990, there was no relationship between the price of cocoa and the extent of planted oil palm areas. The main determinant in the adoption of oil palm seemed to be the implementation of development plans, which meant that there were almost no investment costs. Then the fall in cocoa prices from 1990 to 1994 could have turned farmers away from the crop and towards oil palm, especially after the liberalization of access to seeds in 1995. But this assumes that farmers responded quickly to changes in prices. Furthermore, cocoa prices also exhibited a rising trend starting in 1995.

4.4.2 Aboisso, South-Eastern Côte d'Ivoire

In the 1960s, Aboisso, in south-eastern Côte d'Ivoire (Comoé region), was one of the agricultural frontiers for cocoa and coffee cultivation. With the first development plan of the oil palm sector, this area also became one of the first to host major agro-industrial palm oil production complexes, at Ehania and Toumanguié. This proximity to the processing plants promoted adoption of oil palm by the area's farmers, who until then had mainly cultivated cocoa and coffee.

4.4.2.1 Farmers and Labour Employed on the Farm and Diversification

Neither the age of the farmers, nor their education, nor their professional background, nor their autochthonous or migrant status seemed to have influenced their decision to diversify into oil palm cultivation. At least, a statistical analysis found no statistical difference between these variables. As far as these variables are concerned, farmers who diversified into oil palm cultivation were no different from those who had not adopted oil palm (Table 4.2).

As far as family labour was concerned, an average of 3 men and a little more than 3 women worked full time on each farm. Among farmers who had adopted oil palm, the average was a little over 3 men, whereas it was 2 for cocoa farmers who did not additionally cultivate oil palm. After the privatization of the oil palm sector and the discontinuation of the system for the supply of inputs, farmers seemed to have started relying more on family labour. They limited their paid-labour expenses to site preparation work and planting. However, farmers who had diversified into oil palm cultivation relied more on contractual labour than those who had not (0.7 instead of 0.1 contract worker/farm).

Farmers in Baffia village had the largest farms. The average area of their oil palm, cocoa, coffee and forest reserves were higher than in the other two villages (Table 4.3). Forest reserves were still present in the villages of Baffia and Assouba—

Table 4.2 Distribution of farmers according to their status, age, level of education, possession of a bank account and the adoption of oil palm cultivation in the Aboisso area

Farmer characteristic		Diversification					
		Yes		No		Total	
		Number	%	Number	%	Number	%
Type of resident	Autochthon	40	47.1	16	50	56	47.9
	Ivorian migrant	35	41.2	1	34.4	46	34.3
	Foreign migrant	10	11.8	5	15.6	15	12.8
Age group	23–45 years	32	37.6	13	40.6	45	38.5
	45–60 years	24	28.2	6	18.8	30	25.5
	60+ years	29	34.1	13	40.6	42	35.9
Level of education	Primary	14	16.5	10	31.3	24	20.5
	Secondary/Higher	24	28.2	6	18.7	30	25.6
	Koranic/None	47	55.3	16	50	63	53.8
Bank account	Yes	24	28.2	1	3.1	25	21.4
	No	61	71.8	31	96.9	92	78.6
Profession	None	50	58.8	19	59.4	69	60
	Civil servant	12	14.1	6	18.7	18	15.4
	Other	23	2716	7	21.9	30	25.6

Source survey of 117 farmers in the Aboisso area, 2002

Table 4.3 Average areas (in ha) and numbers of farmers per village and per crop

Villages	Oil palm		Cocoa		Coffee		Forest	
	Average[a]	Number	Average[a]	Number	Average[a]	Number	Average[a]	Number
Koffikro	5.28	32	2.75	24	2.56	25	0.33	38
Assouba	5.15	23	3.36	29	6.46	26	3.87	31
Baffia	6.58	30	4.85	44	5.78	41	8.57	44
Consolidated	5.70	85	3.88	97	5.10	92	4.51	113
Total cultivated areas	485		376		469		510	

Source survey of 117 farmers in the Aboisso area, 2002
[a]Averages of only those farmers cultivating the crop in the village (and not of the total number of farmers in the survey sample)

respectively 8.57 and 3.87 ha on average per farmer—against only 0.33 ha for Koffikro village (Table 4.3). Conversely, in Koffikro, oil palm represented 44 % of the cultivated surface area against 20 % in the other two villages (Fig. 4.4). This difference is almost certainly due to the difference in forest reserves. At Koffikro, the lack of usable forest reserves prevented diversification through the creation of new plots; oil palms were therefore planted to replace cocoa.

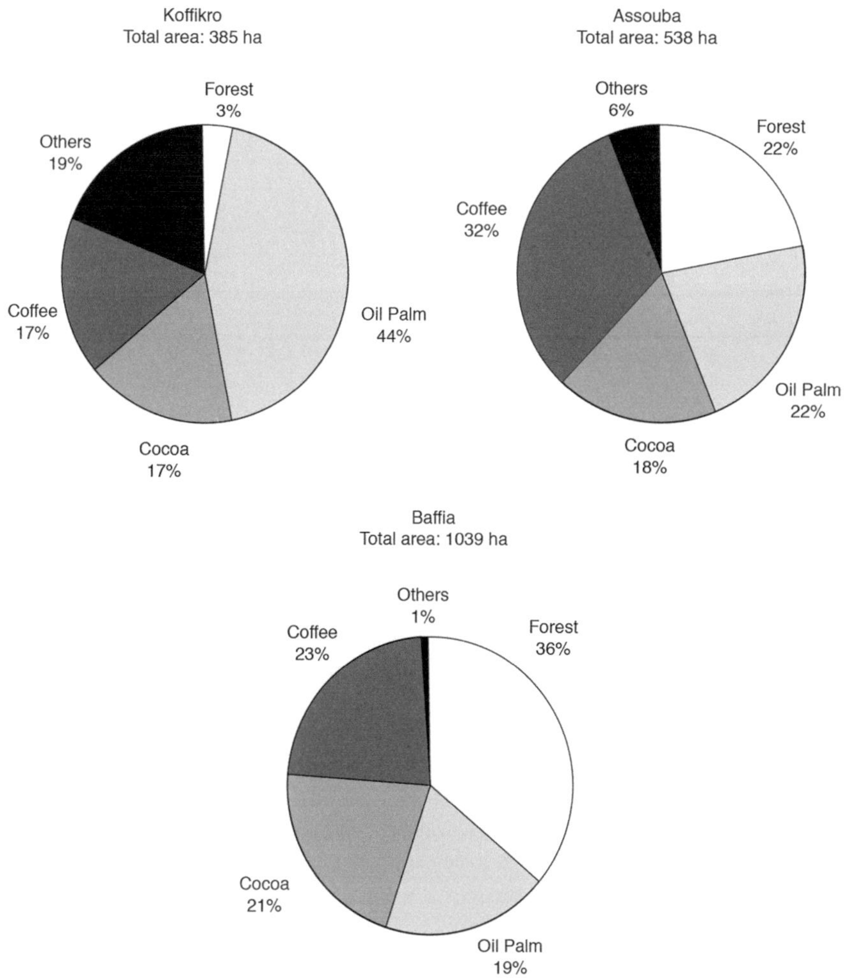

Fig. 4.4 Distribution of the various crops in the three villages

4.4.2.2 The Dynamics of Cocoa Plantations

The plots where cocoa trees were felled still represented a small area. Over the 40-year period that could be recreated by the survey, the average of the area of cut-down cocoa trees is much lower than that of planted areas. From 1960 to 2001, only 191 ha of cocoa plots were cleared against 631 ha of plots planted. Farmers in the area did not, therefore, completely turn away from cocoa cultivation, but its dynamics did weaken. Indeed, 77 % of the cocoa trees were planted before 1990 and only a quarter subsequently. We observe therefore a reversal of the trend of planted and cleared areas: after 1990, the area of new cocoa plantations was lower than that of cleared cocoa plantations (Fig. 4.5). In contrast, 37 % of the cleared

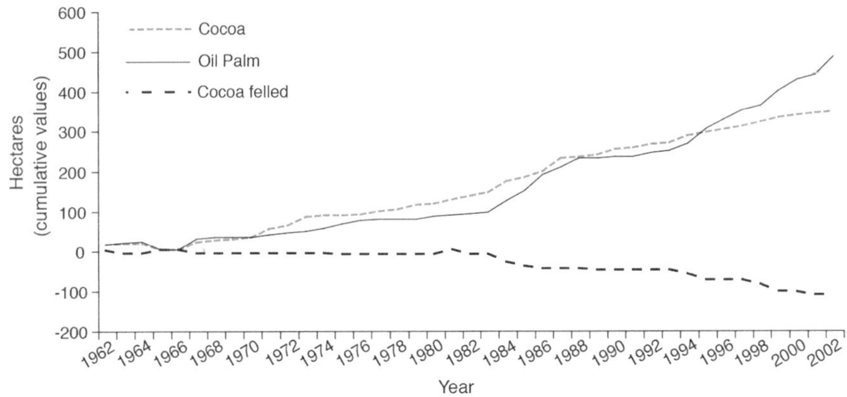

Fig. 4.5 Cumulative areas planted with oil palm and cocoa and areas of felled cocoa trees in the Aboisso area. (*Source* survey of 117 farmers in the Aboisso area, 2002)

plots were cleared before 1990 and 63 % after 1990. Here too, the life cycles of the plantations seemed to be more important than price factors. 83 % of farmers who cleared cocoa reported that they did so because the cocoa trees were old and no longer produced beans. Only 15 % of them declared that they did so in response to lower prices. The distribution of areas of cocoa plots cleared across the three villages was: 63 % by farmers of Koffikro, 29 % by those of Assouba and 8 % by those of Baffia.

As far as the preceding crop is concerned, the results were similar to those in the Dabou area: 76 % of cocoa-coffee plots were planted after clearing of forest. The remaining quarter involved replanting of the crop after old coffee or cocoa. This trend of the creation of cocoa-coffee orchards after forest clearing is less marked in Koffikro where farmers no longer had a forest reserve. At Baffia, forest reserves were still available and 95 % of cocoa-coffee orchards were planted after the clearing of forest.

69 % of the new crops planted after the clearing of cocoa-coffee orchards consisted of oil palm or rubber. Only 19 % of cocoa-coffee plots were replanted with cocoa or coffee. Conversion to oil palm or rubber took place mainly in Koffikro (71 % of plots of oil palm or rubber after cocoa or coffee). Indeed, in the other two villages, the presence of forests meant that oil palm or rubber cultivation could take place on cleared forest land.

4.4.2.3 The Dynamics of Oil Palm Adoption

As in the Lagunes region, the adoption of oil palm happened in three phases in the Aboisso area (Fig. 4.5). The first phase, from 1962 to 1980, corresponded to the first oil palm plan. A more significant phase, between 1985 and 1990, witnessed activity driven by the second oil palm plan. A third phase, in magnitude equal to the previous one, corresponded to the privatization of access to seeds starting in 1995, followed by the sector's privatization in 1997. This phase also coincided with a period of steady increases of palm bunch prices until the end of 1999. During this third phase, 75 % of palm plots were planted with improved planting material, without recourse to intra-sector credit.

In total, approximately 60 % of the oil palm plots were established before 1995. The plots planted post-1995 mainly concerned the village of Baffia (75 %) where the forest reserve was still available. Over the entire sample, 20 % of oil palm groves were created after clearing of forest. Moreover, the areas planted with oil palm were on the whole bigger than the areas of felled cocoa trees. Thus, the plots of cleared cocoa-coffee were mainly replanted with oil palm or rubber (case of Koffikro), indicating a partial substitution of cocoa by these new cash crops. In addition, there was a process of diversification through the creation of new plots without any substitution, which was part of a strategy to diversify and increase incomes.

Of the oil palm plots planted after the clearing of cocoa or coffee, 20 % were established before 1980, 39 % between 1980 and 1995 and 41 % after 1995. This type of diversification, which can be called substitution-diversification, was mainly observed in villages where there was land scarcity. Faced with a lack of land to expand their plantations, farmers preferred to clear, wholly or partially, their cocoa-coffee orchards, in order to replace them with oil palm which offered them an alternative income source. This strategy emerged between 1980 and 1995 and has continued even after that period.

Unlike the Lagunes region, the process of adoption of cocoa was still going strong. But it was mainly due to the village of Baffia (67 % of plots) where forest reserves were abundant.

4.4.3 *Overview of the Determinants of Diversification into Oil Palm Cultivation*

The decision to diversify into oil palm cultivation is due to factors that are primarily structural. These factors are discussed below.

4.4.3.1 Accessibility of Plots and Location Rent

According to the surveys, 79 % of farmers explained their non-adoption of oil palm or rubber cultivation by the difficulty of access to land to plant. The decision to adopt oil palm cultivation is predicated upon access to the plots by the collection trucks in a 30 km radius around the processing plant. 91 % of farmers who did not adopt oil palm and gave this reason were residents of Baffia, which is a remote village and difficult to reach, especially during the rainy season.

4.4.3.2 Forest Rent, Deforestation and Cocoa Tree Life Cycle

The impact of deforestation and the end of the cocoa life cycle on diversification was clearly shown, at least at the village level. Since forest reserves were still abundant, the planting of oil palm after forest clearing was commonplace in Baffia. But there was a greater level of diversification towards the oil palm tree in the other two villages. Oil palm cultivation was in addition to that of cocoa, which still constituted the primary wealth of the area (Fig. 4.4). In contrast, the complete disappearance of Koffikro forests promoted adoption of oil palm after the clearing of aging cocoa and coffee plantations; these two crops were not renewed. The oil palm thus became by far the most significant part of local wealth; it could almost be said that the diversification was close to becoming a conversion. Assouba village, where some forest still existed, was in an intermediate state. The relative degree of diversification—as judged by the percentage of oil palm in comparison with cocoa and coffee—and the principle of diversification by substitution (total or partial) were clearly linked to the aging of cocoa trees and land scarcity.

However, land scarcity remained, in absolute terms, an obstacle to diversification and the adoption of new crops. The creation of new plots requires availability of land. This land must be found within the farm or, failing that, be acquired through lease or purchase. In the Aboisso area, the vast majority of land used for diversification into oil palm cultivation was already available within the farm. It consisted of forest reserves or old plots of coffee and cocoa producing little or not at all. Diversification very rarely took place through the lease or purchase of new plots.

4.4.3.3 Non-agricultural Income and Pensions

In Assouba, there was another factor responsible for the fact that cocoa plots were less systematically cleared than at Koffikro. Many growers were retired. They still had the financial means to buy agricultural inputs to maintain their cocoa plantations. Retirement and pensions thus represent a non-agricultural determinant of diversification.

4.4.3.4 Diversity of Uses and Markets

Oil palm offers various marketing opportunities: sale of palm bunches to a factory, sale to women artisans and home extraction of oil. Thus, although they may have eventually changed their mind, some farmers said that ‘the product from oil palm never remains in the hands of the farmer’. We interviewed 17 women artisans, all close relatives of farmers in our sample. They were artisanal extractors of oil for sale, in addition for family consumption. During periods of high production, these women extracted an average of 536 l per month per woman. There were many of them in the village of Koffikro with this village offering easy access to urban markets.

This activity was especially profitable when the palm bunches came from their husband's or a relative's farm. In such a case, a part of the plantation's profits were ploughed back into the purchase of agricultural inputs and payment of labour (contractual and daily). If, on the other hand, the bunches had to be purchased, the meagre income from this activity only covered the processing expenses. Nevertheless, as all the women producers of palm oil affirmed, these revenues did provide them with a certain degree of financial independence.

4.4.3.5 Effect of Cocoa Prices on the Decision to Clear Cocoa Trees and Replace Them with Oil Palms

The first period of clearing of cocoa trees to replace them with oil palms was around 1985. Though cocoa prices had been rising in nominal terms, they remained stable in real terms—which does not explain the dynamics of the clearing of cocoa plantations (Fig. 4.6). During the second period of clearing of cocoa and planting of oil palm, around 1995, cocoa prices had fallen sharply over the preceding 5 or 6 years. Then they rose again after the clearing of cocoa. This phase probably played a role.

4.4.3.6 Public Policies and Revenues from Cocoa and Oil Palm

The first two phases of oil palm adoption coincided with development projects. These projects, with their initial contribution of capital and information[3] necessary for investment in hybrid oil palm cultivation, were major determinants. As in the Lagunes region, the third phase of the adoption of oil palm, around 1995, did not involve projects. It was a matter instead of individual initiatives of farmers, without access to intra-sector credit. This raises the question of investment capital. For this third wave of diversification into oil palm cultivation, where did the funding for the establishment of plantations of improved oil palms—which are labour- and inputs-intensive during the first 4 years—come from?

Even though there is a lack of data to prove it conclusively, the gradual rise in prices and revenues from palm bunches certainly encouraged investments in oil palm cultivation after falling incomes from cocoa. One should also add the specificity of the oil palm in West Africa: that of obtaining capital from the felling of oil palm trees through the extraction and sale of palm wine, which offers the possibility of reinvesting a part of the capital (Nai Nai 2000; Ruf 2007).

---

[3]More than 60 % of farmers said they received information on the hybrid oil palm through supervisory and managerial personnel (employed by the Palmindustrie company or private acquirers). Others obtained information from relatives. Three decades of integrated development of the sector (including services and information) and the controlled adoption of this crop by the Palmindustrie company led to a significant circulation of information.

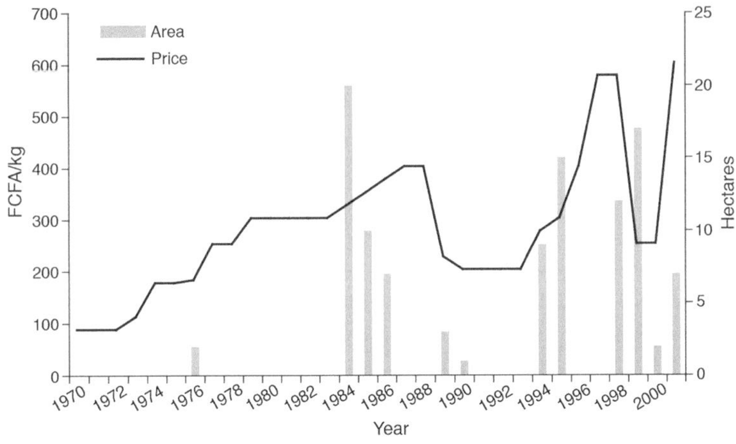

Fig. 4.6 Dynamics of clearing of cocoa plantations and the price of cocoa from 1970 to 2002

Moreover, and as suggested above, even if this variable does not appear to be statistically significant, non-agricultural income, including pensions to some degree, may have played a role in the adoption of oil palm cultivation.

Finally, it is possible that the 1994 devaluation of the currency increased interest in investing in new plantations, including those of oil palm. In fact, with the devaluation reducing labour costs, plantations required more modest investments.

4.4.3.7 Urban Markets and Location Rent for Cassava

The development of urban markets for food products has played an increasing role in the financing of diversification of perennial crops. Therefore, cassava cultivation, especially in Koffikro along the route linking Aboisso to Abidjan, merited some attention. Land prepared for planting oil palm, especially that freed by the clearing of old cocoa, was often leased out to produce cassava, before the farmer started cultivating oil palm. The capital obtained after the sale of cassava was then used to start oil palm cultivation. Once again, the trend towards diversification—here, food crops—was promoted by the development of urban markets and a location rent—proximity to a large city and a road. This seems to be a structural and general trend (Chaps. 2 and 11).

4.4.3.8 Adaptation of the Social Fabric: The Nurserymen

In the Aboisso area and more so in the Lagunes region, we observed the adoption of free or cheap unselected oil palm seeds which are less productive. This was partly explained by the low availability of initial capital and lack of access to credit, combined with a strong artisanal market for palm bunches. Nurseries sprang up and offered hybrid planting material of uncertain quality—which could not easily be checked—and which cost much less than the one available at the formal market (Cheyns et al. 2001).

4.5 Conclusion

Diversification into oil palm cultivation in the Aboisso area and the Lagunes region was the result of favourable public policy in the form of two development plans of the palm oil sector. These plans provided farmers with the capital and the information necessary for taking up hybrid oil palm cultivation. This diversification of crops, desired by the State, originally took place on available forest land. Thus, just like cocoa, oil palm was also a first-cycle crop after forest clearance. However, in a context of massive deforestation in the country, coupled with the problems of replanting cocoa, family farmers quickly grasped the opportunity to extract value from fallow and old plantations by cultivating oil palm.

From a farmer's perspective, diversification into oil palm cultivation made sense because of increased incomes: first through an expansion of his farm with the creation of new plots of oil palms, then through a simultaneous process of intensification and replacement of old cocoa plots. Both processes could take place simultaneously on the same farm (Chap. 2).

The price factor also probably played a role, but one that did not appear to be critical. The rising price of palm bunches between 1995 and 1999 probably encouraged the third wave of oil palm adoption. But the impact of cocoa prices is more difficult to assess, given the frequent changes during the same period. Thus, where forest was available, the volatility of cocoa prices did not prevent a process of investment in cocoa, but where forest was no longer available, all new investment went into oil palm cultivation. There was little or no reinvestment of revenues from oil palm cultivation into cocoa cultivation. This can be linked to the low procurement price of both products: palm bunches and cocoa. Monthly income that the farmers derived from palm remained low, leaving little scope for reinvestment. Nevertheless, the structural role of deforestation and ecological change seemed to be predominant in the diversification process.

The adoption of cassava as a food cash crop was an integral part of the process of diversification, with a determinant that was partially ecological: deforestation. To this ecological determinant were added demographic and economic factors: the growth of cities and the domestic market.

Finally, we must remember the decisive role in the distribution of revenues and cash incomes over the year in motivating farmers to consider oil palm cultivation. Even though the surveys were conducted during periods of low prices, most of the farmers in both Aboisso and Dabou who had adopted hybrid oil palm expressed satisfaction with this diversification. The regular monthly income allowed them to meet their various expenses more easily.

However, the problem of market outlets arose for oil palm growers in the early 2000s. The setting up of new processing plants and the revival of old ones did not keep pace with the widespread adoption of oil palm cultivation. Could women artisans and the informal sector absorb the entire production that was not collected by the factories? In the late 2000s, the answer seemed to be in the negative and thus provided an opening to rubber in several areas.

References

Araujo-Bonjean C, Chambas G (2001) Impact du mode d'organisation des filières agroalimentaires sur la pauvreté: la filière cacao en Côte d'Ivoire. Clermont-Ferrand, 56 p

Cheyns E, Akindes F, Aka Adie F (2000) La filière palmier à huile en Côte d'Ivoire trois ans après la privatisation: état des lieux d'un procès de recomposition institutionnelle. OCL 7:166–171

Cheyns E, Kouamé YS, Verdeaux B, Nai Nai S (2001) Adoption du matériel végétal et itinéraires techniques en plantations villageoises de palmier à huile. In: Assamoi Y, Burger K, Nicolas D, Ruf F, De Vernou (eds). Proceedings of the conference on the future of perennial crops, Yamoussoukro 5–9 Nov 2001. Cirad, BNETD, ESI-VU, Montpellier

Colin JP (1990) La dynamique des systèmes productifs agricoles en basse Côte d'Ivoire. In: Dynamique des systèmes agraires. La dimension économique. Orstom, CNRS, Paris, p 165–190

Léonard É (1997) La reproduction de la société agraire dans la région de Sassandra. In: Guillaume A, Ibo J, Koffi NG (eds) Croissance démographique, développement agricole et environnement à Sassandra, Orstom, p. 137–160

Nai Nai S (2000) Les déterminants de l'adoption du palmier à huile en plantation et en replantation. DAA, Institut national polytechnique, Yamoussoukro, Côte d'Ivoire, 61 p

Naï Naï S, Cheyns E, Ruf F (2000) Adoption du palmier à huile en Côte d'Ivoire. OCL 7(2):155–165

Oswald M (1997) Recomposition d'une société au travers de plusieurs crises: la société rurale Bété (Côte d'Ivoire). Doctoral thesis, INA-PG, 290 p + annexes

Pillet-Schwartz AM (1972) Capitalisme d'État et développement rural en Côte d'Ivoire. La Sodepalm en pays Ebrié. Doctoral thesis, Paris I

Pillet-Schwartz AM (1978) Les grandes entreprises de culture et de promotion des paysans en Côte d'Ivoire. Études rurales 70. http://etudesrurales.revues.org/document686.html

Ruf F (1987) Éléments pour une théorie sur l'agriculture des régions tropicales humides. I - De la forêt, rente différentielle au cacaoyer, capital travail. L'Agronomie tropicale 42(3):218–232

Ruf F (1991) Les crises cacaoyères. La malédiction des âges d'or? Cahiers d'études africaines, 121–122, 31(1–2):83–134

Ruf F (1995) Booms et crises du cacao: les vestiges de l'or brun. Paris, Éditions Karthala. Collection Économie et développement, 455 p

Ruf F (2000) Déterminants sociaux et économiques de la replantation. Dossier: Afrique, plantation et développement. OCL 7(2):189–196

Ruf F (2007) Planned and unplanned agroforestry systems in relation to socio-economic change: cocoa combined with palm-wine in Côte d'Ivoire. In: Second international symposium on multi-strata agroforestry system with perennial crops, 17–21 Sept 2007, Catie, Costa Rica, 12 p

Ruf F, Akpo JL (2008) Étude sur les revenus et les investissements des producteurs de café et de cacao en Côte d'Ivoire. Report to the European Union. Cardno Agrisystems, Abidjan, 90 p

Chapter 5
Development of Oil Palm Plantations and Orange Groves in the Heart of the Cocoa Territory in Eastern Ghana

Isabelle Michel-Dounias, Laure Steer, Emmanuelle Giry, Claude Jannot and Jean-Marie Kalms

This chapter helps us understand the diversification dynamics of family farms in eastern Ghana. The region is not only one of the oldest cocoa growing areas of the country but also one of its most diversified. At the beginning, only cocoa was grown here. But over the past 30 years other perennial crops like oil palm and orange have emerged, with the former gradually becoming the dominant crop. The aim of this chapter is to analyze the factors that led to this diversification, as well as assess its effects at the regional and farm levels. The development of oil palm cultivation was largely influenced by the establishment of an agro-industry that sourced and processed harvests in the region. We, however, believe other factors were also involved, which explain the coexistence of several perennial cash crops in the majority of farms in the same territory. These factors are both ecological and socio-economic in nature.

I. Michel-Dounias (✉)
UMR 951 Innovation, Institut des régions chaudes (Montpellier Supagro), 1 101 Avenue Agropolis, BP 5098, 34093 Montpellier Cedex 05, France
e-mail: isabelle.michel@supagro.inra.fr

L. Steer · E. Giry
Institut des Régions Chaudes (Montpellier Supagro), 1 101 Avenue Agropolis, BP 5 098, 34093 Montpellier Cedex 05, France
e-mail: l.steer@iram-fr.org

E. Giry
e-mail: manugiry@hotmail.com

C. Jannot
Cirad-Persyst, UPR 34, TA B-34/02, Avenue Agropolis, 34398 Montpellier Cedex 05, France
e-mail: claude.jannot@cirad.fr

J.-M. Kalms
UMR 951 Innovation, Cirad-ES, TA C-85/15, 73 Avenue Jean François Breton, 34398 Montpellier Cedex 05, France
e-mail: jean-marie.kalms@cirad.fr

F. Ruf and G. Schroth (eds.), *Economics and Ecology of Diversification*,
DOI 10.1007/978-94-017-7294-5_5

5.1 Mechanism and Method

The factory of palm oil agro-industry GOPDC (Ghana Oil Palm Development Company Ltd, part of the SIAT group) lies about 150 km north-west of Accra, in the tiny region of Kwae, Kade district, and is surrounded by its industrial-scale oil palm plantation. The factory also sources its raw material from farmers, with whom it has contractual agreements, located within a 20 km radius around its factory and its plantation. This area is well serviced by paved roads and is located within the cocoa producing zone. Although GOPDC has long promoted cocoa cultivation in the area, orange cultivation has now also gained favour in the vicinity of palm oil research centres (Oil Palm Research Centre (OPRC), now OPRI) and orange research centres (Agricultural Research Station, ARS). We chose to locate our study area in Abaam village and survey its farmers; this village is located in the east, and is in the heart of GOPDC's supply area.

Abaam village is inhabited by the Akan people of the Akyem group, much like the rest of eastern Ghana. It has been subject to an influx of many migrants in the past, mainly from southern Ghana at a time when cocoa was booming. Migration, albeit mostly seasonal, still takes place from northern parts of the country and Burkina Faso during the cocoa harvests. The population density of the village has risen steeply due to more than 50 years of migration, with an average density of 80 inhabitants/km^2.

The Kwae region has a humid tropical climate with two rainy seasons, from March to June and from September to November. The average annual rainfall, although very variable, is 1400 mm. The average monthly temperature ranges between 23 and 28 °C (GOPDC data).

The results presented in this chapter are drawn from a study based on detailed surveys and observations conducted over 6 months in several stages (Giry and Steer 2003). The physical environment was first analyzed and described in its agricultural context using several transects laid out in the village territory. We adopted common names used by farmers—which were elucidated by GOPDC experts and research (Asanoa and Tenadu 1972)—to help describe the different ecological units we encountered. Various resource persons (GOPDC agents, the village chief, older farmers) were interviewed to recreate the village's agrarian history starting from the period when cocoa cultivation was adopted. We also defined a typology of farms, representing the diversity of existing situations identified from the historical analysis. A total of 77 farms were surveyed in several stages for:

- Analyzing the trajectories of changes and activities;
- Assessing areas under cultivation and agricultural practices, and identifying cropping systems;
- Techno-economic evaluation of productive activities (quantification of yields, inputs and labour time, including for the post-harvest period).

We defined categories of farms based on different combinations of cropping systems. For each major cropping system, we calculated the annual gross value added (GVA[1]) produced per hectare worked on (ha) and per unit of labour needed (md[2]). We then calculated the total farm income generated by workers in the family (TI[3]) for each type of farm. To do so, we created a model for each farm type from our survey data (Ferraton et al. 2002; Ferraton and Touzard 2009). Economic calculations were based on 2003 prices.

5.2 Results and Discussion

5.2.1 Physical Environment and Spatial Distribution of Crops

The topography of the Abaam village area is undulating with gentle slopes (0–10 %). The peaks are rather flat, with altitudes ranging between 130 and 250 m. Soils are ferralitic, with different characteristics based on their location in the toposequence. Figure 5.1 shows a schematic representation of the spatial distribution of soil types along a typical toposequence, according to the way farmers categorize them. They thus distinguish between highland soils (*bekwai*) and soils on slopes (*kokofu*), intermediate soils (*nzema*) and lowland soils (*temang* and *oda*).

Given the region's rainfall pattern, this diversity of soils plays a role in the spatial distribution of crops. Thus, oil palm trees, which require a lot of water all year around, were found in lower, less drained areas, i.e. mainly on *temang* and the lowland soils of *kokofu*. Conversely, cocoa trees and, in particular orange trees, both of which require much less water, could be found on the highlands, mainly in *nzema* and the elevated *kokofu* areas. Habitation was concentrated on *bekwai* soils, where the oldest village cocoa plantations were also found. As for the *oda* hydromorphic soils, they were much less used for cultivation, supporting mainly rice and sugarcane. Food crop plots were found all along the toposequence. Cassava, maize, plantain and cocoyam were grown in association in these plots, in alternation with fallow periods. There were very few livestock rearing activities in the area (sheep breeding).

The physical environment thus appeared to favour a diversity of perennial crops, each occupying a specific ecological unit: the palm on one side, cocoa and orange on the other. Nevertheless, an overlap was possible on *kokofu* soils, which are also the most common in the Abaam territory. However, this spatial distribution of crops varied over time, which is what we will explore now.

[1]GVA = Gross output (GO)—Intermediate consumption (IC).
GO = quantity produced × price, for all products from the plot, regardless of their destination.
IC = inputs used + services consumed and casual labour.

[2]The work unit requested is expressed in days of work performed by a worker or a 'man', i.e., man-days (md). A work day is, on an average, 6 h longs.

[3]TI = sum of GVA – depreciation – fixed costs – staff salaries + income from sharecropping.

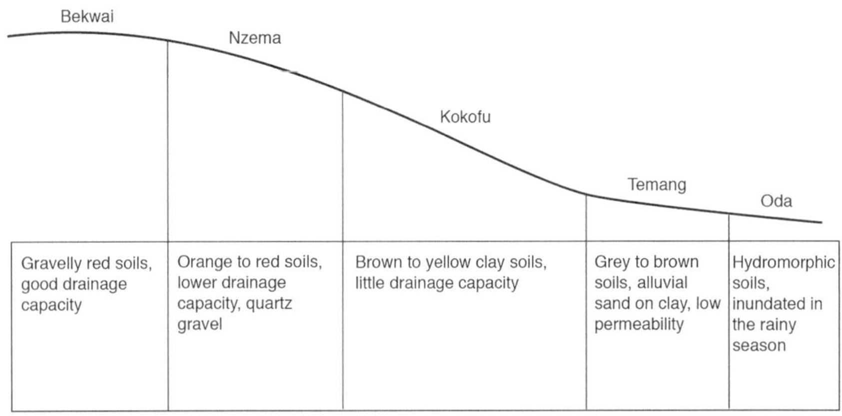

Fig. 5.1 Spatial distribution of soil types in the Abaam area

5.2.2 The Region's Agrarian History

The history of agriculture in this region can be split into seven phases. Each of them can be represented with the help of a diagram showing the cultivated areas and location of crops on the typical toposequence defined above (Fig. 5.2).

5.2.2.1 Phase 1: Prior to 1940

Dense primary forests were dominant. They supported itinerant family agriculture based on slash-and-burn systems mainly used to grow food for home consumption. However, a few cocoa plantations already existed. These were set up by migrants from the country's eastern part. A sharecropping system was established early on to manage these cocoa plantations (Amanor 1999). Lowlands were not cultivated (Fig. 5.2a).

5.2.2.2 Phase 2: 1940–1957

This phase was marked by a significant increase in population due to the influx of migrants, mainly Fanti people from coastal Ghana, and the development of cocoa cultivation. The virus that causes swollen shoot disease was first observed in cocoa in this period. The British colonial administration responded with drastic measures over the next 2 years: uprooting of cocoa trees and introduction of the Upper Amazon variety, hardier and less vulnerable to the disease. Moreover, fluctuations in international cocoa prices and social tensions caused by marketing issues resulted in the establishment of Cocobod in 1947 (Ghana Cocoa Board, which would

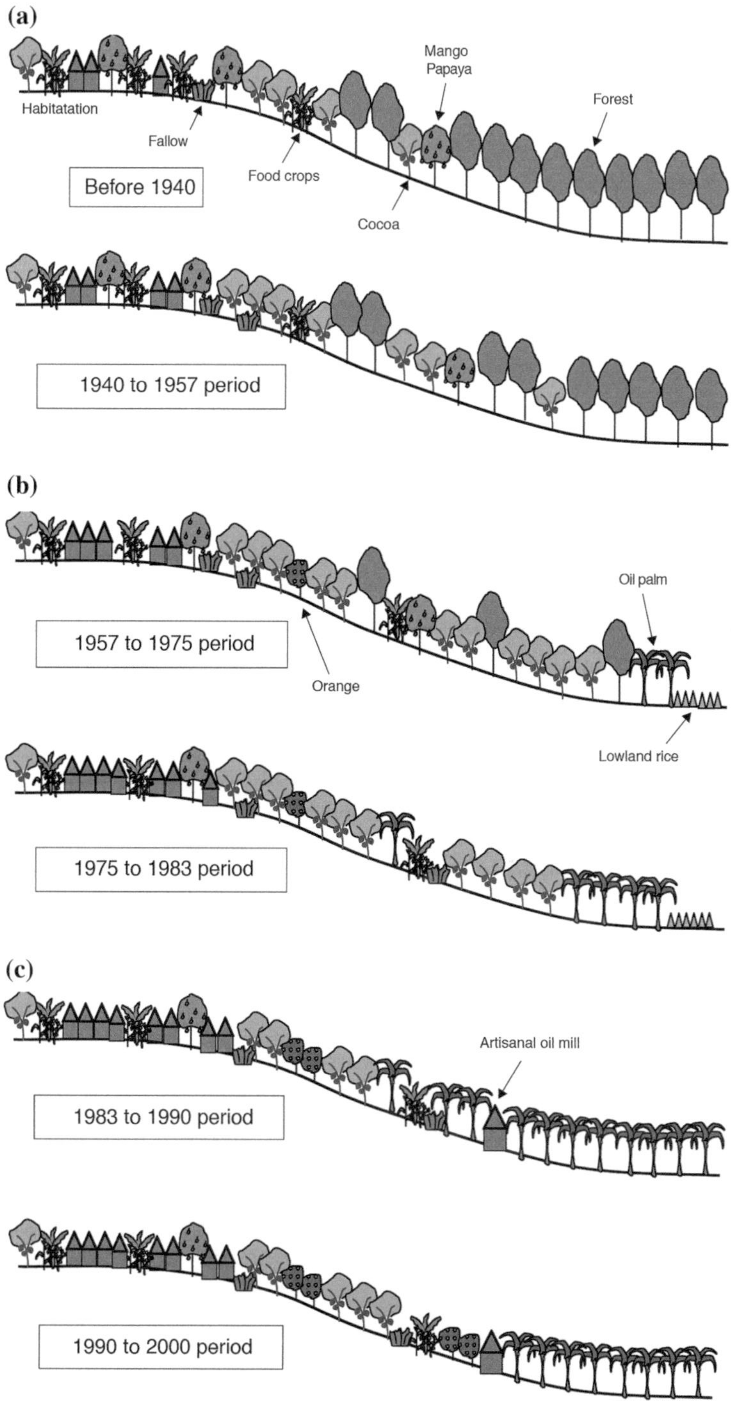

Fig. 5.2 Location of crops along a typical toposequence before 1994 and from 1940 to 1957 (**a**), from 1957 to 1983 (**b**), and from 1983 to 2000 (**c**)

partially be privatized in 1993) by the colonial administration. This official body was made responsible for managing the cocoa sector and exports. This phase was also marked by the limited introduction of rice locally, through the development of the lowlands (Fig. 5.2a).

5.2.2.3 Phase 3: 1957–1975

The swollen shoot disease finally had little impact: cocoa plantations continued to grow. The local history in the period between independence in 1957–1969 clearly reflects the model of the cycles presented in Chap. 2. The development of cocoa continued to be linked to large population increases due to migration, including from neighbouring countries. As an inevitable consequence, expansion of cocoa cultivation resulted in massive deforestation. New cocoa varieties were introduced with the help of research carried out under the colonial administration. These were subsequently promoted by the Cocoa Research Institute of Ghana (CRIG), thus contributing to further adoption of cocoa cultivation as a sole activity. Nevertheless, two new research centres were created to promote agricultural diversification: ARS for fruit trees and OPRC for oil palms. The first oil palm plantations (State farms) and citrus groves were established.

While cocoa cultivation carried on, it faced serious difficulties starting in 1969 (Fig. 5.2b). Unplanned migration, as seen in Kade, was becoming a common occurrence in Ghana and adjacent countries. The political complications that resulted compelled the then government to promulgate a law to regulate foreigners (Aliens Compliance Act) in 1969, which led to the expulsion of many migrants. This policy played a decisive role in the decline of cocoa production in Ghana in the 1970s and 1980s (Ruf 2007a). This national development applied in particular to the Kade region. It was here that the decline of cocoa started, which would mark the next phase.

5.2.2.4 Phase 4: 1975–1983

In a scenario of particularly low cocoa prices in the world market, the government, assisted by the World Bank, encouraged the development of oil palm cultivation. The GOPDC[4] corporation, then owned by the government, was founded in 1975 and consisted of a factory surrounded by a cocoa estate. It received technical support from the Institute for Oil and Oil Plant Research (IRHO). This French research institute had already worked extensively in Côte d'Ivoire before and after that country's independence and had gained a worldwide reputation for the quality of its oil palm hybrids.

[4]GOPDC was privatized in the 1990s, following which the Belgian group SIAT acquired 80 % of it.

GOPDC quickly began to diversify its supply base by promoting village oil palm plantations near the factory. It entered into contractual agreements[5] with farmers to this end. It also extended credit to help them purchase seeds, inputs and harvesting tools, and gave them technical advice, in return for exclusive rights to sell their produce. The cultivation of this crop was taken up by the farmers starting in 1979, and by 1983 it covered 130 hectares (Fig. 5.2b). On the other hand, the cultivated areas of lowland rice and cocoa on the slopes declined.

5.2.2.5 Phase 5: 1983–1990

This period was marked by the great drought of 1983 and by bushfires that destroyed many fields, including ones with cocoa. Once again, local history illustrates the model proposed in this book: ecological change, resulting here from fire damage, promotes diversification. Indeed, it was at this time that farmers converted to oil palm cultivation in large numbers, to the detriment of cocoa. This was the 'oil palm boom' (Fig. 5.2c): farmers under contract with GOPDC planted 5142 hectares of oil palm trees between 1986 and 1992 (Jannot 1998), covering much of the lower slopes.

5.2.2.6 Phase 6: 1990–2000

Villagers who were under contract with GOPDC accounted for 7000 ha of oil palm. However, the period was also marked with unexpected changes in the social environment. Private plantations that had no contract with GOPDC were established throughout the area, as were small oil mills, mainly in proximity to the urban markets in Accra. Farmers were thus able to sell their produce at a higher price than that paid to them by GOPDC since they were not constrained by the repayment of loans. This resulted in a partial diversion of the production from oil palm plantations under contract, resulting in rising tensions between GOPDC and the farmers.

This period was also characterized by a continuation, or even an unexpected incipient recovery, of cocoa, both locally as well as at the national level. Cocoa production began to rise again in the 1990s, partially on the back of a liberalization of the domestic sector which opened cocoa cultivation to competition, albeit tempered by administered prices. This recovery was mainly the result of a hike in the administered price, made possible by the gradual reduction of taxation on the sector (Ruf 2007b).

The 1990–2000 decade saw the development of orange groves on burnt fields of old cocoa on the slopes (Fig. 5.2c). A new orange variety (Valentia Late grafted on

[5]These were quite innovative contracts insofar as GOPDC managers asked farmers to copy the *abusa* sharecropping contract they themselves used with their labourers. A part of the harvest had to be turned over to GOPDC.

Rough Lemon rootstock) helped boost cultivation. This variety was created by the local research station (ARS) and propagated by private village nurseries. Unlike cocoa and the GOPDC oil palm, oranges were destined for local markets where the demand was growing rapidly.

5.2.2.7 Phase 7: 2000 Onwards

The diversification into orange cultivation slowed down in the early 2000s. Farmers lost interest in orange cultivation following a stagnation in market prices, a slump in sales and the non-construction of the orange juice factory promised by the NGO ADRA (Adventist Development and Relief Agency).[6] While orange cultivation continued in Abaam, it became less widespread. Oil palm cultivation was still on the rise (GOPDC 2002). However, farmers demanded greater freedom from GOPDC and continued to divert a part of their harvests to the local micro-mills market (Vandebeek 1999; Jannot 2001). In fact, the domestic market continued to be very buoyant due to the endemic scarcity of edible oil in Ghana (Jannot and Kalms 2003). With prices rising, cocoa again found favour with farmers who again started growing it but in a different manner. While the old cocoa trees that had survived the fires of 1983 were under the shade of large remnant rainforest trees, new cocoa plants were planted in much less shaded areas, even under full sun.

The reconstruction of Abaam's agrarian history illustrates this book's major hypotheses. Historically, we indeed find an interaction between the two major factors: on one hand, the market opportunity (palm bunches) as well as the supply of credit and information by a public (and later, private) entity (GOPDC); on the other hand, a drastic ecological change brought about by the fires of 1983. In this context, the swollen shoot disease in cocoa was not the only significant reason for the diversification process. The supply of information and planting material played key roles in the diversification into both oil palm cultivation and orange cultivation. The development of orange cultivation showed also that diversification can happen more spontaneously, without credit or organized technical support. However, a minimum inflow of information and external expertise is required.

The reconstruction of Abaam's agrarian history also showed, as is the case of most cocoa cultivation histories, that it was closely linked to different forms of migration (Ruf 1994).

5.2.3 Comparative Analysis of Cropping Systems

The analysis of farming practices has allowed us to define a variety of cropping systems. These systems are primarily defined not only by the type of crops grown

---

[6]NGO funded by the United States Agency for International Development (USAID).

on the farms but also by the technical choices made. For cocoa cultivation, these systems varied depending on the plantation's age. In fact, the preceding crops, the planting material, modalities of planting and structure of cocoa plantations have changed over time. In the case of oil palm, we distinguish between plantations that were and those that were not under a GOPDC contract. Farmers under contract routinely used improved planting material, used cover crops during the juvenile phase and used fertilizer. Practices on non-contracted plantations were much more variable. The common point was that perennial crops were systematically interplanted with food crops (except tubers) during the immature stage, including on contracted oil palm plantations. This association, a provisional form of diversification, optimized the use of the workforce through the simultaneous maintenance of both cropping systems.

In order to compare the techno-economic performance of the different cropping systems, we chose to retain, based on our survey results, only one cropping system for each major species. We thus selected four cropping systems from the existing diversity.

5.2.3.1 Cropping System for Food Crops (CSf)

This system was widely adopted by a majority of the farmers in Abaam. It is based on a pattern that combines 4 years of fallow with 2 years of associated food crops. In this system, farmers clear land and then plant cassava, cocoyam and plantain, and sow maize after the fallow period. Maize is harvested in the first year, while the other species are harvested in the second and third years. The plot is weeded in the first 2 years before the new fallow period. The production of this system is primarily intended to feed the family, with the surplus being sold in local markets.

5.2.3.2 Cropping System for Cocoa (CSc)

This system was characterized by recently planted cocoa. Selected planting material was purchased and planted on plots that were used earlier to grow food crops or on which old cocoa was burnt. In the juvenile phase, cocoa was interplanted with a cycle of food crops (cassava, corn, taro and plantain in association). In the productive phase, under light shading, an average of three pesticide treatments, three weedings and a sanitary harvesting were carried out in a year. Cocoa trees were regularly pruned. The observed planting densities were of the order of 1500 cocoa trees per hectare. Cocoa trees started providing yields after 4–5 years. The production increases and, according to our survey, levels off at around 412 kg of fermented beans per hectare per year between 14 and 35 years. It starts declining thereafter. We estimated, in concurrence with the farmers, that the cocoa tree needs to be replanted after it becomes 50 years old.

After the harvest, the beans were extracted from the pods, fermented, dried, bagged and sold to a Licensed Buying Company (LBC).

5.2.3.3 Cropping System for Oil Palm (CSp)

The cropping system for oil palm was typical of palm plantations under contract with GOPDC. On plots that had a different preceding crops (secondary forest, food crops, rice monoculture, burnt cocoa), improved palm saplings (*Tenera* hybrid) provided by GOPDC were planted. During the juvenile phase, prior to the planting of a cover crop (*Pueraria phaseoloides*), an annual crop cycle (rice or maize) was undertaken. During the production phase, weeding was done an average of three times a year. Fertilizer was applied only to juvenile plants.[7] The planting density of oil palm was 150 trees per hectare. The oil palm trees start producing after 3 or 4 years. Production increases steadily thereafter until it levels off between 10 and 20 years at 11 tonnes of fruit per hectare per year (estimated from our survey), after which it starts declining. We estimated, in concurrence with the farmers, the productive life span of an oil palm tree to be 25 years. It is then cut down and further value derived by producing alcohol, mainly palm wine, from it.

We estimated that in such an oil palm plantation, 25 % of the fruits were, on the average, diverted to be sold on the markets.

5.2.3.4 Cropping System for Oranges (CSo)

The preceding crop was mainly old cocoa that was burnt. Selected planting material was purchased and planted. During the juvenile phase, orange trees were interplanted with a food crop cycle (cassava, maize, cocoyam and plantain). Although there was a substantial amount of maintenance work (weeding and pruning of orange trees), no inputs were used. During the production phase, weeding was done an average of three times a year. The planting density was 280 orange trees per hectare. Orange trees start production in the 3rd year. Since the orange trees in our sample farms were all at the early stage of production, we studied plantations in the research centre to establish reference points for this cropping system. The survey showed that orange yields increase steadily until they level off between 10 and 20 years at 220,000 oranges/ha/year, after which they started declining. Since cultivation conditions were less favourable in orange groves in villages, we estimated the maximum production there to be 80,000 oranges/ha/year. We also estimated the productive life of an orange tree to be 30 years.

Oranges were sold to itinerant buyers from all over Ghana as well as from neighbouring countries (Burkina Faso and Togo). They either bought the production immediately after harvest, or the standing crop itself.

[7]GOPDC adopted a strategy of organic production. The farmers surveyed, even those under contract, no longer used fertilizer on oil palm plantations which had entered the production phase.

5.2.3.5 Comparison of the Four Cropping Systems

We calculated the Gross Value Added (GVA) generated over the duration of each of these four cropping systems (e.g., over 50 years for CSc, taking into account different production phases). We thus arrived at an average annual GVA that allowed us to compare the systems. In order to compare labour and land productivity of these four systems, we expressed the GVA per hectare and per unit of labour required, in labour hours per day (Figs. 5.3 and 5.4). The values are expressed in Ghanaian Cedis (1 Euro was equivalent to 10,000 Cedis in 2003).

Cocoa cropping systems had lower labour and land productivities than the other systems in spite of its price beginning to increase, an increase that was clearly insufficient in 2003. Labour productivity was particularly low due to the high workload. In fact, we estimated an average labour requirement of 88 md/ha/year for cocoa, as against 20 and 21 respectively for the oil palm and orange systems.

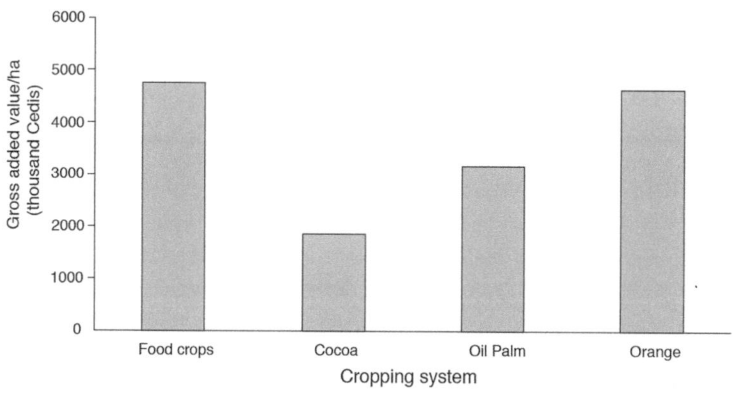

Fig. 5.3 Gross value added of cropping systems per hectare in the Abaam territory

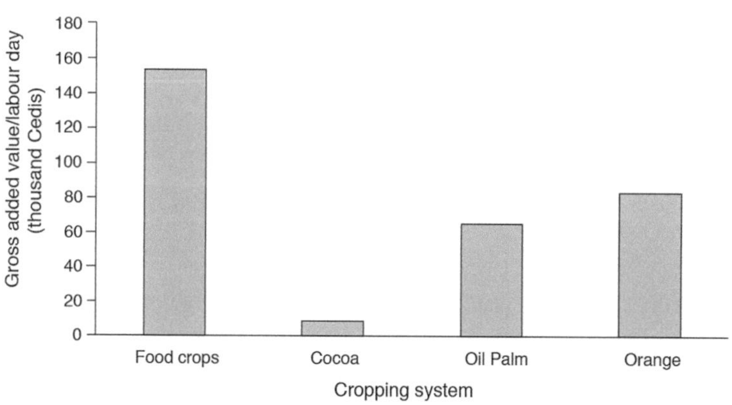

Fig. 5.4 Gross value added of cropping systems per labour day in the Abaam territory

Table 5.1 Maximum area for a worker per cropping system

	CSc—cocoa	CSp—oil palm	CSo—orange	CSf—food crops
Maximum area per worker (ha)	1.2	3.1	1.8	1.0

Source surveys

According to our results, the orange cultivation system derives the best value from the land and labour, based on prices prevalent in 2003.

With the caveat of all the inherent difficulties in estimating yields and work time for interplanted crops, we observed that cultivation of food crops was an efficient use of land and an even more efficient use of labour. The production of these food crops were almost exclusively destined for home consumption. However, the development of commercial food crops could be an interesting option in the region.

Faced with the constraints of the cocoa system, we looked more closely at labour time. An analysis of the monthly distribution of labour time for each of the four cropping systems was used to estimate the maximum surface area that a single worker could cultivate (Table 5.1).

One farmer can cultivate only a small area with food crops or cocoa. Our surveys indeed confirmed that farmers rarely cultivated more than one hectare of food crops per year. In addition, we estimated that a 0.4 hectare plot in production can feed two or three persons for a year. The only crop that allows a single worker to maintain a large agricultural surface area is the oil palm, as the work required for it is spread out over the entire year. In reality, however, a farmer rarely works alone, and he combines several cropping systems on his farm. Even though, these figures for the maximum surface area under a cropping system did not therefore reflect the reality on the farms, they helped us understand why and how farmers diversified, or why they did not. The actual surface area of farms in this region was often more than four hectares for two workers. The use of hired labour was common during peak work seasons, especially during cocoa harvests.

The comparative evaluation of cropping systems thus showed that the techno-economic performance of cocoa was still low in 2003. The amount of work it required limited the surface area under cultivation. Nevertheless, this crop did not disappear completely from the Abaam territory. In fact, it even enjoyed a revival.

Our analysis will now continue by ascertaining the determinants in the choice of species cultivated at the farm level. Who are the farmers who can benefit from combining several cropping systems?

5.2.4 Diversification and Complementarity of Income

Figure 5.5 shows the periods of crop production. Under climatic conditions prevalent in the Kwae region, oil palm and cocoa production takes place over the entire year, albeit with seasonal peaks. Oil palm production dips between May and

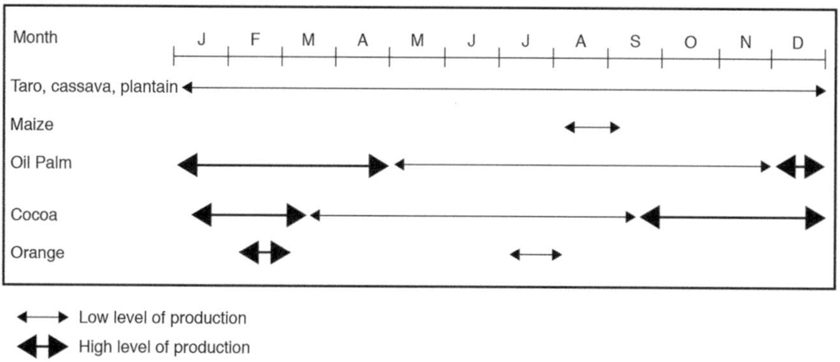

Fig. 5.5 Crop production periods

November and that of cocoa between April and August. We found that the combination of different crops within one farm could help shorten the periods of reduced production due to this seasonality of perennial crops:

- in April, while there is little cocoa production, oil palms are still producing normally;
- in July, there is a small harvest of oranges;
- in August, it is time for the maize harvest, just before cocoa begins producing again;
- in May and June, the most difficult period, it is possible to harvest tubers and plantains.

We can thus highlight the important role of food crops in the livelihood of families during periods when perennial crops produce little or not at all.

It should be noted that the January-February period is when the work on preparing new plots (clearing and felling) takes place. But this is the time when the three perennial crops are simultaneously in full production. Farmers who already owned productive plantations encountered no major difficulties in financing the preparation work for new plantations. However, the time of sowing perennial crops is in May and June, corresponding to the non-income period, which made the purchase of plants difficult.

A diversification of crops in a farm thus allows a better distribution of income over the year, as also improved food security. We shall now analyze how and why farmers combined these different crops on their farms.

5.2.5 *Typology of Farms and the Dynamics of Diversification*

Our survey on the functioning of farms showed that farmers combined various crops differently. The only common point was that farmers interplanted food crops with perennial ones. Depending on the case, this could, however, involve one, two or three perennial crops covering different areas and proportions. The maintenance of the cocoa plantations we observed varied from farm to farm: on some it was good, on others not so good.

Other elements, like the mode of cultivation of perennial crops, also varied: either directly by the farmers (with or without employees) or by sharecroppers, with similar variations on land tenure status. In this manner, we identified twelve types of farms (Table 5.2).

We noticed that the age of the head of the family was decisive in such a typology. In fact, the age was linked to the size and composition of the family, and thus to the size of the family labour force (initially without children or with under-aged ones, later with children of working age and, finally, with children who leave to undertake some other activity). In the case of inheritors, age was also related to the prospect of access to land; it defined the position in family succession. However, other factors like the farmer's origin were also significant. In fact, migrants did not obtain access to land in the same way that members of long-established families did. This had consequences on cultivation areas and the possibilities of establishing perennial plantations. The period when farms were set up, in addition to their condition, also influenced the choice of perennial crops.

Over time, access to land became increasingly difficult and complex in Abaam. To explain it in a simplified manner, access to land could be obtained through inheritance in the old families, or through sharecropping (*abunu* and *abusa*[8]) or marriage (Amanor 1999, 2001, 2005). The *abunu* and *abusa* contracts were also used between different generations within the same family in cases when the legal transfer of land to descendants took time (Amanor and Diderutuah 2001). Migrants who had settled in a place for a long time could, to an extent, be beneficiaries of plots gifted to them.

Our study on the way farms evolved clearly showed that existing farms all followed a particular development course. Their path of development differed depending on the starting point and the period when it happened. Based on current situations and their retrospective analysis, it became possible to identify major types of trajectories linking and better characterizing the different types of farms identified above.

[8]The *abunu* contract is used for cocoa and, more recently, orange trees. The owner entrusts the land to a sharecropper, who clears the plot, sets up the plantation and maintains it until it starts production. The owner then gets half the production until the time the trees are felled. Similar contracts are also common for cocoa in Ghana (Ruf 2010) and were also coming up in Côte d'Ivoire (Colin and Ruf 2009). The *abusa* contract is used for oil palm trees which cost more to set up. In this case, the harvest is divided into three parts, one for the land owner, one to pay for the costs of setting up the plantation, one for the labourer.

Table 5.2 Types of farms and crop combinations in Abaam

Type no.	Description of the type	Characteristics of a representative farm
Farms with little or no diversification		
T1	Farmers who only cultivated food crops by themselves, and let out their cocoa plantations through sharecropping. This was a typical situation for very elderly land-owning farmers (including widows)	1 worker, 0.5 ha of food crops in production (+1.6 ha of old cocoa plantations in sharecropping)
T2	Farmers who cultivated several cocoa plantations with their families; these plantations were owned or entrusted to sharecroppers, with renewal. The head of the families were young to middle-aged	2.5 workers, 0.8 ha of food crops in production; 2.8 ha of well-maintained cocoa with renewal, in ownership
T3	Older farmers with descendants, who continued to work in self-owned old and poorly maintained cocoa plantations. They also grew food crops	2.5 workers, 0.8 ha of food crops in production; 2 ha of poorly maintained old cocoa
T4	Young farmers who mainly grew food crops for home consumption with sale of the surplus, and who recently planted orange trees on self-owned land. These were rare cases	2 workers, 1.2 ha of food crops in production; 1.2 ha of young orange groves in ownership
T5	Young farmers who grew food crops and palm trees on land they owned or entrusted to sharecroppers; they started with help from GOPDC	2 workers, 0.8 ha of food crops in production; 2 ha of oil palm in ownership (1 ha in production with GOPDC, 1 ha not under contract, in immature phase)
Diversified farms		
T6	Older farmers with descendants and with lots of land. In addition to food crops, they continued to grow old cocoa, and had palm and/or orange plantations in sharecropping	3 workers, 1.2 ha of food crops in production, 4 ha of poorly maintained old cocoa (+5 ha of old palm in sharecropping)
T7	Retired people returned to the village, who still owned cocoa plantations which were entrusted to sharecroppers. After their return, they set up oil palm plantations with the help of GOPDC	2.5 workers, 1.2 ha of food crops in production; 4 ha of self-owned oil palm (half with GOPDC) (+4 ha of old cocoa in sharecropping)
T8	Young to middle-aged farmers who farmed both cocoa and palm with the help of their families on land they owned. They employed seasonal labour	2.5 workers, 0.8 ha of food crops in production; 2.4 ha of cocoa with renewal, 3.2 ha of oil palm (half with GOPDC, half immature)
T9	Young to middle-aged farmers with poor access to land; they resorted to sharecropping to develop plantations. They started with oil palm with GOPDC help and continued with orange	2 workers, 0.8 ha of food crops in production; 2.8 ha of oil palm (2/3 with GOPDC, half in sharecropping), 1.2 ha of orange (half in immature phase)

(continued)

Table 5.2 (continued)

Type no.	Description of the type	Characteristics of a representative farm
T10	Young to middle-aged farmers who cultivated all three perennial crops on small self-owned plots and with sharecropping. They started with oil palm with GOPDC help and continued with cocoa and orange	3 workers, self-owned plantations; 1.2 ha of food crops in production; 1.6 ha of GOPDC oil palm; 0.8 ha of young cocoa, 0.8 ha of young orange
T11	Middle-aged farmers who cultivated all three perennial crops on large self-owned plots and with sharecropping. In addition to family workers, they also employed permanent and/or seasonal labour	4.5 workers including 2 employed labourers, self-owned land; 0.8 acres of food crops in production; 6 ha of cocoa with renewal; 6 ha of palm (2/3 GOPDC), 0.8 ha of oranges
T12	Middle-aged farmers who cultivated all three perennial crops on large self-owned plots and with sharecropping, but with a strong emphasis on oil palm. The few cocoa trees cultivated were all old, with no renewal	2.5 workers, 1.2 ha of food crops in production, 1.2 ha of old cocoa in ownership, 7.2 ha of oil palm (half with GOPDC, half in sharecropping), 1.2 ha of orange (in sharecropping)

Sources surveys

5.2.6 *Typical Trajectories of Farms*

Ease of access to land at the time of establishment, in conjunction with the farmers' origin, allowed us to identify two main types of trajectories:

- The trajectory for young and elderly members of old families (Fig. 5.6);
- The trajectory for young farmers, and even many older ones (mostly migrants) who had poor access to land (Fig. 5.7).

In addition, we identified a specific trajectory (Fig. 5.8) corresponding to city dwellers who returned to the village when they retired. These were absentee owners of old cocoa plantations they had inherited and entrusted to sharecroppers. It is important to distinguish between sharecroppers who were involved in establishing the plantation and those who were entrusted with already established ones. In the first case, we considered that the farmer actively promoted plantation activities while, in the second, we consider him merely a sharecropper.

Young farmers, regardless of whether they were born into old families (but were yet to inherit land) or were migrants, had no easy access to land and had little capacity for financial investment. The situation was similar in the past as well. Thus, young people normally started with food crops while developing other activities, including sharecropping. Presently, most young people who want to start farming enter into an agreement with GOPDC (type T5) to cultivate oil palms and, less frequently, orange trees (type T4) (Figs. 5.6 and 5.7).

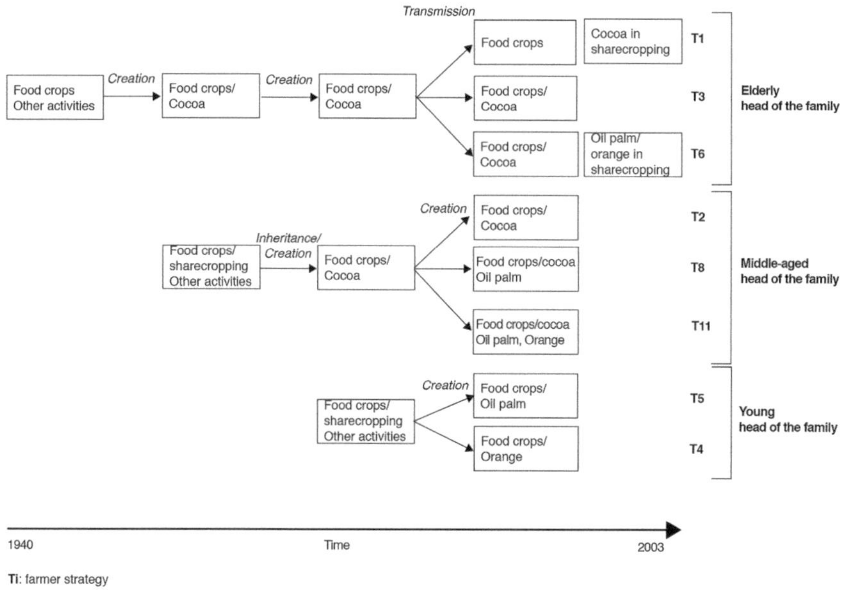

Fig. 5.6 Cropping systems and cultivation strategies of landed farmers

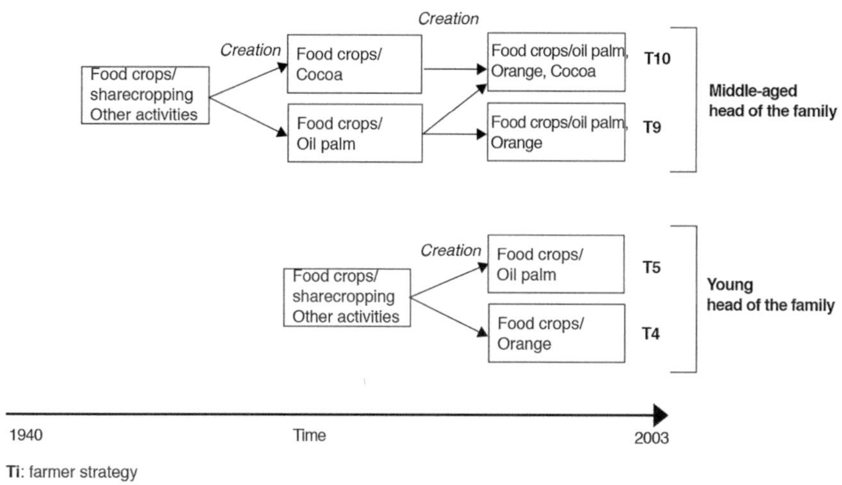

Fig. 5.7 Cropping systems and cultivation strategies of migrant farmers

It was common for young migrants to thus start with palm trees, in keeping with the trajectory reconstructed for type T9 (Fig. 5.7). In this age and social-origin category, type T10 shows that some of these farmers were willing to also plant cocoa since some time, in addition to oil palm and orange trees.

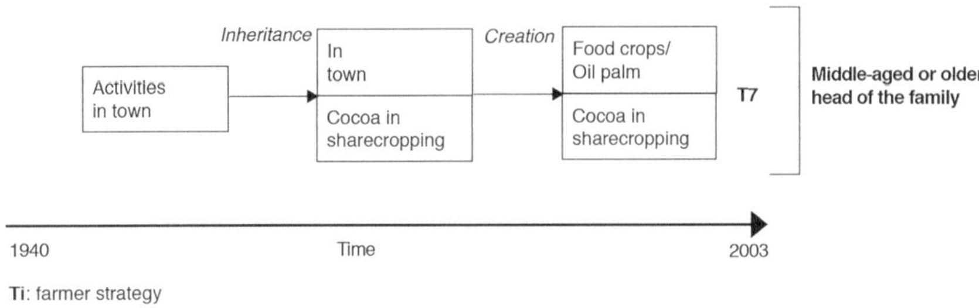

Fig. 5.8 Cropping systems and cultivation strategies of city-dwelling farmers who returned to the village

The trajectories of middle-aged to old farmers of old families were very specific for members; they were defined by cocoa plantations, as shown in Fig. 5.6. These farmers all started with cocoa plantations. The old farmers found themselves with cocoa plantations they had developed over time. They were now too old and their children had all left. These old farmers gradually either passed on these cocoa trees and plots to their descendants (type T3) or entrusted the cocoa trees to sharecroppers (type T1). Some had sharecroppers who set up oil palm and orange plantations (type T6). Middle-aged farmers, in interaction with them, pursued a somewhat different trajectory. They had received already established cocoa plots from their ancestors which helped strengthen their cocoa cultivation. They also pursued more diversified trajectories: some remained exclusively cocoa farmers (type T2), while others (type T8) planted oil palm with or without orange trees. The three perennial crops were relatively well balanced (type T11) or exhibited a clear dominance of oil palm (type T12). The trajectory of type T7 clearly shows this cocoa 'domination': even though people who had left the village and returned for retirement invested in oil palm trees, they preserved old cocoa plantations they had inherited and entrusted to sharecroppers.

We have thus highlighted various processes by which farms evolved in Abaam, with a special place accorded to cocoa cultivation. The degree of diversification of perennial crops also varied. It was not surprising that young farmers primarily chose to grow oil palm and orange trees, while the older farmers who chose to grow these crops did so with more resources. Farmers who presented the most varied trajectories were middle-aged members of old families (types T2, T8, T11 and T12). We continue our analysis by assessing the economic impact of these production choices.

5.2.7 Indicators of the Economic Performance of Production Choices

We estimated agricultural farm income by taking into account the income from land entrusted to sharecroppers and expenses, including those of temporary and

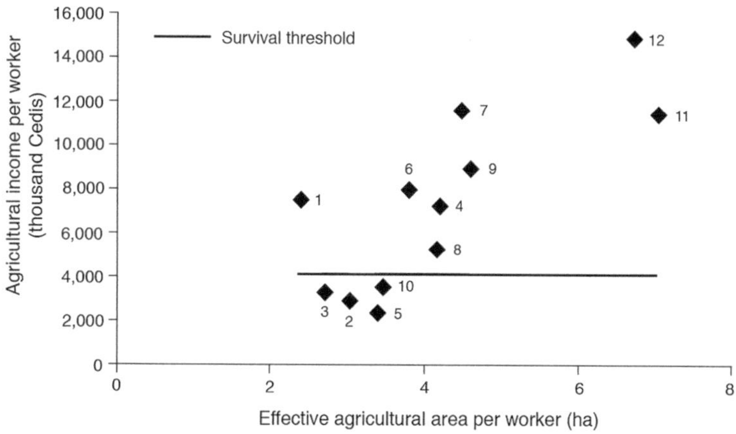

Fig. 5.9 Change in economic performance indicators

permanent labour. A survival threshold was estimated, in other words a minimum income that a worker must generate to allow for his continuation (Ferraton and Touzard 2009). In 2003, this was estimated at about 4 million Cedis per worker per year in Kade. Figure 5.9 shows the change in these indicators of economic performance. The y-axis represents the income for each family worker. The x-axis indicates the surface areas cultivated by each worker.

The first result (not surprisingly) shows that the annual farm income changed depending on the area cultivated per worker. The second result shows that the types of farms that obtained the highest agricultural income per worker were the ones that were most diversified. However, type T1 was an exception; it consisted of older farmers who generated little wealth, but benefited from sharecropping income, which could be regarded as a kind of land rent. The performance of type T12 seemed to be the highest; the farmer grew the three perennial crops over a large area with a preponderance of palm trees and neither employed sharecroppers nor a permanent workforce, but used casual labour during peak work periods.

In contrast, one third of the farms were below the survival threshold, and thus in a difficult situation; they were the ones whose operations were the least diversified (types T2, T3 and T5). Added to this was type T10 which, although clearly diversified with the three perennial crops, had small land surfaces to work on.

Should we therefore conclude that diversification is economically efficient? In part, what was seen here was primarily the economic efficiency of a farm with a minimum size of about 3.6 hectares per family worker. Efficiency was achieved subsequently, when the plantations entered into production. In fact, excluding elderly farmers of type T3, who no longer invested, the others below the poverty line, especially those in type T5, were actually in the investment phase, with a high proportion of immature plantations.

Nevertheless, in terms of return on investment, the benefits of diversification were clearly more remunerative than merely growing one perennial crop. The performance of type T12, highlighted above, illustrates a success mainly linked to two key factors:

- good complementarity with the agricultural calendar, which allows the development of a large area;
- good income distribution to be able to cope with a specific need for seasonal labour at any time of the year.

5.3 Conclusion

Our results clearly show that the majority of the farmers in Abaam started their plantation activities with cocoa. It was only the newer farms that started out by planting oil palm and orange trees, regardless of whether the farmer was descended from an old family or not. Cocoa remained the primary crop of the older farmers of the region. While some of them practiced a cocoa mono-culture, others diversified their farms by planting oil palm or orange groves.

The reconstruction of farm trajectories highlights the key role played by GOPDC and its contract-based support in developing oil palm cultivation. Young farmers took to planting palm trees more readily when they had access to credit for planting and inputs. Once the first palm plantation became productive they often developed other non-contract palm plantations. The development of orange groves, albeit more limited, was the result of a promising national and regional market, as well as access to a variety developed by a nearby research station.

In conformity with the overall hypothesis of this book, ecological factors seemed to play a clear role in triggering the development of oil palm in this cocoa-dominated region, with the fires of 1983 only strengthening these factors. We must, however, also highlight the role of prices and their impact on remuneration of labour. In the current scenario, notwithstanding the increase in the procurement price paid to cocoa producers, the economic performance of oil palm and orange plantations are found to be better.

However, diversification did not amount to a conversion at the regional level: cocoa continued to retain its overall place in the Abaam territory as well as on most farms. We even observed a revival of interest in cocoa, including by young farmers.

In fact, these three perennial crops have become complementary in the Abaam territory by deriving value from different ecological units. Furthermore, their combination allowed for a better mobilization of the labour force during the year, as well as a higher income which was also more regular over time.

Only a small minority of farmers in the region were satisfied with cultivating only one perennial crop. They were mainly older farmers, close to the end of the life cycle (T1 and T3). There were also some younger farmers who were just beginning their farming activities (T2) but they were already on a trajectory towards diversification. Through their practices, the vast majority of farmers in Abaam seemed to exhibit a preference for a polyculture of perennial crops.

References

Amanor KS (1999) Global restructuring and land rights in Ghana :forest food chains, timber and rural livelihoods. Report from the research programme:The political and social context of structural adjustment in Africa. Nordic Africa Institute, Sweden, 149 pp

Amanor KS (2001) Land, labour and the family in southern Ghana:a critique of land policy under neo-liberalisation. Report from the research programme:the political and social context of structural adjustment in Africa. Nordic Africa Institute, Sweden, 127 pp

Amanor KS (2005) Jeunes migrants et marchandisation de l'agriculture au Ghana : jeunes ruraux. Afrique Contemporaine 214(2005):85–101

Amanor KS, Diderutuah MK (2001) Contrats fonciers et contrats de travail dans la zone de production du palmier à huile et d'agrumes du Ghana. Gret, IIED, 50 pp

Asanoa GK, Tenadu DO (1972) Report on the semi-detailed soil survey of the world bank oil palm project area at Kwae, near Kade. Technical Report, Soil Research Institute, 35 pp

Colin JP, Ruf F (2009) The ant & Share contract in Côte d'Ivoire. Incomplete contracting and land conflicts. In: 13th Annual conference of the international society for new institutional economics. University of California at Berkeley, Walter A. Haas School of Business, Berkeley, 18–20 June 2009

Ferraton N, Cochet H, Bainville S (2002) Initiation à une démarche de dialogue : étude des systèmes de production dans deux villages de l'ancienne boucle du cacao (Côte d'Ivoire). Dossier pédagogique Agridoc :observer et comprendre un système agraire. Éditions du Gret, France, 134 pp

Ferraton N, Touzard I (2009) Comprendre l'agriculture familiale. Diagnostic des systèmes de production. Agricultures tropicales en poche, CTA, Presses agronomiques de Gembloux, Éditons Quae, Versailles, 123 pp

Giry E, Steer L (2003) Diagnostic agraire et dynamique de plantation en région Est du Ghana. Mémoire Cnearc, Montpellier, 95 p + annexes 23 p

Jannot C (1998) Assessment of the Ghana oil palm development corporation, 14–19 Décembre 1998. Cirad-CP, document 899. Montpellier, 55 pp

Jannot C (2001) Les déterminants historiques du commerce de l'huile de palme : entre filière artisanale et agro-industrie, quel avenir pour le palmier en Afrique ? Communication présentée à la conférence internationale sur l'avenir des cultures pérennes, Yamoussoukro (Côte d'Ivoire) Berkeley, Cirad-Tera, Montpellier, 5–9 November 2001, 28 pp

Jannot C, Kalms JM (2003) Les cultures pérennes au Ghana. Cirad-Tera, n°Tera 20/03, Montpellier, 32 pp

Ruf F (1994) Le cacao: cycles, coûts, externalités. Quelle place pour le politique. In: Benoit-Cattin M (ed) Économie des politiques agricoles dans les pays en développement. Revue française d'économie 3:149–198

Ruf F (2007a) Boom du cacao au Ghana, fruit de la libéralisation? Grains de sel (38):5–6

Ruf F (2007b) The new Ghana cocoa boom in the 2000s. From forest clearing to green revolution. European and Cirad, n°9 ACP GH 6/23. Cirad, Montpellier, 40 pp

Ruf F (2010) You weed and we'll share. Land dividing contracts and cocoa booms in Ghana, Côte d'Ivoire and Indonesia. Cirad and GTZ, Montpellier, 53 pp

Vandebeek H (1999) Case study of an outgrower project in the oil palm sector in Ghana. Thesis of master in agricultural science: Katholieke universiteit Leuven, The Netherlands, 74 pp

Chapter 6
Rubber in the Kingdom of Cocoa. The South-West of Côte d'Ivoire in the 1990s

Armand Yao and Kouame Fiko

At the time of independence of Côte d'Ivoire in 1960, the south-west of the country was still a vast virgin forest with a population density of only 0.5 inhabitants per km^2. In the late 1960s, President Houphouët-Boigny implemented a cocoa-cultivation policy encouraging Ivorian and non-Ivorian migrants to settle in the forested south. The south-west was included in this plan. Boignykro was its iconic village; it was created to accommodate Baoule farmers whose villages were disappearing under the waters of Lake Koussou, formed by the country's first hydro-electric dam. At the same time, Houphouët-Boigny formulated a real policy of diversification, primarily to oil palm and, to a lesser extent, coconut and rubber. For this diversification, state-owned estates as well as public-private ones found favour. In the south-west, which it saw as a free land for investment, the State planned for thousands of hectares of agro-industrial plantations of oil palm, coconut, rubber and trees for paper pulp.

Some schemes were implemented effectively, but others never saw the light of day because, at the same time, tens of thousands of migrants crossed the newly constructed bridge over the Sassandra and swept over the south-west to clear the rainforest and plant cocoa (Schwartz 1979, 1982; Léna 1979; Ruf 1985; Chauveau and Léonard 1996). It is partly due to this development in the south-west, following on from that in the centre-west, that Côte d'Ivoire became the world's largest producer of cocoa in 1978. But this massive deforestation by migrants would also lead to considerable ecological change.

Starting in 1989, cocoa farmers were confronted with a drop in prices, then the removal of the price guarantee to producers. As the cocoa migration had extended into the early 1980s, many farmers still managed to escape the crisis. Their plantations were young and production was rising. But for the first migrants and autochthons, the crisis was severe, with aging plantations and falling incomes. Labour

A. Yao
Ecole Nationale Supérieure de Statistique et D'économie Appliquée, Abidjan Ivory Coast, Côté d'Ivoire

K. Fiko (✉)
BNETD, Abidjan, Côté d'Ivoire
e-mail: Kouamwill_fr@free.fr

F. Ruf and G. Schroth (eds.), *Economics and Ecology of Diversification*,
DOI 10.1007/978-94-017-7294-5_6

constraints also started making themselves felt all around the country, especially in the south-west (Léonard and Oswald 1995; Léonard and Vimard 2005; Ruf 2002).

During the 2000s, agricultural landscapes changed rapidly, at least along the historical cocoa-cultivation axis: the asphalt road from Soubré to San Pedro. There were rapid conversions to rubber plantations (Chap. 2). How did rubber make an entry into the cocoa farms of south-western Côte d'Ivoire? Who were the farmers who seized this opportunity? What were their reasons for doing so? What was the connection between this diversification into rubber and the structural problems of cocoa cultivation? Did the falling price of cocoa and the liberalization of the sector play a role in the process of diversifying into rubber? Were ecological change and the aging of cocoa plantations as important as this book makes them out to be?

The objective of this chapter is to analyze the decisions of diversification into rubber and, in particular, to test the hypothesis of the role played by ecological change—in interaction with economic change—in this diversification in the 1990s.

6.1 Concepts and Hypotheses

Diversification is seen here as introduction of a new activity, in addition to existing ones. However, diversification can go all way up to conversion: the substitution of one activity by another. In our context, adopting the cultivation of rubber in a farm initially dedicated to cocoa cultivation may lead to an association of the two crops, or a total conversion of the cocoa plantation into a rubber plantation. This diversification or conversion can be driven by two requirements: increase in income and reduction in the risk of undertaking an activity. What are the factors that can influence, one way or the other, the decision to diversify?

6.1.1 Age of the Farmer

There is abundant economic literature which links innovation to youth. Young people are more willing than the old to make changes in their habits and adopt new crops. This is what Ndabalishye and Keli (2000) confirm in the case of rubber in the centre-west of Côte d'Ivoire; they note that older farmers are less open to innovation and learning than the new generation.

6.1.2 Level of Education

We can hypothesize that a crop as technical as rubber will be better accepted by growers with a minimum level of education. According to our survey on diversification and replanting of perennial crops by rubber growers in the Dabou region

(50 km west of Abidjan) and in the Bettié region (in the country's east), 66 % of farmers who have taken up rubber cultivation are literate).

6.1.3 *Origin of the Farmer*

According to the study undertaken in Dabou and Bettié, it is for the most part (92 %) autochthonous farmers who have taken up rubber cultivation. Similarly, rubber was more or less a speciality of autochthonous farmers in the 1990s in the centre-west (N'Dabalishye and Keli 2000). But migrants emerged as the new farmers in the 2000s (Ruf 2012; Chaps. 2 and 7 of this book). Since the 1990s in the south-west, migrants have been adopting rubber faster than the autochthons (Brouzro 1999).

6.1.4 *Net Farmer Income and Access to Credit*

The question of capital, and therefore of credit, a priori based on existing income, arises whenever a perennial crop is adopted, but to varying degrees depending on the crops and systems. For rubber, capital plays a major role from the very first crop cycle, especially because of the relatively high cost of clonal planting material and because of the crop's long period of immaturity. Thus, Brouzro (1999) confirms that access to credit is a driving force in the expansion of rubber groves. From 1994 to at least 2000, the suspension of credit, and therefore the lack of capital, put brakes on the expansion of areas planted with rubber trees (Chap. 3 of this book; Ndabalishye and Keli 2000).

6.1.5 *Forest Rent and Ecological Change*

For cultivating cocoa, farmers require little capital in the pioneering phases because they rely on the forest rent. On the other hand, capital becomes necessary for replanting. Therefore, the disappearance of the forest slows down cocoa expansion, but it can orient the farmers' investments towards other crops (Ruf 1987). Indeed, rubber plantations seem to come up primarily on degraded fallows and old coffee plantations (Ndabalishye and Keli 2000; Chap. 2 of this book).

6.1.6 *Land Access*

Ruf and Konan (2001) note that access to land is the first condition for an individual to invest in a perennial crop. The studies cited confirm that scarcity of land slowed down the expansion of rubber in Côte d'Ivoire, at least during the 1995–2000 period.

6.1.7 Labour Availability

Problems of labour availability arose naturally with the expansion of cocoa plantations, which provided employment to an increasing number of workers, absent any decisive increase in labour productivity (Léonard and Oswald 1995; Léonard 1997).

6.1.8 Family Life Cycle and Marital Status

The family life cycle and the aging of the farmer, with the concomitant loss of his own physical strength, impact the level of investment in cocoa (Ruf 1991). The phenomenon is also beginning to affect rubber. The farmer's marital status is an indicator of the level of his social responsibilities. A married farmer has a priori a greater amount of family responsibility than a single one. But bearing a heavier family burden also means having access to family labour, which can facilitate diversification.

6.2 Study Area and Methodology

The département of San Pedro is located in south-western Côte d'Ivoire. There is quite a lot of economic activity here, with its autonomous port from where hundreds of thousands of tonnes of cocoa are exported. In fact, the region's economy is based on and revolves around the port. San Pedro also has some areas of oil palm development. Food crops—and now increasingly rubber—are grown almost everywhere.

Historically, rubber cultivation was brought to the area by agro-industrial complexes: the Grand Béréby Rubber Company (SOGB) with 15,700 ha of plantations, the Go rubber company (Hevego) with 1,537 ha, and the SAPH rubber project with its 500 ha. These companies then embarked on a mission to promote rubber cultivation in family farms. These companies buy the entire village production to process in their factories.

Our survey dates from 2002. Nine villages were identified initially. In each village, at least one farmer was identified to serve as an interlocutor (representative) for the purposes of the survey. During the inventory phase, six villages were selected based on their accessibility. In view of the survey's limited budget we chose sites easily accessible by bush taxi. Given the limited working time (4 weeks), the size of the questionnaire and statistical processing planned, the sample size was set at 100 farmers, selected from amongst the population of farmers identified in the six villages in proportion to the workforce in each village (Table 6.1).

For the sampling, the number of cocoa growers who had diversified into rubber was increased to be able to analyze the diversification process. The sample thus broke down as follows: 2/3 of cocoa farmers who had diversified into rubber; 1/6 of

Table 6.1 Number and distribution of farmers identified and surveyed

Farmer	Village	Number of farmers identified	Initial sample	Number of farmers surveyed
1	Gnity-Touaguy	42	16	15
2	Gnake	35	13	14
3	Touih	21	8	8
4	Gabiadji	73	28	20
5	Fahe	72	27	23
6	Africa no.1	20	8	10
Total		263	100	90

cocoa growers who had not diversified into rubber; and 1/6 of rubber farmers who grew no cocoa but who had done so in the past (in this latter case, we move away from the concept of diversification and towards that of conversion at the level of the family farm).

6.3 Results and Discussion

6.3.1 Areas Devoted to Various Crops on Farms

Despite the relatively high number of farmers who had already diversified into rubber in 2002, cocoa plots still represented 20 % of the total cultivated area (in addition to 27 % of reserve lands, forests and fallow). These results reflect the reputation of the south-west region as a major producer of cocoa. The proportion of coffee, still existing but already in decline in 2002, partly reflects the "all cocoa" strategy of thousands of migrants in this region. It also reflects the beginning of the trend to clear old coffee plantations to convert them to rubber plantations (Fig. 6.1).

Adopted for the first time in 1989 thanks to a project undertaken by SAPH, rubber ranked second after cocoa. Then, other crops developed, but much more hesitantly, such as cola, teak and oil palm. Their adoption suffered from the lack of a robust system of collection and of processing. This was particularly the case of oil palm, whose bunches were processed in the traditional way by women. Unlike for rubber cultivation, there was no formal support for farmers for these crops.

Food crops included cassava, banana, yam, maize and rice, which was the dominant culture. Food crops continued to occupy a respectable position (13 %) for two main reasons:

- they were essential because of home consumption requirements;
- they represented a significant extra income. This income is beneficial in the lean season, especially for farmers who do not grow rubber and thus have irregular incomes.

Fig. 6.1 Areas devoted to various crops on farms

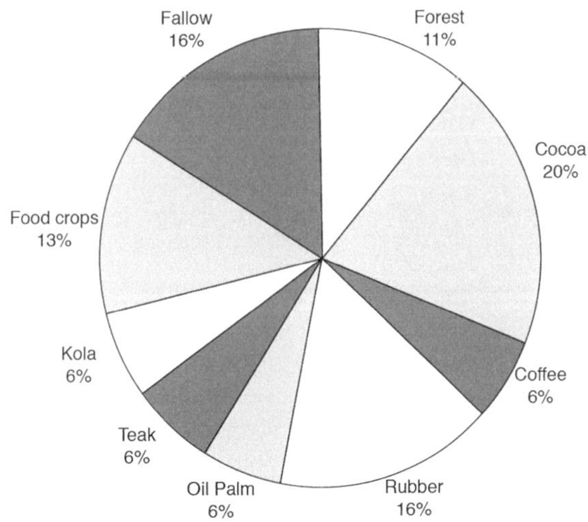

The fact that forest and fallow occupied a significant area (27 %) shows that the pressure on land, which farmers complain about, was only relative and still left room for investment. Nevertheless, this extra space was already in competition with food crops. They need a fallow period, which is an integral part of the food crops system. Furthermore, the soil types were not all equivalent. A part of the forest reserves were located in the lowlands, a priori not suitable for rubber cultivation.

6.3.2 Dynamics of Adoption of Rubber

Rubber cultivation was thus introduced to farmers in the late 1980s, at a time when the agricultural landscape was marked by the cocoa-coffee combination. It is thanks to an official project that farmers adopted this new crop, which until then had only been cultivated in large SAPH and SOGB plantations. This village rubber project really took off in 1989/1990 but then stopped abruptly in 1991 (Fig. 6.2).

From 1994, there was a gradual recovery of areas planted with a new peak achieved in 2001. The official project no longer existed but the proximity of operators of the agro-industrial sector seems to have contributed to this recovery. Two categories of actors were part of it:

- farmers who had already adopted rubber during the project. They increased their areas under cultivation, encouraged to do so by the first income from rubber, around 1994–1995;
- some farmers who had not joined the project. They took the initiative to plant rubber, convinced by the successes of their neighbours.

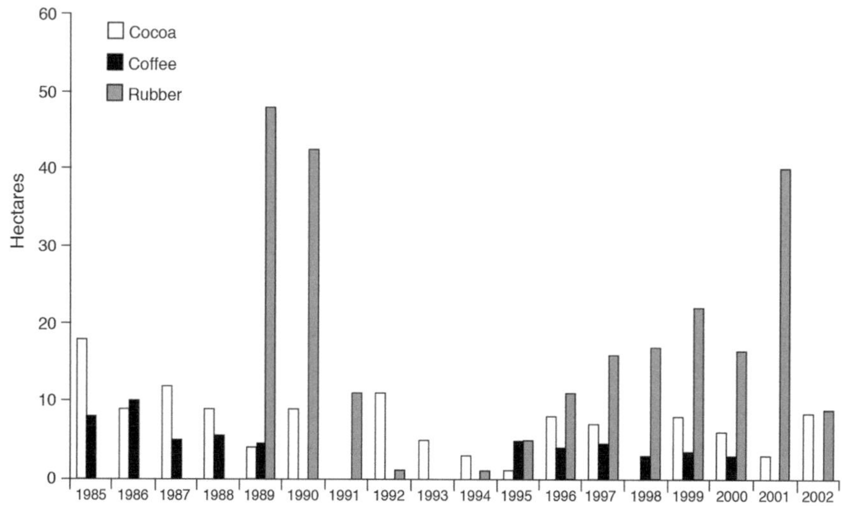

Fig. 6.2 Annual changes in areas of perennial crops cultivated by 90 farmers

This revival in the form of the creation of new rubber groves happened before the liberalization of the cocoa sector in 1999. It peaked 2 years later, in 2001. This was also the year of very low cocoa prices. This first indication suggests that the setbacks of the cocoa sector worked in favour of the adoption of rubber.

6.3.3 What Motivations and Means were Involved in the Process of Diversification?

6.3.3.1 Increase in Incomes and Regular Monthly Revenue

More than half the farmers mentioned the regular monthly income as the reason for diversification into rubber. Indeed, the regularity of income—however small—provided by rubber helps farmers cope more easily with their daily expenses.

The second reason cited was the increase in gross income. It is similar to the first reason insofar as it still revolves around the satisfaction of a financial need (Fig. 6.3).

6.3.3.2 Opportunity Offered by the Project, Ease of Maintenance of the Crop and Life of Rubber Trees

The relatively easy maintenance of rubber trees during production, as compared to a cocoa plantation at the same stage, was the third reason cited by 8 % of growers.

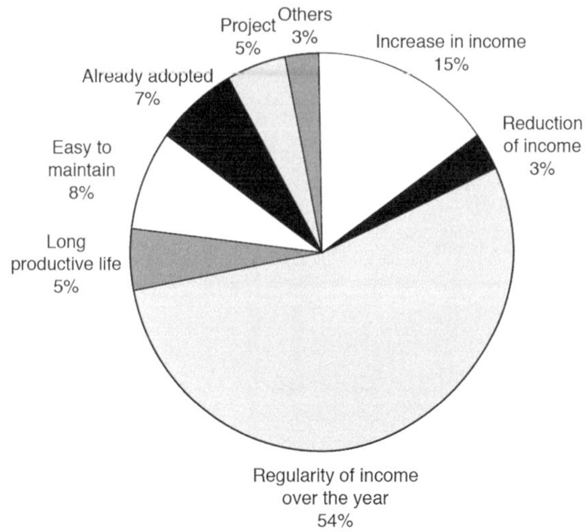

Fig. 6.3 Reasons for adopting rubber cultivation by growers in San Pedro *département*

The long life of the rubber tree, at least 40 years, was mentioned with the same frequency. Project aid influenced the motivations, even if it was only mentioned as a secondary reason by the village farmers.

Finally, the low cocoa prices and the price risk associated with its fluctuations were almost never mentioned by the farmers.

6.3.4 Constraints to Diversification

6.3.4.1 Lack of Income, Credit and Labour Availability

The constraints cited by farmers confirmed the observations described in the recent literature and initial hypotheses. The lack of credit remained the main obstacle to diversification. It was cited by 56 % of the farmers surveyed. Furthermore, these farmers complained about the cost or unavailability of labour (Fig. 6.4). Indeed, many labourers, mainly of foreign origin, had become plantation owners and had obtained land as compensation for their work.

6.3.4.2 Lack of Suitable Land

Land scarcity was a constraint to diversification for 20 % of the farmers interviewed. As suggested above, much of fallow lands were lowlands unsuitable for rubber cultivation.

Finally, support and availability of planting material were concerns for only a tiny minority of farmers (2 %). This is explained by the proximity of private organizations providing these goods and services in the surveyed area.

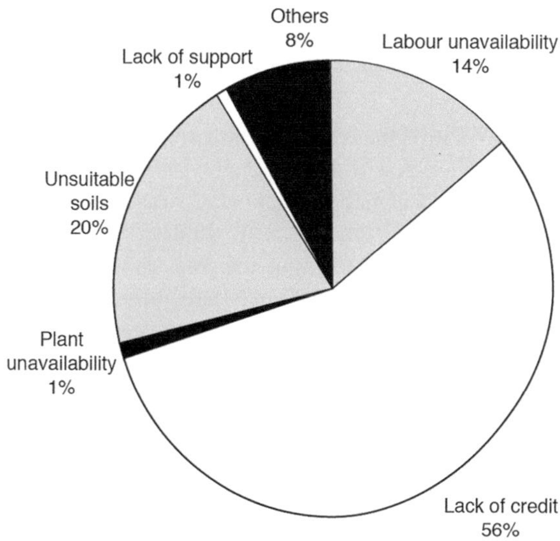

Fig. 6.4 Constraints to adoption of rubber cultivation

6.3.4.3 Income of Rubber Farmers and the Success of Diversification

The range of net income per hectare was very wide (Table 6.2). The negative minimum income shows that some farms were still in the investment phase.

The difference in incomes is significant between growers who undertook diversification and those who did not (Fig. 6.5). The mean income of farmers who adopted rubber cultivation was almost double that of cocoa farmers who had not diversified.

Table 6.2 Net income of farmers (FCFA/ha) in 2002

Characteristics	Minimum (FCFA/ha)	Maximum (FCFA/ha)	Mean (FCFA/ha)	Standard deviation (FCFA/ha)
	−15,143	732,000	182,861	177,930

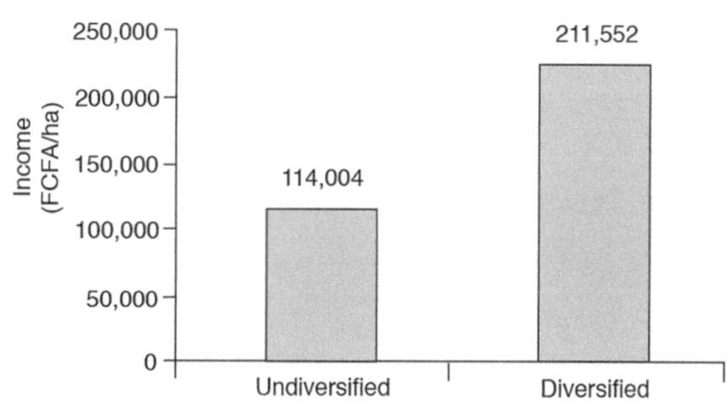

Fig. 6.5 Diversification of cocoa plantations into rubber cultivation and net income of farmers

6.3.5 Project SAPH, Liberalization of the Cocoa Sector and Imitation Effect

We saw above that the farmers hardly ever explicitly mentioned the low price of cocoa as a factor of diversification into rubber.

In 1989, the initial adoption of rubber by farmers was triggered by a project undertaken by SAPH (African Rubber Plantations Company). It provided information, credit and planting material to village farmers. Nevertheless, the fall in cocoa prices in 1989/1990 may have been an incentive to diversify into rubber.

But from 1994 to 1998, the procurement price paid to cocoa producers in current CFCA increased but it did not deter the decision to diversify. Instead, the curves of cocoa prices and annual investments in rubber plantations are almost parallel (Fig. 6.6). A similar situation prevailed in the centre-west, with a possible explanation: growers were investing revenue from cocoa cultivation into rubber plantations, even and especially when the price of cocoa rose (Ruf 2012).

In 1999, the liberalization of the cocoa sector brought about a further fall in cocoa prices. The prices recovered somewhat in 2001–2002 due to the military-political crisis in Côte d'Ivoire. At the same time, new areas planted with rubber trees reached their peak in 2000 and started decreasing in 2001. It is difficult to interpret this evolution over a time interval of 2 or 3 years. It is possible that liberalization and what it entailed—high volatility in procurement prices paid to cocoa producers—had a positive influence on the diversification process. However, the revival of investments in rubber in 2000 also stems from the impact of harvests and income from rubber cultivation obtained by its initial adopters who had participated in the SAPH project.

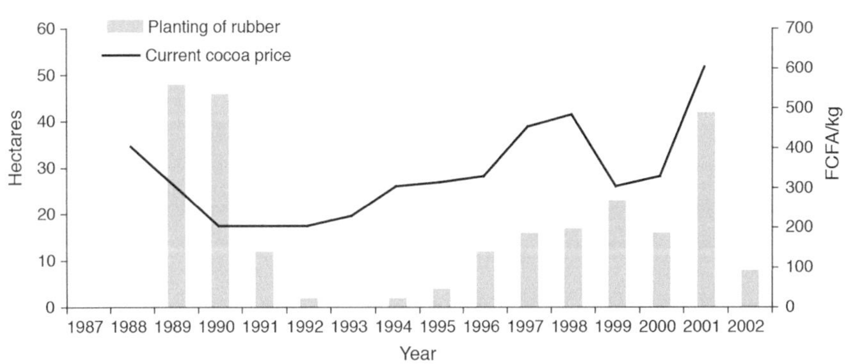

Fig. 6.6 Diversification and liberalization of the cocoa sector

Table 6.3 Results of the Logit[a] model of diversification

Variables	Estimated coefficients	Wald test	Associated probabilities
Constant	−2.7815	4.0989	0.0429
Labour availability	−0.3815	0.3631	0.5468
Access to credit	2.2040	7.0954	0.0077****
Access to land	0.4954	0.5448	0.4604
Farmer's age	0.2074	0.1118	0.7382
Net income	0.5807	2.7450	0.0976**
Educational level	−0.8456	1.3522	0.2449
Ethnicity	2.1073	2.6374	0.1044*
Marital status	1.8982	3.4820	0.0620**

Sample size: 90 farmers
(χ^2): 29.830 **** Percentage of good prediction: 76.67 %; **** Significant at a threshold of 1 %S; *** Significant at a threshold of 5 %; ** Significant at a threshold of 10 %; * Significant at a threshold of 15 %
[a]The results of the Logit model are consistent with a test of χ^2 significant to 1 %. The good prediction coefficient signifies that the model explains the behaviour of respondents to 76.67 %

6.3.6 Results of the Econometric Analysis

6.3.6.1 Credit, Income and Autochthony

Consistent with the hypotheses and the descriptive analysis, an econometric model —in this case, a Logit model—confirms that the determinant variables of the diversification process are access to credit, net income and ethnicity, as well as the farmer's marital status. They are all positively correlated with diversification (Table 6.3).

6.3.6.2 Access to Credit

Access to credit is significant at a threshold of 1 %, the highest level of significance. More the farmer has access to credit, the more he can diversify into rubber cultivation. The result is not surprising and perfectly consistent with the history of the adoption of rubber. Indeed, the SAPH project launched rubber cultivation in the region in 1989. All these farmers received government funding from this project.

6.3.6.3 Farmer's Net Income

Net income is significant at a threshold of 10 %. It is positively correlated with diversification. We are unable, however, here too, to distinguish between two possible directions of causality. But the first cause, the relatively high incomes facilitating diversification, is the one that is most likely. The higher the income of the farmer, the more he is likely to diversify. The long period of immaturity of rubber requires a minimum level of financial resources since the payback period is long. Moreover, during the surveys, we saw that the farmers at a certain social level viewed diversification as a means of asserting their position. Most of the major cocoa growers who were interviewed clearly expressed the desire to cultivate rubber.

6.3.6.4 Ethnicity and Autochthony Status

Ethnicity, or rather the autochthony status—as opposed to the migrant status, is significant at a level of 15 %. This is a low level of significance, but nevertheless real. Everything indicates that the autochthons were more willing to diversify into rubber.

This is explained largely by the activities of the SAPH project. The project's feasibility study had revealed a particular desire on the part of autochthons to grow rubber. Favoured by the project, the autochthons were quick to grasp the issues involved in rubber cultivation and the advantages that this and other projects could provide. They took the opportunity to catch up with migrant populations, who had left them far behind as far as cocoa cultivation was concerned (Chaps. 2 and 7 of this book; Ruf 2012).

6.3.6.5 Marital Status

A significance threshold at 10 % seems to indicate that the married farmers diversify more than the single ones. This result is explained by the selection criteria of the diversification project. The membership requirements stipulated the need to have sufficient family labour for various activities pertaining to the setting up and maintenance of the rubber plantation.

6.3.6.6 Not Significant: Access to Land, Labour Availability, Educational Level

The variables 'labour availability', 'access to land', 'educational level' and 'farmer's age' do not show statistically significant differences in this model. Educational level and labour availability even show a negative effect on diversification.

6.3.6.7 Quasi-elasticities

A calculation of quasi-elasticities can provide some additional information on the role of diversification-promoting factors. The results are shown in Table 6.4, by the

Table 6.4 Quasi-elasticities of the probability of diversification

Relevant variables	Estimated coefficients	Quasi-elasticities (μ)
Access to credit	2.2040	0.31
Net income	0.5807	0.08
Ethnicity	2.1073	0.30
Marital status	1.8982	0.27

use of the following equation: $\mu = [P_i(1 - P_i)] \times \beta$, where P_i is the sample adoption rate; β is the vector of estimated coefficients.

The slope of each relevant variable is calculated from the following equation: $(0.83) \times (0.17) \times \beta$.

Access to credit has the largest impact because its availability can increase the possibility of diversification by 31 %. Ethnicity is of similar importance; indeed being autochthonous increases the probability of diversification by 30 %. Marriage increases the probability of diversification by 27 %. The determinant that has the lowest impact is net income, since it increases the chances of diversification by only 8 %.

6.4 Conclusion

The study conducted in south-western Côte d'Ivoire had the goal of analyzing the decisions taken on cocoa-based farms for crop diversification towards rubber.

The first factor, chronologically as well as hierarchically, was the intervention of public policy in the form of help from companies specializing in rubber cultivation and from funding entities. Indeed, this diversification initially happened because of the implementation of a project to create village rubber plantations around the Rapides-Grah industrial plantation complex. The launch of this project in 1989 marked the beginning of a plantation dynamic that persists even today.

Without having been able to prove it conclusively here, the second factor pertains to the concept of the life cycle of cocoa, the depletion of forest rents and difficulties of replanting cocoa. This factor is explored in detail in Chap. 2. To begin with, rubber plantations were established on available spaces, and then went on to replace old coffee and cocoa plots. Fortunately for these farmers, the possibility to diversify into rubber cultivation did not come 'too late' for them. Indeed, even though cocoa plantations were aging on soils that were unfavourable to cocoa, they still managed to produce enough to be able to reinvest. Rubber tree groves, other than those planted as part of the project set up by SAPH (African Rubber Plantations Company), were made possible due to revenues from first cocoa and later rubber itself.

As is the case of the oil palm (Chaps. 4 and 5) and rubber in other Ivorian regions or other countries (Chaps. 7 and 8), the 'monthly' income also played a very important role in motivating farmers to adopt rubber. Even though slowed down somewhat by the lack of credit, diversification into rubber in south-western Côte d'Ivoire was motivated by a desire of obtaining monthly income. In this regard, almost all the rubber-cultivation adopters were satisfied with their decision. Farmers who had not yet opted for rubber also had a very positive perception of this diversification.

The liberalization of the cocoa sector in 1999, which resulted in a collapse of procurement prices paid to cocoa producers, seems logically to have accelerated a process of diversification initiated a decade earlier in this region. The prestige of growing cocoa somewhat tempered the farmers' enthusiasm for the adoption of rubber, at least for a few years. But we should emphasize the decisive role played by projects—and the credit they made available—in launching the diversification into rubber cultivation.

References

Brouzro H (1999) Analyse des déterminants socio-économiques de la pratique de l'hévéaculture en Côte d'Ivoire. End of study thesis ESA. Yamoussoukro, Côte d'Ivoire, 89 pp

Chauveau JP, Léonard É (1996) Côte d'Ivoire pioneer fronts: historical and political determinants of the spread of cocoa cultuvation. In: Clarence-Smith WG (ed) Cocoa pioneer fronts since 1800: the role of smallholders, planters and merchants. Macmillan, London, pp 176–194

Léna P (1979). Transformation de l'espace rural dans le sud-ouest ivoirien. Thèse de doctorat en Géographie. Orstom, Abidjan, Côte d'Ivoire, 345 pp

Léonard E (1997) La reproduction de la société agraire dans la région de Sassandra. In: Guillaume A, Ibo J, Koffi NG (eds) Croissance démographique, développement agricole et environnement à Sassandra. Orstom, pp 137–160

Léonard E, Oswald M (1995) Cocoa smallholders facing a double structural adjustment in Côte d'Ivoire: responses to a predictable crisis. In: Ruf F, Siswoputranto PS (eds) Cocoa cycles. The economics of cocoa supply. Woodheap Publishing, Cambridge, pp 125–150

Léonard É, Vimard P (2005) Crises et Recompositions d'une agriculture pionnière en Côte d'Ivoire. Karthala, Paris

Ndabalishye I, Keli J (2000) Diversification des systèmes de production à base de plantes pérennes dans le Centre-Ouest de la Côte d'Ivoire: la place de l'hévéa. Survey Report. CNRA, Abidjan, 27 pp

Ruf F (1985) Fonction d'un projet de développement en zone pionnière: l'exemple du sud-ouest ivoirien. In: Cirad (ed) États, développement, paysans. Colloquium Proceedings, Cirad, Montpellier, pp 184–197

Ruf F (1987) Éléments pour une théorie sur l'agriculture des régions tropicales humides. I - De la forêt, rente différentielle au cacaoyer, capital travail. L'Agronomie tropicale, 42(3):218–232

Ruf F (1991) Les crises cacaoyères. La malédiction des âges d'or? Cahiers d'études africaines, 31 (1–2):83–134, 121–122

Ruf F (2002) Changes in labor and non-labor inputs in Côte d'Ivoire cocoa farms:1997–2001. What impact on cocoa supply? Cirad, Montpellier, 60 pp

Ruf F (2012) L'adoption de l'hévéa en Côte d'Ivoire. Prix, mimétisme, changement écologique et social. Économie rurale 330–331:103–123

Ruf F, Konan A (2001) Les difficultés de la Replantation. Quel avenir pour le cacao en Côte d'Ivoire? OCL 8(6):593–598

Schwartz A (1979) Colonisation agricole spontanée et émergence de nouveaux milieux sociaux dans le sud-ouest ivoirien: l'exemple du canton de Bakwé dans la sous-préfecture de Soubré. Cahiers de l'Orstom. Série sciences humaines 16(1–2):83–101

Schwartz A (1982) Grands projets de développement et pratique foncière en Côte d'Ivoire. L'exemple de l'opération de San Pedro. In: Le Bris E, Le Roy E, Leimdorfer F, Grégoire E (eds) Enjeux fonciers en Afrique. Karthala, Paris, pp 293–300

Chapter 7
Rubber: Natural Rent, Capitalization Rent? West-Central Côte d'Ivoire and Southern Thailand

François Ruf, Bénédicte Chambon and Chaiya Kongmanee

Chapters 2 and 6 presented the process of diversification towards rubber cultivation in cocoa farms in Côte d'Ivoire. The first determinant, chronologically speaking, was the combination of the efforts of the private sector and government policy (actively supported by one bilateral aid agency) to provide information and capital to encourage the adoption of clonal rubber. Then, with attractive prices and the revenue generated by the rubber sector, as well as the convenience of a monthly income from growing this crop, the market emerged as a second determinant in the smallholders' interest in rubber. As far as procurement prices were concerned, government policies helped the rubber industry by imposing a minimal tax on it, even as the heavily taxed cocoa sector remained the proverbial 'goose that laid the golden eggs' for the Ivorian government. We also discussed the effects of learning and imitation between farmers at the village level, especially among farmers of the same ethnic or geographical origins. This proximity was also a factor in linking upper middle class investors (UMCI[1]) from Abidjan with the villagers for the management of large plantations. We observed an example of a manager of Senufo origin interact mainly with villagers of the same ethnic origin to help them adopt rubber.

Finally, we also confirmed that biological and ecological change—the aging of cocoa and environmental degradation—was an important structural determinant

[1]For a definition of UMCI, see Footnote 3 in the Chap. 1.

F. Ruf (✉)
UMR Innovation, Cirad-ES, TA C-85/15 (Bât. 15, Bur. 223), 73 rue Jean-François Breton, 34398 Montpellier Cedex 5, France
e-mail: francois.ruf@cirad.fr

B. Chambon
UPR Systèmes de Pérennes, Cirad-Persyst, 34 398 Montpellier, France
e-mail: benedicte.chambon@cirad.fr

C. Kongmanee
Prince of Songkla University, Hat Yai, Thailand
e-mail: Chaiya.k@psu.ac.th

F. Ruf and G. Schroth (eds.), *Economics and Ecology of Diversification*,
DOI 10.1007/978-94-017-7294-5_7

since it increased the difficulty of replanting cocoa. This ecological change and its consequences were briefly described in terms of differential rents in Chap. 2.

The objective of this chapter is to again look at this trilogy of 'government policy and private initiatives; ecological change; and markets and imitation', and to expand the concept of differential rent, based on the comparison of two regional dynamics.

Rubber cultivation became part of the dynamics of the region of Gagnoa (west-central Côte d'Ivoire) in the 2000s, in the heart of this old cocoa-cultivation zone. The process of regional and national diversification was well underway, even though there were tendencies observed in some farms and even villages to convert to rubber—rather than diversify into it—which posed a risk of reverting to a single-crop specialization.

In the agricultural landscape of southern Thailand, rubber has, for decades, been the dominant crop except in the lowlands, which have always been reserved for rice cultivation. Southern Thailand, however, is witnessing a growing process of specialization in rubber cultivation. At the farm level, the diversification of rice fields often results in the conversion of rice to rubber. On the other hand, the diversification process is persisting in north and north-eastern Thailand. Rubber trees were introduced in the wooded landscape consisting of fields of cassava, sugarcane and lowland rice. In this area too, the trend of rural labourers migrating to find work as sharecroppers in rubber plantations in the south plays a key role in diversification. It is often they who are the pioneers of transmitting the concept and know-how of rubber cultivation to farmers in the northern and north-eastern provinces.

These two regions are following different paths and dynamics of the adoption of rubber cultivation. But they both raise the issue of the integration of technical advances in clonal rubber cultivation.

Rubber has a good capacity to grow on poor and acidic soils. From this point of view, it would be akin to obtaining a natural rent from rubber, which other perennial crops that are more demanding in terms of soil quality cannot provide (Djkman 1951; Chap. 2 of this book). But is it not the progress made through the creation and breeding of rubber clones which gives a decisive edge to this plant?

Although not essential, we could recall the existing research work done on rubber. In the Francophone world, the French Rubber Institute and its offshoot, the Indochina Rubber Research Institute, were the first specialized organizations to be established in the tropics in 1936 and 1940 respectively. They were very close to the colonial rubber industry, which owned large plantations, and were thus directly interested in technical advancement. It was not until 1957 that an institute for coffee and cocoa was established in Côte d'Ivoire (Volper 2011). This institute played an important role in disseminating the Upper Amazon cocoa planting material in the country, but it benefitted from neither the resources of, nor the pressure from, an upstream industry. The chocolate industry owned no plantations and until the 1990s exhibited little interest in production. It was very recently—over 50 years after research began on rubber cultivation—that research on cocoa started focusing on clones.

The hypothesis on the effects of progress in rubber plantations states that the human capital invested in research and creation of clones, which incidentally results in a relatively higher cost of planting material, provides a rent of capitalization (Box 7.1). To illustrate the concept of capitalization rent, we can draw a parallel with French agriculture which reduced its dependence on natural-fertility rent in favour of capitalization rent by, for example, investing in drainage (Reboul 1977; Cottet and Perrier-Cornet, 1980, cited by Lifran 1985). As far as tropical perennial crops are concerned, the tree is often the main capital of a production system, and can be seen as an investment representing a capitalization rent. In this perspective, any work for improving or enhancing the value derived from the tree's genetic potential can be viewed as capitalization rent.

Box 7.1. Capitalization rent

The Ricardian principle is to apply different 'doses' of capital and labour to homogeneous lands (Gigou 1982). While it is true that Ricardo used it to obtain a rent for a single product, namely wheat, we can, however, apply a similar reasoning for a set of different agricultural products—for example, rubber in comparison with cocoa—to mobilize Ivorian farmers' resources of land, labour and capital.

An illustrative example is when a farmer decides not to spend 75,000 FCFA annually on labour and capital for 1 ha of cocoa, but instead invests 150,000 FCFA per year on 1 ha of rubber (including a depreciation value of the clonal planting material which constitutes a large proportion of the invested capital). Whereas an investment in cocoa generates an annual yield (here cocoa beans) worth between 350,000 and 600,000 FCFA, that of rubber is worth between 400,000 and 2,000,000 FCFA, depending on the price of rubber.

The investment per unit area is the capital accumulated in the high-performing planting material. This performance is partly explained by a favourable world rubber market in the late 2000s (lasting until 2011–2012), as well as by the combined human and financial investment in scientific research, by the physiology of rubber and by the product extracted (latex). Indeed, when we double the quantity tapped from an incision on the tree, the amount of labour required remains the same. This is not the case with cocoa or other trees grown for their fruits. If we wanted to double the number of fruits per tree, the harvesting and post-harvest labour required will almost double.

Has the work on rubber by geneticists and agronomists led to a capitalization rent? Even though the investigative methods used in Thailand and Côte d'Ivoire are different, this is a question worth asking here.

7.1 Methodology

The analysis of the rubber adoption process in west-central Côte d'Ivoire is based on a survey conducted in 2008 in 350 farms, whose basic income came from cocoa cultivation. Nearly half of these farms had adopted rubber, partly under contract with the rubber agro-industry. Based on a survey conducted earlier by Ndabalishye and Keli (2000), we selected four out of the five villages which were the most dynamic in adopting rubber. The results of the 2008 survey pertain to 170 farms which had already planted rubber, with most doing so after 2000. All the surveyed farms were located in the four selected villages.

In southern Thailand, our investigations are at a nascent stage (Delarue and Chambon 2012). We thus relied on available literature (limited, but very focused on the subject), statistics (on the whole, well maintained in the country) and pre-surveys on some twenty rice farmers.

7.2 The Facts

7.2.1 In Côte d'Ivoire

During the 2000s, there was a significant increase in smallholder rubber production. According to the association of rubber professionals, Côte d'Ivoire produced 200,000 tonnes of rubber in 2008, up from 100,000 tonnes in 2000. In 2008, our surveys led us to estimate that the threshold of 300,000 tonnes could well be surpassed in 2012 or 2013. This threshold was eventually reached in 2013 (Fig. 7.1).

Côte d'Ivoire owes this doubling of the production in 9 years—or tripling in 13–14 years—in large part to smallholder rubber farms, whose annual yields jumped every year. In 2004, they had risen to match yields of rubber estates (Fig. 7.1) and from 2005 onwards, surpassed them; ever since, smallholder production accounts for more than 50 % of the national output.

This amazing jump in performance in the 2000s caught the sector's foremost experts by surprise; none of them had predicted in the late 1990s. Given the high investments that clonal rubber plantations require, the complete withdrawal of the State—with a resulting abrupt end to credit and projects—gave rise to fears of an end to the rural dynamics of the 1980s. For example, our interaction in 1999 with the APROMAC (Côte d'Ivoire Association of Natural Rubber Professionals, http://www.apromac.ci) revealed the disillusionment of its members following the drying up of the funding. This perception, widespread at the time, was highlighted in sections of the report by Hirsch (2002), one of the leading experts on the Côte d'Ivoire rubber sector: 'For the rural sector, the freezing of major programmes has, without a doubt, ruptured the dynamics of the 1980s. A few projects stand out as exceptions, however, and challenge the assumption that the sector is likely to perish without long-term financing.' What happened between this gloom of the early 2000s and the doubling of production in 2008?

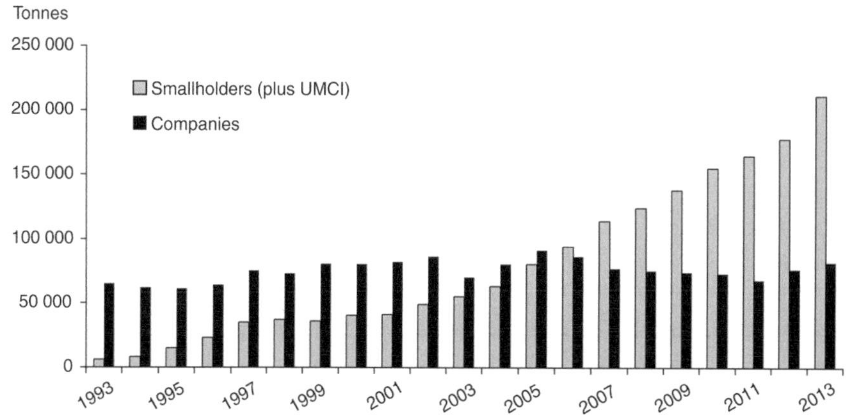

Fig. 7.1 Rubber production in Côte d'Ivoire from 1993 to 2013 (in tonnes). (*Sources* Apromac, 2014 and SAPH, personal communication, 2014)

Our survey highlighted three big waves of investment in west-central Côte d'Ivoire:

- In the late 1980s, corresponding to the innovation phase through the project;
- In the 1990s, and the increases in non-project investment in rubber, especially in 1998 and 1999;
- In the 2000s, with the resumption of an exponential growth in investments in 2006 and 2007 (Figs. 7.2 and 7.3).

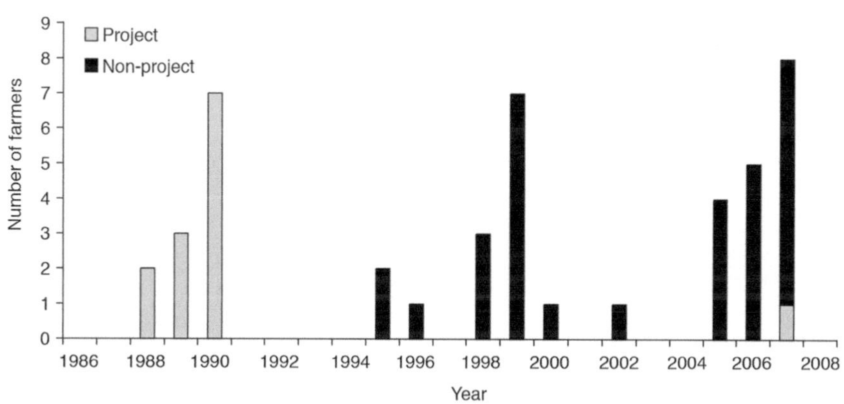

Fig. 7.2 Year of first adoption of rubber cultivation in Gnaliepa village (former Fromager region, west-central Côte d'Ivoire), for the 1986–2007 period. (*Sources* Surveys by CIRAD and A&C-Vie, 2008)

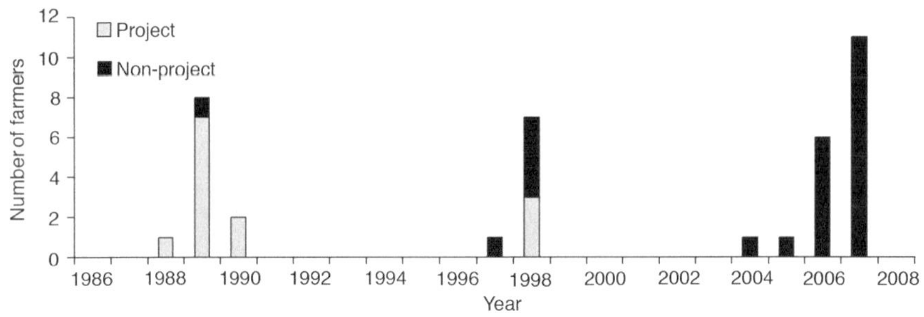

Fig. 7.3 Year of first adoption of rubber cultivation in Tehiri village (former Fromager region, Côte d'Ivoire), for the 1986–2007 period. (*Sources* Surveys by CIRAD and A&C-Vie, 2008)

7.2.2 In Southern Thailand

Rubber cultivation has dominated the landscape of southern Thailand for the past several decades. However, the lowland regions, with its rice paddies, have always contributed a little bit of diversity. Even though these rice fields evoke a picture of perfectly irrigated rice on alluvial plains, they represent, in fact, either a relatively extensive cultivation of rainfed rice in flooded fields or poorly managed irrigation. The farmers' cropping practice consists of one cycle per year (Robert 2008), using a local variety that is well adapted to the environment and is disease-resistant, but whose yield is low (Simien and Penot 2011). Nevertheless, all available information indicates that this system is on the decline.

National statistical data shows that the area covered by rice fields fell sharply in southern Thailand between 1993 and 2009, from about 500,000–300,000 ha (National Statistical Office 2003; Office of Agricultural Economics 2010).

At the local level, according to a study conducted between 1976 and 1990 in the watershed areas of Phatthalung province in southern Thailand, nearly 25 % of the lowland rainfed rice fields were converted to rubber plantations. The conversion rate was even higher between 1990 and 2006: more than 50 % of the rice fields were replaced by rubber plantations (Pensuk and Shrestha 2008). This trend seemed to continue, since there were reports of a sub-district in the same province witnessing the conversion of half of its rice fields to rubber between 2008 and 2010 (Anon 2011). Two recent studies undertaken in southern Thailand refer to the conversion of rice fields into rubber plantations (Simien and Penot 2011). Following a process of expansion of rubber from the hills to the mountains, these plantations are colonizing the plains and, more recently, the lowlands (Robert 2008). While rainfed paddy fields began to be converted into rubber several years ago, the conversion of irrigated fields is more recent.

Some farmers have converted all their lowland rice fields into rubber farms, while others have retained a small rice field (2–3 rai, equivalent to 0.3–0.5 ha) to provide for family consumption. This is also a way to ensure they obtain quality rice, free from pesticide pollution.

7.3 Prices, Incomes and Imitation

7.3.1 In West-Central Côte d'Ivoire

7.3.1.1 Relative Prices and Imitation

The tripling of the procurement price per kg of rubber paid to producers between 2001 and 2008 has certainly been a pivotal factor in the frenzy of investments that were made in rubber across the country. However, in the second stage, particularly between 1997 and 1999, the high price could not, by itself, explain everything. For example, the adoption of rubber in the Gagnoa region reveals a combined effect of the price and the imitation effect, clearly discernible in Gnaliepa in particular (Fig. 7.4).

Under the impact of devaluation, increased international prices and a very low taxation of the sector, the rubber prices obtained by farmers jumped in 1994. The first rubber farms were established between 1986 and 1988, and it was only 6–8 years later that the first non-project farms came up. This lag was equivalent to the period required for rubber to enter into production: theoretically 5–6 years if maintenance is perfect or 6–7 years in practice. The imitation effect took hold independent of rubber prices. Neighbours realized the potential of rubber when they saw that rubber growers obtained a high and monthly income, unlike cocoa farmers who went for months without income. For the coffee and cocoa farmer, this was a revolutionary aspect of growing rubber. Such farmers were suddenly willing to plant rubber trees. Subsequently, this imitation effect was given a further boost by the price rise—which translated immediately into increased revenue (Ruf 2012).

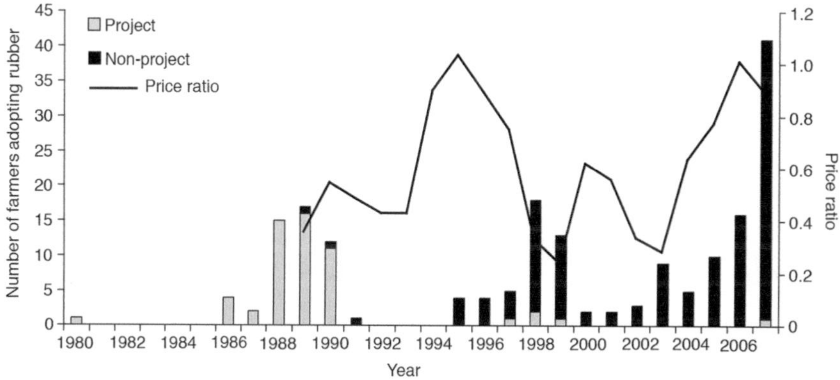

Fig. 7.4 Year of first adoption of rubber and the rubber-cocoa price ratio during the 1980–2007 period. (*Sources* Surveys by CIRAD and A&C-Vie, 2008)

7.3.1.2 Antagonistic Effects of Relative Prices

Cocoa prices experienced cyclical increases in 1998 and 2003, which resulted in a decline of the ratio between prices of rubber and cocoa. Yet, investments in rubber plantations continued unabated. How can this be explained? In villages like Liliyo, where rubber had become the dominant crop, and where every house had its own rubber nursery, all extra income from cocoa was partially reinvested into rubber. Côte d'Ivoire remains a cocoa country but this is a very indicative sign of the nascent crop revolution spreading across the country.

7.3.1.3 Expected Income

A look at the total income on the basis of average incomes from 1 ha of cocoa and rubber removed any doubt of the financial advantages of cultivating rubber, especially from 2000s onwards (Fig. 7.5).

7.3.2 *In Southern Thailand*

7.3.2.1 Relative Prices

In the 1990s, the 'dry rubber/rice' price ratio stagnated between 4 and 5, and reached 8 between 2003 and 2004. Since 2000, increases in the price of natural rubber (USS grade 3) in comparison to the price paid to farmers for rice have favoured rubber cultivation, and inevitably supported the decision to convert to this crop (Fig. 7.6).

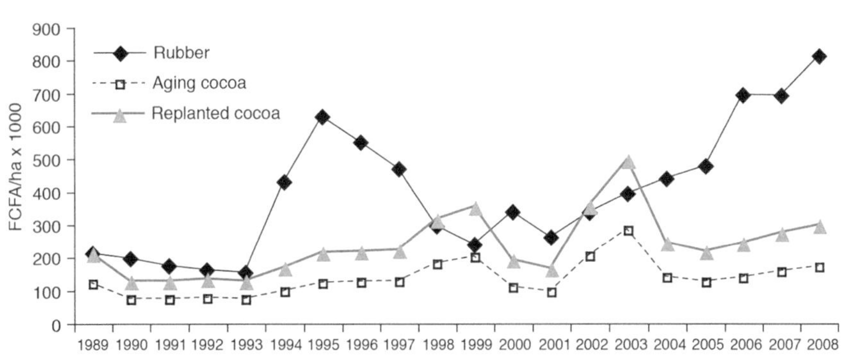

Fig. 7.5 Estimation of gross average income from 1 h of cocoa and rubber in Côte d'Ivoire in the 1989–2008 period. (*Source* Ruf 2012)

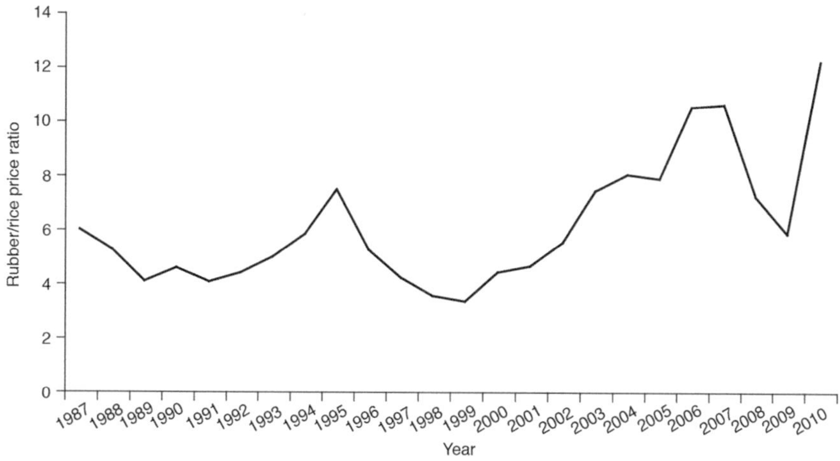

Fig. 7.6 Rubber/rice price ratio in southern Thailand. (*Source* Office of Agricultural Economics 2011)

During the 2000s, the price-cost squeeze (difference between the prices of rice and inputs required for its cultivation) became so acute that more than half of the farmers surveyed in an area in Songkhla province stopped cultivating rice. Food security was, however, not an issue as some families had put away up to 10 years of stock. Plots in the lowlands were first turned into pastures or left fallow. From 2004 onwards, there was an accelerated conversion of these erstwhile lowlands rice fields into rubber farms (Robert 2008). To summarize: a low price for rice coupled with rising rubber prices led to this apathy towards rice.

7.3.2.2 Expected Incomes

It was easy for farmers, including those who cultivated rice, to estimate the incomes obtained by a neighbour who already had diversified into rubber. By estimating the average gross income from 1 ha of rice and of rubber on the basis of national statistical data on yields and prices, there is no doubt about the income gap, especially starting in the 2000s (Fig. 7.7). This situation is similar to the cocoa-rubber combination in Côte d'Ivoire.

Under conditions prevalent in 2008, the annual income from 1 ha of rice was poor, and amounted to less than 300 €/ha, whereas even under unfavourable conditions, rubber generated an income of close to 1400 €/ha. Labour productivity in rice fields was also much lower than in rubber plantations: 19 €/workday in rice fields as against at least 43 €/workday in rubber plantations (Robert 2008). The decreasing profitability of rice farming forced farmers to opt for more remunerative crops such as rubber (Simien and Penot 2011).

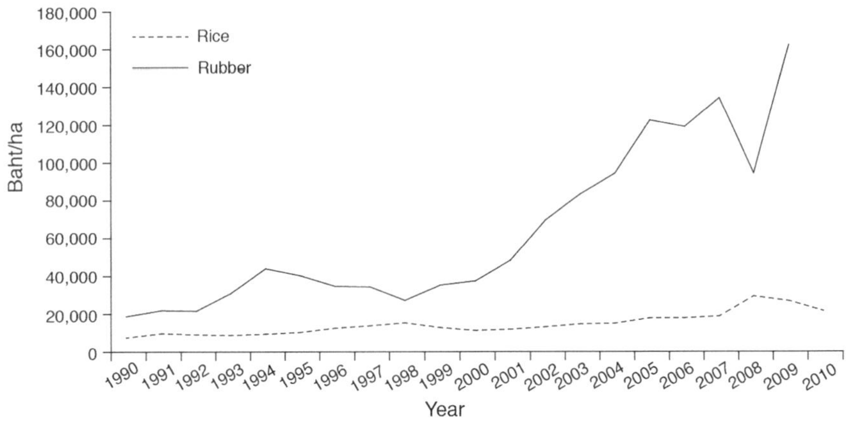

Fig. 7.7 Estimation of gross average income from rice and rubber in southern Thailand. (*Source* Office of Agricultural Economics 2011)

Most farmers who decided to plant rubber in their lowland rice fields were aware that the agronomic performance of the new crop in these conditions would be lower than in areas that were not flooded: in lowland areas, higher seedling mortality and slower growth delays the start of production of rubber trees. The average yield was estimated at 1400 kg/ha/year in the lowlands, as against 1600 kg/ha/year in non-flooded area (Pacheerat and Somboonsuke 2010). To compensate for this discrepancy, some farmers use more fertilizer on their plots to boost yields, thus increasing their production cost. Nevertheless, they reason that income from rubber will still be higher than from rice. Rubber was also preferred over fruit trees which were already abundant in the area and required higher investments than rubber plantations.

7.3.2.3 Future Income from the Sale of Timber

The expected income from the sale of latex is not the only remuneration farmers expect when they make the conversion from rice to rubber. Planting rubber trees, even in unfavourable areas, also allows them to capitalize on the value of timber (Robert 2008). The price of rubberwood has actually surged with the rise in the price of natural rubber. In fact, expectations of higher incomes cause farmers to delay felling and replanting activities on their farms, resulting in a reduced supply of rubberwood. It was estimated in 2010 that the income from the sale of timber exceeds the annual income derived from latex for a farm that is in full production and where tapping is done using family labour. The soaring price of rubberwood is also related to government-imposed restrictions on the use of forest species.

7.4 Public Policies: Conflict Between Two Strategies

7.4.1 Promoting Cocoa-Rubber Diversification in Côte d'Ivoire

In 1988, Gagnoa, in west-central Côte d'Ivoire, was among the last few regions to benefit from a rubber village programme. This programme was undertaken between 1988 and 1990. Rubber cultivation was introduced by SAPH (African Rubber Plantations Company), which was then a semi-public company with public capital, at the rate of 1 ha per farmer per year. A number of village heads ignored the project as rubber was still largely unknown in the region. The main reason advanced for this rejection was the perceived lack of a market to sell the produce. Some of the farmers who did accept to become part of the project were better informed; they had heard of rubber grown in villages in other regions, especially in Angédédou, where it was introduced in 1968.

The main reason, however, that led these pioneers of rubber cultivation in their village to join the project was that it would fund the entire investment. Nevertheless, save for a few exceptions, the projects all stopped after 1991. Farmers who wished to grow rubber were obliged to do so without project help. SAPH took the decision, however, of maintaining its team of extension workers, a decision considered vital by the first rubber adopters. Without the support of these advisors, the first smallholdings in this region would have been abandoned (Ruf 2012). We therefore see, on the one hand, a very strong support for the promotion of rubber farms by government policies and private companies through projects and, on the other, an abrupt discontinuation of these projects, even though the support system was maintained.

7.4.2 Replanting Rubber and Opposition to Rice-to-Rubber Conversion in the Lowlands of Thailand

A consensus is emerging in the literature recognizing the commendable effectiveness of ORRAF (Office of Rubber Replanting Aid Fund) in the replanting of rubber plantations in Thailand. The Rubber Research Institute of Thailand (RRIT), on the other hand, opposes the establishment of rubber plantations in the lowlands, which it considers unfavourable for rubber. The agronomic performance of rubber plantations (growth and production) in lowlands does remain limited. Any initiative to convert to rubber in the lowlands is not eligible to receive support from ORRAF, whose main focus remains on replanting in the south. Such plantations are also not eligible for aid meant for new plantations for the same reason: unfavourable zones for rubber cultivation. The innovation to 'convert rice fields to rubber plantations' was therefore undertaken without any support of public policies. It was solely an initiative by farmers, who were able to self-finance these plantations using other agricultural or non-agricultural income and sometimes through bank loans.

However, in areas where the conversion of rice to rubber plantations happened long ago, some farms have reached the replanting stage. In such cases, ORRAF can finance the replanting, even if the plots are located in unfavourable areas. However, an official from ORRAF noted that under such conditions, farmers have to self-finance about 30 % of the cost of establishing the plantations, while the financial burden borne by farmers in non-flooded areas, who are eligible for ORRAF's support, is only 10 %. Planting or replanting rubber trees in the lowlands therefore requires farmers to have access to some funds.

7.5 Ecological Change and Environmental Exploitation

7.5.1 In Côte d'Ivoire

All these villages involved in rubber cultivation have done so in response to ecological change and the aging of cocoa trees—and, of course, to rising rubber prices. In terms of the surface areas planted, we can clearly perceive the relationship between the discontinuation of all investment in coffee and the gradual decrease in investments in cocoa, on the one hand, and the spectacular increase (in three distinct waves) of investments in rubber (Fig. 7.8), on the other.

In a village like Gnaliepa, many coffee and cocoa plantations of the 1970s have disappeared from the landscape. There are two reasons for these disappearances: accidental mortality resulting from the great West African drought of 1983 or,

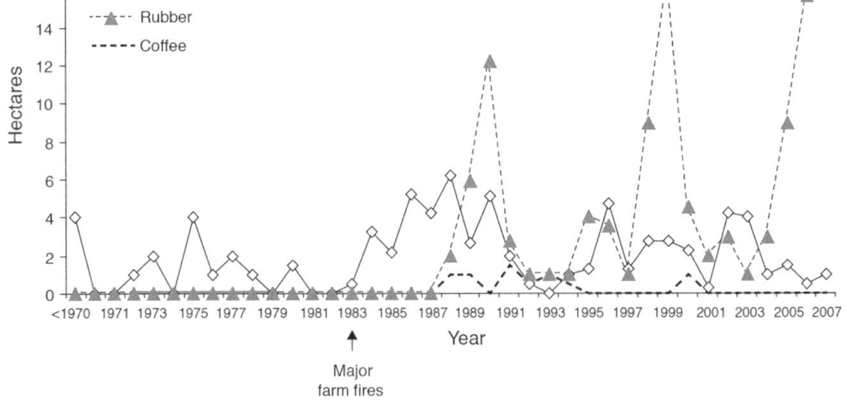

Fig. 7.8 Area planted annually with cocoa, coffee and rubber in Gnaliepa during the 1970–2007 period. (*Sources* Surveys by CIRAD and A&C-Vie, 2008)

simply, natural mortality. Farmers tried to re-establish their cocoa plantations following the natural disaster of 1983, but encountered several difficulties, including seedling mortality and a lengthening of the unproductive phase. These attempts at replanting cocoa were, however, disrupted by each successive wave of adoption of rubber.

Farmers also ran a number of risks in cultivating cocoa. According to them, the variation in yields from 1 year to the next is greater for cocoa than for rubber. More importantly, the risk of mortality of cocoa seedlings is much higher than for rubber tree saplings when replanted on soils used to grow cocoa for two or three decades.

7.5.2 *In Thailand, the Production Risk Is a More Significant Determinant Than the Price*

Our initial investigations suggested that the belief that conversion was primarily determined by the rubber-rice price ratio required qualification. A number of farmers affirmed that the price of natural rubber did not influence their decision to convert rice fields to rubber plantations.

In fact, the first reason that the farmers spontaneously advanced to explain the conversion of their lowland rice fields to rubber plantations was the decline in their rice yields. It was a matter more of production risk than of market risk. To begin with, the yield of lowland non-irrigated rice varies from 1 year to the next much more than that of rubber (Fig. 7.9). The risk that stems from this variation can be anticipated and managed, for example, by non-agricultural activities. But after several years of bad harvests, farmers give up cultivating rice and turn instead to rubber. According to the farmers, this decrease in rice yield is mainly due to water

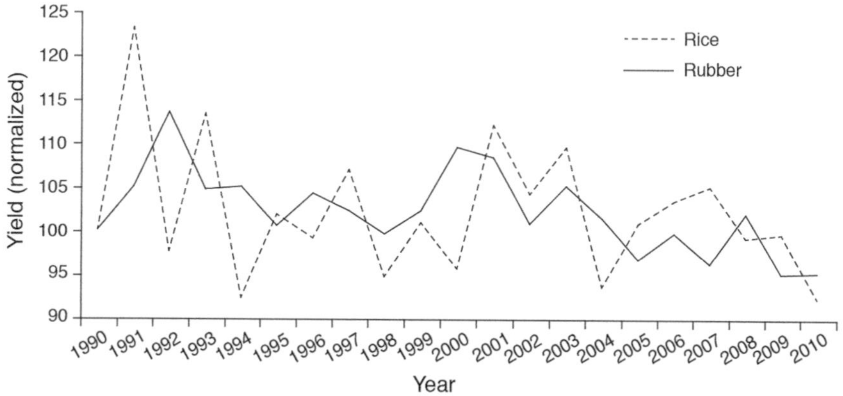

Fig. 7.9 Changes in normalized yields (1990 = 100) of rice and rubber in southern Thailand during the 1990–2010 period. (*Source* Office of Agricultural Economics 2011)

availability: the water level in the paddy fields is insufficient to ensure a good harvest.

Surveys suggest that the first conversions (more than 20 years ago) were made by farmers who owned only lowlands. Faced with declining yields and, consequently, declining incomes, the only solution was to innovate by planting rubber trees.

The phenomenon of conversion then took hold. In fact, the water supply to some paddy fields had become limited. Channels that brought water were clogged due to lack of maintenance by upstream farmers who had already converted their rice fields to rubber plantations. It was also observed that rice fields sandwiched between two rubber plantations were quickly converted to rubber, since the shade of rubber trees hindered rice cultivation. Finally, the pest pressure in existing rice fields increased since many surrounding farmers had stopped cultivating rice. All these factors led to reduced rice yields in the lowlands and thus encouraged a conversion of the plot to rubber.

Following the decline in rice yields, farmers chose rubber as the alternative crop for several reasons:

- The existing biophysical conditions were not considered favourable to other crops (fruit trees);
- The tendency was to copy what neighbours did, especially when the innovations were seen to be successful;
- At the time the decision was taken, farmers were not aware of other alternatives (for example, oil palm).

7.6 Role of Labour in the Conversion

7.6.1 In West-Central Côte d'Ivoire

Labourers mentioned that 'there were virtually no farmer-rubber tappers here'. Indeed, smallholders rarely tapped their own rubber trees. Many of them entrusted almost all the work to hired labourers, some of whom had learned the work while being employed by rubber estates.

As we saw in Chap. 2, this heavy reliance on labour introduced an initial concept of rubber rent, partly based on low labour costs. Benchmarks of the agricultural labour market are still based on social relations created by the country's dominant sector: cocoa. Rubber farmers are benefitting greatly from this. But even though rubber plantation owners draw advantage from this form of rent, their level of income from rubber allows them to pay a higher and more regular compensation for labour than cocoa grower are able to.

We observed cases of labourers migrating from Burkina Faso, more motivated by opportunities in rubber and oil palm plantations than of cocoa. This new orientation of labourers may also encourage farmers to diversify their investments.

7.6.2 *In Southern Thailand*

By the late 1990s, a little under half of the rubber plantations were being tapped by labourers, often through a sharecropping agreement. Labour requirements for rubber tapping were considerable. According to some calculations, a 1 ha rubber plantation required twice as much labour as 1 ha of rainfed rice (Robert 2008). Some farmers, however, thought differently and believed that rice paddies required more labour than rubber plantations. It is true that rice cultivation requires a larger number of workers, but this is only for a limited period. Moreover, it is difficult to find casual labourers for rice fields as they always prefer to tap rubber instead because of the higher remuneration.

Finally, in both situations, the impact of income on the availability of labour encourages farmers to reduce or abandon their initial crop—rice or cocoa—in order to adopt rubber. In the mean time, technical innovations also became a factor.

7.7 Technical Progress, Entrepreneurial Farmers and Capitalization Rent in Côte d'Ivoire and Thailand

For each country, we draw attention to the decisive progress made by national and international research efforts in the field of clonal rubber cultivation. These developments are followed and implemented by some of the more innovative and entrepreneurial of the farmers.

7.7.1 *In Côte d'Ivoire*

Family farming benefitted directly from research conducted by the erstwhile IRCA (Rubber Research Institute), the know-how of agro-industrial plantations, as well as the support and detailed advice provided to them during the initial years of conversion. In the decade between 1980 and 1990, these factors allowed IRCA to ensure the best rubber yields (in the order of 2,500 kg of wet rubber per hectare) in smallholdings in all of Africa. In the course of our most recent survey, in 2011, we often came across cases of recorded yields that were greater than 3,000 kg/ha of wet rubber. Similar results were also obtained in Ghana (Chap. 8). Such performance, among the best in the world for smallholder plantations, had an impact on income and, consequently, on the effect of imitation amongst villagers; especially at a time when cocoa yields were on the decline. It should, however, be noted that rubber yields per hectare are also expected to decline.

The rapid growth in smallholder rubber plantations makes it increasing difficult for support and extension systems to provide the same quality of advice to each farmer. At the same time, Ivorian farmers are developing and implementing indigenous innovations, like making cuttings from rubber branches to create their own low-cost clonal planting material. It is very likely that such techniques will reduce the per-hectare rubber yields, but they also help drastically reduce investment costs. As a result, techniques like these local cuttings increase the number of farmers who adopt rubber, as well as the areas devoted to rubber cultivation.

7.7.2 *In Southern Thailand*

7.7.2.1 Introducing Clones

In southern Thailand, the gains in productivity per hectare generated by the widespread introduction of clones were all the more obvious and convincing since this innovation was introduced in a country with existing rubber plantations. More than the prices, it was the increase in rubber productivity which led to a sharp rise in incomes (Fig. 7.7). Planting material grafted on selected clones had been introduced gradually in southern Thailand in the early 1960s. Compared to seedlings grown from seed, this new material helped double average rubber yields in the late 1980s, and quadrupled yields within 20 years (Fig. 7.10). While it is likely that the rejuvenation of plantations also contributed to higher yields per hectare, the technical progress represented by 'clonal rubber' capital remained the most significant factor in this increase in productivity.

We have mentioned earlier the changes that took place in Phatthalung province: between 1976 and 1990, 25 % of the flooded rice fields were converted to rubber, and more than 50 % between 1990 and 2006 (Pensuk and Shrestha 2008). It seems certain that the progress made because of the clones and their considerable impact on yields and incomes had played an important role in the decision to convert rice fields to rubber. But for this to happen, it is necessary for the farmers themselves to have an innovative spirit.

7.7.2.2 Innovation and Entrepreneurship in Family Farming

Rubber is vulnerable to the risk of flooding. The first innovative farmers developed and implemented a technique to prevent rubber plants from getting inundated: they constructed raised beds that were more than a metre high. This technique requires

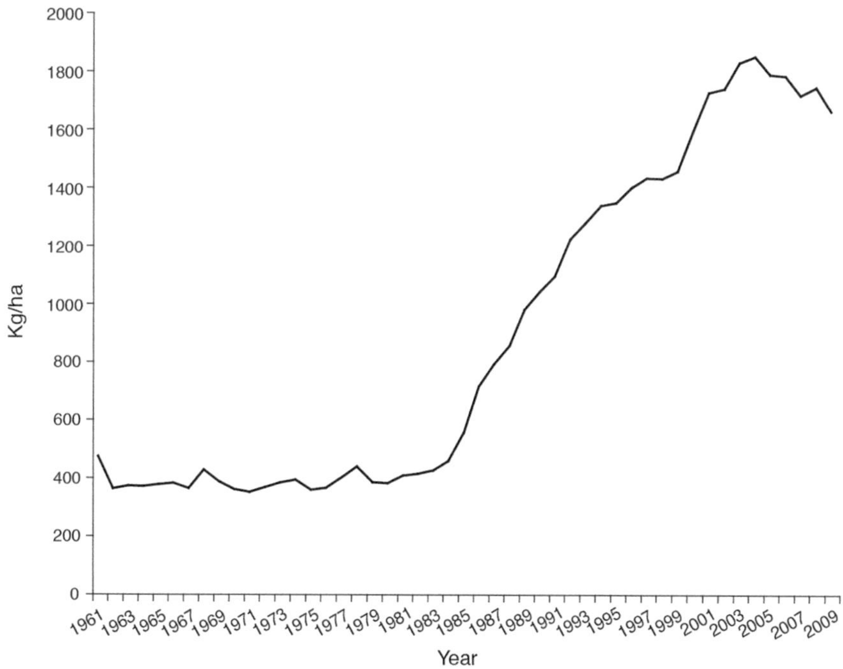

Fig. 7.10 Average annual rubber yields in southern Thailand in the 1961–2009 period. (*Sources* RITT and FAO)

adequate availability of financial and material resources (extra expense of around 25 %), institutional arrangements and genuine entrepreneurship. Farmers in southern Thailand possess all these advantages; for example, tractors and disc ploughs are very common in the region and can easily be rented.

7.8 Conclusion: Differential Rents and Schumpeterian Evolution

We observed two major determinants in the adoption of rubber in west-central Côte d'Ivoire and southern Thailand: on the one hand, prices and incomes, and, on the other, ecological change and deterioration of conditions for growing the initial crop. A third factor, government policies and private initiatives, played an essential role in Côte d'Ivoire. In the case of Thailand, the expansion of rubber cultivation also benefitted indirectly from the introduction of clones and the overall support provided to the rubber industry by ORRAF, even though public institutions were dismissive of farmer initiatives in the lowlands.

This comparison of two regional dynamics confirms therefore the key role played by prices, taxation, subsidies and the logistics necessary to provide selected plants in ensuring the success of rubber cultivation.

Above all, this chapter and this comparison illustrate well the role of ecological change and allow a prior deliberation of the rubber-adoption process based on the concepts of differential rent. Through the adaptation of the Ricardian approach to the tropical environment in terms of natural-fertility rent and capitalization rent, this comparison suggests an analogy between the soil and trees. Wild rubber obviously grows better on rich soils than on poor and acidic soils. But it is also one of those rare perennial tropical crops that can thrive on poor soils or those impoverished by decades of cultivation. Referring to the known hardiness of cassava, rubber has been called 'the cassava of perennial crops' (Jean Pichot, pers. comm.).

We can thus refer to an advantage of the 'natural rent of the rubber tree' over other crops or systems (cocoa cultivation in Côte d'Ivoire and flooded rice fields in southern Thailand in the examples described here). This natural rent from rubber plays a fundamental structural role in the growth of smallholder rubber plantations. Indeed, rubber today has a virtual monopoly on the colonization of degraded fallows in Côte d'Ivoire.

We can also refer to a capitalization rent by applying to trees the Ricardian approach used on soils. In fact, investments made on the soil, like drainage, can generate capitalization rents that exceed natural-fertility rents. We can also consider research in tree breeding as a capitalization rent. In the case of rubber, this rent resulted in a tripling, or even a quadrupling, of the yield per hectare and of labour productivity. It is a matter of a rent that has been harnessed by smallholders, new actors attracted by the opportunity and the rubber industry at the expense of other sectors such as of cocoa and rice.

This capitalization rent, which is partly responsible for the advantages of the rubber industry, stems partly from the fact that entities undertaking research are closely linked to the industry and its large plantations.

On the agricultural and land front, since planting material is relatively mobile,[2] the capitalization rent from trees can be harnessed by other countries and actors. But it takes several years of adaptation since a new agricultural sector normally takes time to establish itself. In the case of Côte d'Ivoire, it took almost 50 years to get from the idea of establishing a national rubber industry, which was first thought of

[2]As an example: 'In early 1955, drawing on its extensive experience of cultivating rubber in Vietnam and Cambodia and at the initiative of the Indochina Rubber Plantation Company (SIPH), several French companies explored Côte d'Ivoire in order to identify suitable sites for rubber cultivation. The first nurseries were set up in 1955 using three million seeds airlifted from Vietnam. These nurseries would provide the saplings to set up the first rubber plantations of SAPH.' This extract from Wikipedia (French) corroborates the firsthand accounts of former SIPH executives collected by B. Losch.

in the 1960s,[3] to a production level of 250,000 tonnes in 2012 and an overcapacity—albeit temporary—of rubber processing factories.

Finally, these innovations and reconversion processes in both countries refer logically to Schumpeter. It is worth recalling that according to this Austrian economist, innovation acts as a driving force in economies and the transformation of production systems. While it can create some new activities, innovation can also prove destructive for others. An innovation also creates a 'temporary monopoly' for the entrepreneur who introduces it: until his competitors imitate him, he retains pricing power and can impose prices.

Schumpeter's theory finds application in tropical agriculture and we encounter there many of these structural processes. The 'clonal rubber' innovation is contributing, at least temporarily, to the decline of old cocoa in the cocoa cultivating regions of Côte d'Ivoire and of rice cultivation in southern Thailand. Moreover, by changing the scale—no longer considering individual entrepreneurs but instead comparing the status of the rubber and cocoa industries—, we find this form of the lead taken by the more innovative sector, that of rubber: clonal material, innovation in the logistics of the distribution of clones, upward tapping, and stimulants.

The rubber innovation, encompassing and reinforced by all the many innovations that this crop has seen in recent decades and years, now exerts a temporary quasi-monopoly on soils acidified and degraded by decades of cultivation. Smallholders and upper middle class investors able to plant rubber are benefitting from this monopoly. The clonal rubber innovation is playing a major role in the linkages between regional economic cycles, namely the cocoa cycle and the rubber cycle in parts of Côte d'Ivoire, and the rice cycle and rubber cycle in southern Thailand.

In Thailand, public policy took the step of price protection for rice. In Côte d'Ivoire, there was an initial move to reduce taxes on cocoa, while slightly raising those on rubber. Even though these measures were useful, particularly in Côte d'Ivoire, experience has shown that they are of limited use in the long run. Since innovation is unavoidable, it is a matter of regaining the upper hand over the temporary monopoly of rubber on degraded lands.

In Côte d'Ivoire, the chocolate industry, feeling that its supplies are threatened, clearly understands this. It has started contributing to research efforts on cloning and somatic embryogenesis as well as on all possible innovations pertaining to cocoa cultivation and soil management. Realization is dawning that the rubber entrepreneurs should not be the only ones to have the quasi and temporary monopoly of degraded lands and plantations while cocoa entrepreneurs remain dependent on the monopoly—necessarily temporary—of forests.

[3]In March 1965, the *Institut français d'organisation et productivité* (IFO) published a study titled 'Étude sur l'implantation éventuelle d'une industrie du caoutchouc en Côte d'Ivoire' (Dollfus et al. 1965).

In association with this process that seesaws between diversification and re-specialization, a race towards innovation is underway. As we saw have seen in this chapter, farmers are, for the most part, committed to the way forward. It is in the interest of all competing agro-industries to come to know and understand better the innovation processes of family farms in order to create their own innovations.

References

Anon (2011). Expansion of rubber plantations in Thailand benefit only some parties. http://www.therubbereconomist.com/The_Rubber_Economist/News/Entries/2011/6/6_Expansion_of_rubber_plantations_in_Thailand_benefit_only_some_parties.html. Accessed 11 Nov 2011

Cottet C, Perrier-Cornet P (1980) L'impact du drainage sur les conditions de développement des exploitations en Côte-d'Or. Dijon, Inra ESR, 31 p

Delarue J, Chambon B (2012) La Thaïlande: premier exportateur de caoutchouc naturel grâce à ses agriculteurs familiaux. Économie rurale 330–331:191–213

Djkman MJ (1951) Hevea: thirty years of research in the far East. University of Miami Press, Coral Gables, Miami

Dollfus D, Crouzet E, De Cazenove A (1965) Étude sur l'implantation éventuelle d'une industrie du caoutchouc en Côte d'Ivoire. IFO, Paris, 263 p

Gigou JL (1982) La rente foncière, les théories et leur évolution depuis 1650. Economica, Paris, 954 p

Hirsch R (2002) L'hévéaculture ivoirienne après les privatisations et la libéralisation de la filière. Essai de Bilan. AFD, Paris

Lifran R (1985) Les systèmes de production et la division sociale du travail. In: Systèmes de production et transformation de l'agriculture, vol. 1. Inra, Paris, pp 39–48

National Statistical Office (2003) Agricultural census, Southern Region. MOIAC Technology, Bangkok, Thailand, 21 p

Ndabalishye I, Keli J (2000) Diversification des systèmes de production à base de plantes pérennes dans le Centre-Ouest de la Côte d'Ivoire: la place de l'hévéa. Survey Report. CNRA, Abidjan, Ivory Coast, 27 p

Office of Agricultural Economics (2010) Production information of agricultural yield. http://www.oae.go.th/main.php?filename=agri_production. Accessed 9 Nov 2011

Office of Agricultural Economics (2011) Price of unsmoked sheet rubber (grade 3). http://www.oae.go.th/oae_report/price/price_month_result.php. Accessed 9 Nov 2011

Pacheerat K, Somboonsuke B (2010) Comparison of rubber smallholding technology utilization between montain, rolling and plain ecological zone in Thachamoung sub-district, Rattaphum district, Songkhla province. Hat Yai, Faculty of Natural Resources, Prince of Songkla University, 246 p

Pensuk A, Shrestha RP (2008) Effect of paddy area conversion to rubber plantation on rural livelihoods: a case study of Phatthalung watershed, southern Thailand. GMSARN Int J 2 (4):185–190

Reboul C (1977) Déterminants sociaux de la fertilité des sols. Actes de la recherche en sciences sociales 17–18:85–102

Robert Y (2008) Dynamiques agraires et évolution des exploitations hévéicoles dans le sud de la Thaïlande péninsulaire, province de Songkhla. Thesis IRC, Montpellier, 99 p

Ruf F (2012) L'adoption de l'hévéa en Côte d'Ivoire. Prix, mimétisme, changement écologique et social. Économie rurale 330–331:103–123

Simien A, Penot E (2011) Current evolution of smallholder rubber-based farming systems in Southern Thailand. J Sustai Forest 30:247–260

Volper S (2011) Une histoire des plantes coloniales. Editions Quae, Versailles, 144 p

Chapter 8
From Firestone to Michelin, a History of Rubber Cultivation in a Cocoa-Growing Country: Ghana

Emmanuel Akwasi Owusu and François Ruf

Although the rubber tree (*Hevea brasiliensis*) would only appear on the former Gold Coast much later, bush rubber (*Funtumia elastica*), or *funtum* as it was locally known, was commonly used to make slingshots and balls for children. A few rubber plantations were established in the colonial era, before the Second World War, but they had no economic impact. It was only after independence, in 1957, at a time when Ghana had already been the world's leading cocoa producer for over 40 years, that its first president, Kwame Nkrumah, and his government launched a countrywide programme to promote the cultivation of other perennial crops. This programme, of which rubber was part, however, was never completed. President Nkrumah was overthrown in 1966.

The Ghanaian government launched an agricultural diversification programme in the late 1980s, with rubber cultivation once again occupying a key role in it. This time, the programme was rolled out in the south-western part of the country. By 2007, the number of farmers from all over Ghana who wanted to join the rubber plantation programme had reached such proportions that GREL (Ghana Rubber Limited), the entity that managed the programme, could not keep up with the demand.

We thus see that public policies played a major role in this diversification process. What were the others factors in this success? This chapter offers the experiences of an actor of development of the rubber sector in Ghana.

E.A. Owusu
Ghana Rubber Estates Ltd, P.O. Box 228, Takoradi, Ghana
e-mail: eaoaccamail@yahoo.com

F. Ruf (✉)
Cirad-ES, UMR Innovation, 73 Rue Jean-François Breton,
TA C-85/15 (Bât. 15, Bur. 223), 34398 Montpellier Cedex 5, France
e-mail: francois.ruf@cirad.fr

F. Ruf and G. Schroth (eds.), *Economics and Ecology of Diversification*,
DOI 10.1007/978-94-017-7294-5_8

8.1 Method

Emmanuel Akwasi Owusu, one of the authors of this chapter, was part of the project team for the promotion and development of rubber cultivation. The presentation of this project's historical aspect and the description of the outgrower[1] programme here are drawn mainly from his personal experience and documents from GREL's archives. The chapter's fourth section explores the smallholders' reasons for deciding to diversify. It is based on the same experience as that of the developer, combined with François Ruf's investigations, conducted in the form of surveys in the cocoa-cultivation area of Manso Amenfi and in the former coconut-growing area of Yediyesele.

8.2 The Place of the Diversification Project in Smallholder Rubber Cultivation

8.2.1 History of Rubber in Ghana

Rubber was introduced in Ghana in 1898 as an ornamental tree in the botanical gardens in Aburi, near Accra. In the 1930s, the UAC (United African Company) established some trial plantations in the Western region. In 1957, the year that the former Gold Coast got its independence and was renamed Ghana, the East Asiatic Company—a Danish company better known in Ghana as RT Briscoe—established a plantation of 923 ha at Dixcove, also in the Western region. RT Briscoe held a 51 % stake in this venture, while the remaining 49 % was retained by the Ghanaian government.

This plantation was, however, nationalized 3 years later, in 1960, and its management was taken over by a new entity, the Agricultural Development Corporation (ADC), especially created to oversee the plantation and promote the crop. ADC had just enough time to develop rubber cultivation up to an industrial level around the city of Abura, in the western region, before it was merged in 1962 with a larger entity, the State Farms Corporation (SFC). President Nkrumah had a far-sighted vision of agricultural diversification. SFC was assigned the task of developing all perennial crops, including cocoa—of which Ghana was by then the world's largest producer—rubber, oil palm, orange and coconut in any favourable forest area in the

[1]The term outgrower or 'external grower' is part of the semantics of projects implemented by an agro-industry that entered into contractual agreements with smallholders. From the point of view of the 'industrial plantation + rubber factory' combine, the smallholders formed a 'periphery'. Smallholders who joined the project and entered into a contract participated in the regional rubber production programme as 'external growers' to the 'industrial plantation + rubber factory' combine. We have decided to retain the term 'outgrower' in order to conform to this representation.

country. SFC managed to develop some rubber cultivation in the western region and the western part of the Eastern region.

In 1961, President Kwame Nkrumah organized, in the spirit of African socialism of the time, a large gathering of traditional chiefs and farmers at Nsein in the Western region. He 'proposed' to them in this meeting that rubber production be increased rapidly by setting up cooperatives. This project was not limited to the Western region, but also included the Centre, Eastern and Ashanti regions. Farmer's planted over 3500 ha of rubber trees under the 'cooperative' banner, without being sure of what the term actually meant. These plantations were all grown with seedlings.

President Nkrumah was overthrown in 1966 and his socialist ideology disappeared from the national scene. In 1968, the US firm Firestone Tyre Company, based in Akron, Ohio, signed an agreement with the new Ghanaian government. Firestone built a tyre factory in Bonsaso, in the Western region, and the Ghanaian government placed its plantations as capital in the joint venture. The 'plantation' part of the venture was named Ghana Rubber Estates Limited (GREL). Firestone owned 55 % of the venture, with the government of Ghana holding 45 %.

This arrangement was annulled in 1981, following a coup d'état by Jerry Rawlings, and Firestone left the country. GREL and the cooperatives who were selling their production to Firestone were literally abandoned. The cultivation of rubber and bush rubber ground to a halt in the same year. This state of affairs continued till 1988. Between 1988 and 1990, the Ghanaian government and a French company, Sodeci, got together to undertake the first phase of rehabilitating GREL on 3000 ha.

From 1991 to 1996, the second phase of GREL's rehabilitation encompassed 6175 ha of rubber plantations. In addition, 2700 ha were planted and a new factory was built. Simultaneously, a study was undertaken to identify the former cooperatives in the Western region.

In May 1992, GREL established its Rubber Outgrower Purchases Unit (ROPU), the entity that would purchase rubber from the 'external growers'. These external rubber and bush rubber farms were owned mainly by smallholders settled in villages north of the former Firestone factory. GREL offered them assistance in the form of advice and inputs. It also sent a lorry to do the rounds of the villages every month to buy their rubber.

This initiative from GREL restored confidence among erstwhile rubber growers, smallholders who had never really been integrated into the cooperatives. It also played an important role in revitalizing the rubber sector in several ways. For example, this new development also attracted entrepreneurs from the city in a context of the impoverishment of Ghana at the time. These entrepreneurs (called agents) entered into agreements with the owners of abandoned rubber plantations. They rehabilitated these plantations, sold the rubber to GREL and paid a rent to the farmers. The production by agents totalled 159 tonnes of dry rubber in 1992 and rose to 1000 tonnes per year in 2006 and 2007.

In November 1994, smallholders and agents came together to form an association called Rubber Outgrower and Agents Association (ROAA) with the support of GREL. The association became a true partner of the country's rubber industry. Its representatives were elected farmers. The main objectives of the association were to negotiate the price of rubber with GREL, improve the living conditions of its members and establish a true network of smallholders. To this end, the association organized a meeting with all stakeholders in June 1998, including the French Development Agency (AFD), the World Bank, the Agricultural Development Bank (ADB) and GREL to establish a pricing mechanism.

The first phase of the outgrower project was launched in 1995. Farmers from several villages began setting up clonal rubber plantations. Even though the first plantations were only just beginning to produce rubber, their success encouraged funding entities to roll out the second phase after 6 years, from 2001 to 2005. In 2005, the Michelin group invested in GREL. The third phase began in 2006 and concluded in 2010. A fourth phase, started in 2010, is currently underway and is scheduled to run till 2015.

However, the Ghanaian government had drafted a development plan in June 2001 that envisaged covering 50,000 ha by 2020 with the help of AFD, with production expected to reach about 70,000 tonnes/year. Land suitable for rubber cultivation, the dynamics of the project and of the farmers all suggested that this goal could be met or even exceeded. The area covered by rubber in 2007 was about 23,000 ha (Table 8.1) and by the end of 2011, this figure had already gone up to 38,000 ha (Table 8.2).

In summary, it is clear from this brief history that public policy and intervention were unable on their own to develop smallholder rubber cultivation in Ghana. In fact, the Nkrumah regime did not get enough time to implement the cooperatives project, given the unstable political environment of the 1960s.

Private companies too experienced failures in the prevailing state-controlled context. However, it was ultimately a private company that succeeded in promoting rural rubber smallholdings in the late 1980s. And it must be noted that it was these village plantations that helped Ghana grow its rubber sector. We observe here the

Table 8.1 Number of hectares planted with rubber in Ghana, figures as of October 2007

Plantation	Area in production (ha)	Area of immature plants (ha)	Total (ha)
GREL	9,756	2,174	11,933
Wilson estates	530	10	540
Rubber plantations—Topease	105	500	605
Cooperatives with part of the area planted with seedlings rented out to agents	3,000		3,000
Family and private plantations with non-project funding		250	250
Outgrower family plantations. Phases I + II + III	1,401	5,653	7,054
Total	14,792	8,587	23,379

Table 8.2 Number of hectares planted with rubber in Ghana, figures as of December 2011

Plantation	Area in production (ha)	Area of immature plants (ha)	Total (ha)
GREL	8,000	5,000	13,000
Wilson estates	500	0	540
Rubber plantations—Topease	200	500	700
Cooperatives with part of the area planted with seedlings rented out to agents	3,000		3,000
Family and private plantations with non-project funding		2,200	2,200
Outgrower family plantations. Phases I + II + III + IV	4,696	14,274	18,970
Total	16,396	21,974	38,370

principle of interaction between government policies and initiatives of private groups, with long-term support from funding entities.

8.2.2 *Outgrower or Smallholder Plantation Projects*

Outgrowers are smallholders located within GREL's ambit of operations. Although they are 'exterior' to GREL's plantations, they are part of the rural environment and of family farming.

8.2.2.1 Phase I

The first phase of the project started on 1 January 1995, with the following objectives:

- Establishment of 1200 ha of rubber plantations in the Western region by 400 farmers over a period of 6 years;
- Rehabilitation of 1300 ha of existing plantations (individual or part of a cooperative) over a period of 3 years;
- Purchase of 620 tonnes of rubber in the first year, 930 tonnes in the second and 1000 tonnes from the third year onwards.

At the end of this phase, the results largely exceeded the initial objectives:

- 400 farmers were helped to plant 1200 ha over a period of 5 years;
- 1200 ha of old trees grown from seedlings were rehabilitated;
- 2500 tonnes of production was bought from farmers every year;
- 41 km of roads were repaired.

Table 8.3 Breakup of funding for the first phase of the project, 1995–1999

Institution	Amount (Euros)	Type of aid
AFD	823,000	Credit for agricultural activity
AFD	211,000	Grant for infrastructure
IDA-World Bank	609,000	Grant for overheads
Total	1,643,000	

The project was funded by the French Development Agency, the World Bank and the Ghanaian government (Table 8.3).

8.2.2.2 Phase II

The second phase was launched on 14 September 2001 with the following objectives:

- 2800 ha to be planted over 5 years;
- 500 new outgrowers to be selected to join the programme;
- 40 km of roads to be repaired;
- Support to be extended to the outgrowers' association;
- Research for supporting outgrower plantations.

The results obtained at the end of 2005 were:

- 2855 ha were planted;
- 500 additional outgrowers were selected;
- 20 km of roads were repaired.

The outgrower association, supported for 3 years by the 'Institution and Development' consultants, consolidated its position as a stakeholder in Ghana's rubber sector.

Participatory research was carried out by CIRAD in smallholdings and included 83 trials with 60 farmers and 10 trials in the factory's plantations.

The total funding of 6.093 million Euros during the period 2001–2005 was allocated as follows:

- 2.733 million Euros for agriculture, including credit to farmers.
- 3.360 million Euros for technical assistance (buildings and roads).

8.2.2.3 Phase III

The third phase was launched on 14 May 2006 with the following objectives:

- Plantation of 7000 ha by 1750 outgrowers in the Western and Centre regions;
- Provision of credit to farmers for planting rubber;

- Undertaking applied research in smallholdings to maintain or increase productivity;
- Strengthening the Rubber Outgrowers and Agents Association (ROAA);
- Improving 70 km of roads.

The funding for this plan included a German contribution (Table 8.4).

In October 2007, figures from the villages revealed that more than 2000 farmers had adopted rubber and 7000 ha of rubber trees had been planted (Table 8.5).

The increase in cultivated surfaces in 2011–2012 was exponential. By the end of 2012, GREL had helped set up 22,000 ha of smallholdings (Table 8.6). More

Table 8.4 Funding plan for the third phase, from 2006

Participant	Amount (million Euro)	%	Notes
French Development Agency (AFD)	8.615	3.3	
Kreditanstalt für Wiederaufbau (KfW)	5.420	27.2	
Government of Ghana	1.785	9.0	Support to the project after 2010
GREL	2.070	10.4	Costs for coordination and management
Outgrowers (farmers)	2.015	10.1	CESS, Mandays, ROAA
Total	19.905	100	

Table 8.5 Number of farmers who benefitted from the project and number of hectares of rubber planted, figures as of October 2007

Project	Surface (ha)	Number of outgrowers
Phase I (1995–1999)	1,200	400
Phase II	2,855	500
Phase III	2,999	1,221
Total	7,054	2,121

Source GREL

Table 8.6 Number of farmers who benefitted from the project and number of hectares of rubber planted, figures as of December 2011

Project	Surface (ha)	Number of outgrowers	Notes
Phase I	1,200	400	
Phase II	2,855	500	
Phase III	7,855	1,800	
Phase IV	7,060	2,750	3,440 ha were planted in 2012
Total	18,970	5,450	22,410 ha at the end of 2012

Sources GREL

importantly, these plantations were maintained properly and their productivity was exceptional: average yields in 2011 were about 2900 kg/ha of wet rubber, i.e., 1700 kg of dry rubber.

8.2.2.4 Project Benefits

According to GREL's advisers, many benefits have accrued to outgrowers:

- An increase in the incomes of farmers and their families, and an improvement in their quality of life;
- Creation of jobs by the project which helps link urban and rural centres;
- Participation of women in the project (see below);
- Access to inputs and credit. In fact, GREL provided fertilizer and phytosanitary products on credit under specific conditions (10 % to be repaid in the first year of production, 15 % the second and 25 % over the next 12 years). A small sum of money was given as cash every month, or once in 3 months. This was also in the form of a 15-year credit;
- A guaranteed rubber market;
- Transparent pricing mechanism;
- Development of roads. The multiplier effect of roads resulted in an increase in economic activities that were unrelated to rubber production;
- Introduction of the banking system among farmers and access to credit for all types of farming;
- Acquiring knowledge on accessing credit and its repayment;
- Assistance to farmers for procuring land certificates.

Outgrowers were selected on the basis of the following criteria:

- Land title;
- Plot area;
- Suitability of the land for cultivating rubber;
- Distance from the factory;
- Slope of the plot;
- Assessment of the farmer's willingness to follow technical advice provided by GREL.

8.2.2.5 Technical Support for Farmers

The selected farmer signed an agreement with GREL and the bank, following which a bank account was opened for him. The system could be viewed as paternalistic, but each outgrower benefitted from technical monitoring of quality at least ten times a year. Three types of visits were planned:

- Routine visits. These were designed to ensure that the outgrowers were following the recommended farming practices in a timely and correct manner (soil preparation, planting, application of fertilizer);
- Technical field visits. The status of the plot was assessed using an observation checklist in order to provide focused advice and address any weak points. Each outgrower received at least two technical visits every year. For example, if a root disease was observed, immediate action was taken. Technicians would direct the farmer to dig around the base of the rubber trees and apply the fungicide provided by GREL;
- Visits to monitor adherence with recommendations. If the state of the plot was found to be clearly deficient, a monitoring visit was scheduled within 2 weeks of the technical visit. In case of non-implementation of the recommendations made during the technical visit, the bank was informed and the farmer's access to grants for maintaining his young plantation was blocked.

As a summary of this second section on projects and outgrowers, we can highlight the quality of supervision that led to a remarkable productivity and the role of funding entities in diversification projects. More than a two-actor 'government policy and private sector' initiatives, the case of Ghana shows that success was often due to a trilogy of 'government policy and private sector initiatives and initiatives of funding entities'. These latter could be international organizations such as the World Bank or national organizations such as AFD or KfW. In this case, 'government policy' was influenced, at least in part, by countries which were consumers of the promoted crops. In 2006, during an interaction with the association of rubber farmers, Mr. Pierre Jacquemot, Ambassador of France, had expressed this aspect well, to the delight of his audience, 'I experienced feelings of gratitude towards you during my journey from Accra to Takoradi, because it was you who make the material that creates the tyres that guarantee our driving safety.'

8.3 Diversification: Why Did Farmers Participate in the Project?

The 'outgrower' project to promote rubber in family farming covered seven districts in south-western Ghana: Nzema East, Wassa West, Mpohor Wassa East, Jomoro, Ahanta West and Wassa Amenfi (East and West). There was little diversification in family farming prior to the advent of the project. For example, although Jomoro district did have a few cocoa and oil palm trees towards the interior, the entire economy of the zone was very similar to one of coconut monoculture (Chap. 2). All these districts had, and still have, a very dominant perennial crop: cocoa plantations in inland regions and coconut groves or, more rarely, oil palm groves, near the coast (Table 8.7).

Clonal rubber trees seemed to offer an opportunity for diversification in areas, in each district, where one perennial crop, such as cocoa or coconut, dominated all

Table 8.7 Districts and crops before the introduction of clonal rubber and number of growers involved in the project

District	Dominant crop	Secondary crop	Marginal crop	Number of farmers in the rubber project	Percentage of farmers
Ahanta West	Coconut	Oil palm	Orange	404	23
Nzema East	Coconut	Oil palm		408	23
Wassa West	Oil palm	Cocoa	Orange and old rubber	409	23
Mpohor Wassa East	Cocoa	Oil palm	Orange	155	9
Jomoro	Coconut	Oil palm	Cocoa	141	8
Wassa Amenfi West	Cocoa			199	11
Wassa Amenfi East	Cocoa		Oil palm and old rubber	49	3
Total				1765	100

economic activity. Yet in 1998, when GREL restarted its operations on the basis of a Sodeci-government partnership, the entire South-west region wanted nothing more to do with rubber. As mentioned earlier, rubber from plantations was no longer purchased following the collapse of the agreement between Firestone and GREL in 1981. The few old rubber plantations from that time had been abandoned. Why then did the farmers decide to participate in the project?

8.3.1 *Restoring Confidence by Purchasing the Production and the Objective of Income Generation*

By resuming the regular purchase of rubber from cooperatives and individuals in 1992, GREL regained the confidence of the farmers. We noted that several abandoned plantations were rented out to agents for revival. When the first phase of the project was launched in 1995, farmers who already had such old plantations, and agents who had rented them were among the first to join the project.

8.3.2 *Imitation Effect*

The rapid change in the lifestyle of these agents, with some even being able to purchase vehicles, drew the attention of smallholders and led to more farmers signing up for the project.

8.3.3 Ecological Change and the End of Local Cocoa Cycles

Kuwornu et al. (2011) statistically demonstrated that the history of cocoa cultivation in Ghana was related to this crop's dependence on forests. Cocoa yields declined with the depletion of forests. Thus, the average yield in several districts, such as Wassa Amenfi, was usually around, or below, 320 kg/ha. In addition to the relative aging of cocoa plantations, several decades of continued cultivation and heavy rainfall had together led to increased soil acidity. Here too, the diversification into rubber appears to be a response to ecological change. In the large village of Manso Amenfi, investments in cocoa cultivation began to decline as far back as in the 1980s and 1990s, partly due to an increasing interest in oil palm cultivation following the establishment of a palm oil factory in the region. Having attracted the attention of the remaining farmers who still had rubber plantations planted in the early 1960s, GREL offered an opportunity to cocoa farmers who faced difficulties in replanting their crop (Fig. 8.1).

8.3.4 Ecological Change or Accident in Coconut Groves

In districts dominated by coconut cultivation, the lethal yellowing disease destroyed monocultures of this crop in several coastal villages (Chap. 2). Some farmers adopted precautionary measures even though their own coconut trees were not affected. Investments in new coconut plantations saw a considerable decline.

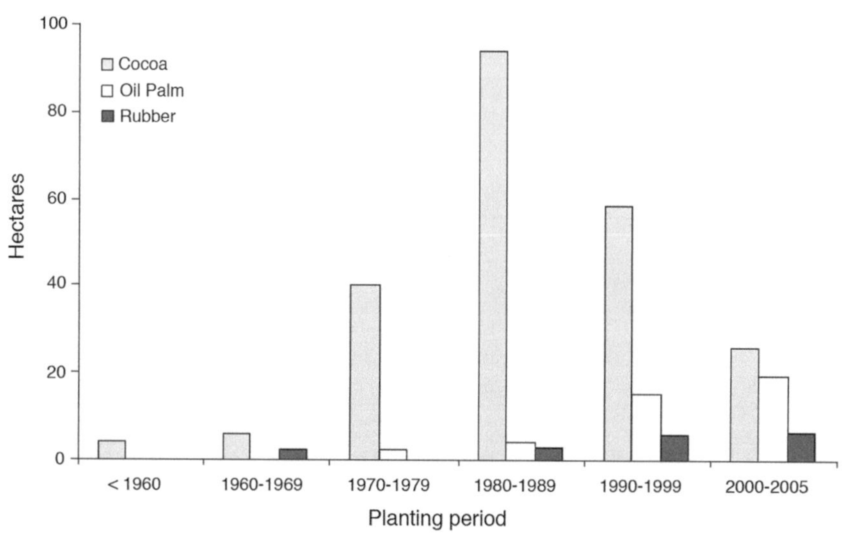

Fig. 8.1 Age structure of plantations in Manso Amenfi in 2005. (*Sources* surveys by CIRAD and A&LC, 2005)

Farmers diversified into oil palm cultivation by partially substituting their coconut trees, and also established new cocoa and rubber plantations further inland. Surveys conducted along the lower coast and in Yediyesele clearly demonstrated this trend (Figs. 8.2 and 8.3).

Yediyesele is one of the villages where rubber was adopted particularly rapidly, thanks to a direct transfer of information from the factory to the village through villagers working in the rubber factory. This kind of information transfer by rural labourers is a very common occurrence in the history of adoption of perennial crops (Gérard and Ruf 2001).

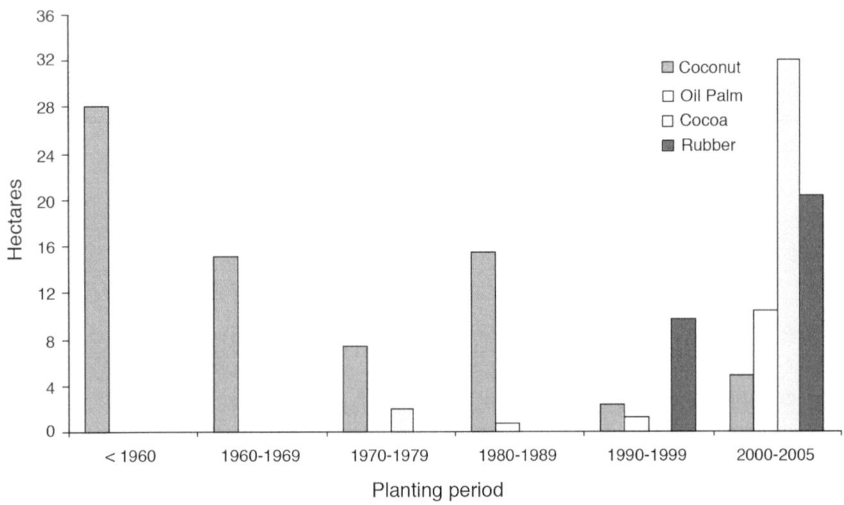

Fig. 8.2 Age structure of plantations along the lower coast in Nzema district in 2005. (*Sources* surveys by CIRAD and A&LC, 2005)

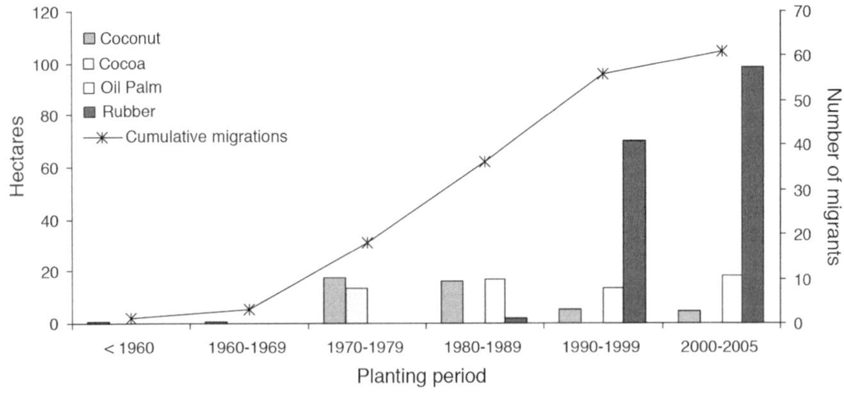

Fig. 8.3 Age structure of plantations in Yediyesele village, Nzema district, in 2005 (*Sources* surveys by CIRAD and A&LC, 2005)

8.3.5 Credit

The project's 'credit' component was also an attraction since farmers, on their own, would not have been able to finance activities such as forest clearing and purchases of planting material, fertilizer and inputs.

8.3.6 Long-Term Investment Suited to the Family Life Cycle

Compared to oil palm trees, which are usually cut down after about 25 years and often between 15 and 20 years due to difficulties in harvesting the bunches, clonal rubber trees indisputably represent a long-term investment: they have a priori an economic lifespan of 30 years, maybe even more. The very first interview we conducted with a rubber farmer in Ghana, in June 2002, was with a village chief who himself illustrated an ideal diversification. At age 55, he already possessed a wonderful heritage of several hectares of oil palm, rubber, orange, teak, fish farming ponds and a sheep farm. He represented the typical case of diversification made possible with the help of abundant access to land and to information by virtue of being village chief. The most important crop at that time was the oil palm. Rubber was first introduced as a diversification crop in 1995. Why rubber? His answer to this was very clear in June 2002: 'When I cut down my palm trees in 1995, I could easily have replanted the same crop. I was then 47 years old. I thought ahead and realized that when the time for the next felling of the trees comes in 20–25 years, I will probably be too old to replant them. So I took the decision to plant rubber, whose life span will ensure that I will not need to do any replanting when I get old.' Similar responses were recorded in Côte d'Ivoire 2 years earlier (Ruf 2000).

8.3.7 Relative Land Security

Since a migrant has few rights on the land he occupies, rubber appeared as an opportunity to safeguard land tenure. For example, when a migrant cut down a 15–20-year-old palm plantation, there was a very high risk of the owner reclaiming the land. With rubber, this situation was not likely to arise for at least 30 years.

In the case of Yediyesele, the role of rubber in land tenure security was even more pronounced because it was a new crop supported by a project. As far as coconut and oil palm trees were concerned, all the farmers of the village who were regarded as migrants by the 'king' of Axim—the holder of all land rights—were generally accorded a quasi-status of sharecroppers. In principle, they turned over a third of their income to the village chief, who then passed some of it on to the 'king' of Axim. Rubber was, however, regarded as a 'government thing' and became a tool of emancipation against the king and his control over land. This emancipation was thus also one of the factors in the rapid adoption of rubber. Compared to two other projects which promoted two other perennial crops (hybrid coconut and hybrid oil palm), the result was unequivocal:

- Hybrid coconut project: 11 farmers;
- Hybrid palm project of 2004: 20 farmers.

The results of these projects are poor compared with those of the clonal rubber project: 100 farmers, i.e. all the families in the village.

The most recent adopters felled their old coconut trees, as well as palm trees that were only 5–6 years old, in order to plant rubber. At the village level, this posed a risk of the diversification turning into a conversion to another monoculture.

8.3.8 Rubber as Life Insurance and Retirement Fund

A life insurance agent who observed the characteristics of rubber plantations said it best: 'Here is a life insurance product which is better than anything we have to offer.' As in Côte d'Ivoire and Thailand, rubber attracted upper middle class investors (UMCI[2]) too to invest in its cultivation. Out of the 1765 rubber farmers who participated in the project in the Western region, 5 % were doctors, lawyers, teachers, accountants or traders. One of the reasons why the project accepted them was because smallholders actually regarded them as a knowledgeable elite class. They became symbols of the desirability of investing in this crop.

8.3.9 Regular Income

As is evident to anyone who has observed the adoption of rubber by cocoa farmers, a regular monthly income is an unmatched incentive and thus makes rubber a good complementary crop in the low-yield cocoa growing areas of Manso Amenfi (Ruf 2007; Chap. 5 of this book).

8.3.10 Steadily Rising Prices and Incomes

Rubber prices started increasing again from 2000 (Fig. 8.4). In November 2007, the price reached 616 Ghana Cedis[3] per tonne, or 670 US dollars. At this price, a farmer

[2]For a definition of UMCI, see footnote no. 3 in the Introduction.

[3]In July 2007, the new unit of the Ghanaian currency became the Ghana Cedi equivalent to 10,000 old Cedis. The value of 1 new Cedi was equal to about 1 US dollar at the time. All prices are in Ghana Cedis, at the rate prevalent in 2007.

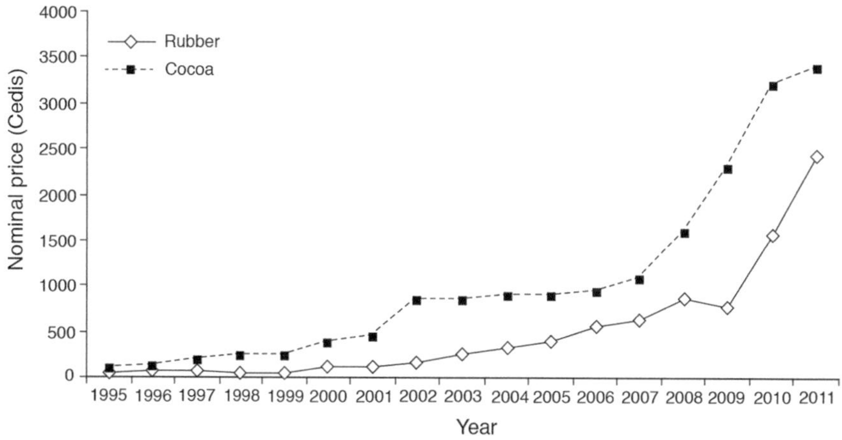

Fig. 8.4 Nominal prices of rubber and cocoa in Ghana, for the 1995–2001 period. (*Sources* for rubber: GREL; for cocoa: information obtained from farmers by CIRAD)

with 4 ha of rubber plantation in production could expect a gross income of about 410 Ghana Cedis or 450 USD per month. In November 2011, a combination of rising international prices and increased yields raised this gross monthly income to between 850 and 1000 USD.

8.3.11 Public Policy and Relative Prices

In the 2000s, the government of President John Kufuor gave a stimulus to the cocoa sector by establishing cocoa processing facilities (mass processing). It also distributed subsidized fertilizer and increased prices paid to farmers. This policy has been very effective and helped revitalize cocoa production in the country in the 2000s.

Furthermore, because of a decrease in taxes on the cocoa sector and an increase in the international rubber price, which was anyway not highly taxed, the prices paid for these two crops to Ghanaian farmers continued to increase in tandem.

However, because of very high rubber yields, the gross income from rubber is significantly higher than that from cocoa (Fig. 8.5). This is not only because of an increase in world rubber prices, but also due the remarkable productivity of smallholdings that now yield 2500 kg of wet rubber per hectare.

Relative income is the ratio of income from rubber divided by the income from cocoa. It is calculated from price and yield estimates: it is around 2500 kg/ha of wet rubber and 500 kg/ha of cocoa grown without fertilizers; it goes up to 1000 kg/ha in fertilized cocoa plantations. This ratio is therefore lower when cocoa yield increases due to the use of fertilizer.

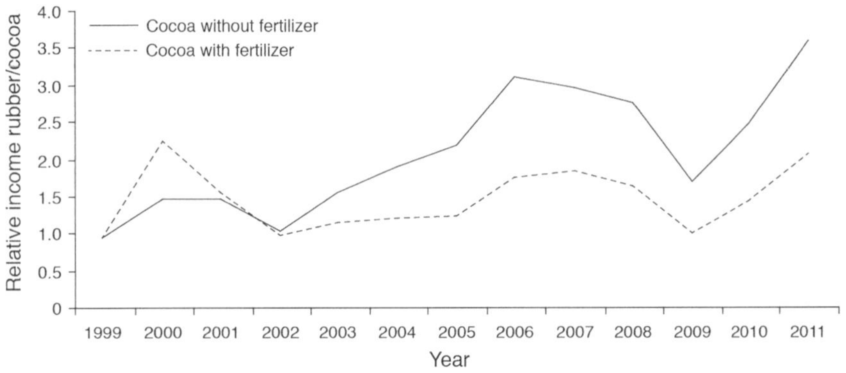

Fig. 8.5 Modelling of relative incomes from rubber and cocoa cultivation in the Manso Amenfi area in Ghana

However, the comparison of relative income, with or without fertilizer in cocoa plantations, clearly illustrates the choices open to Ghanaian cocoa farmers and, by extension, to neighbouring countries: diversification into rubber or intensification of cocoa cultivation.

8.4 Participation of Women and Young People

In 2007, out of the 1765 outgrowers involved in the project in the Western region, 344 were women, i.e., 19.5 %. This proportion was high here compared to other sectors, and especially so when compared to other countries. In principle, one must be able to prove his or her access to land to be selected by the project. But of the 344 women, only 22 owned land (6.4 %). Apart from some 'Queen Mothers', the majority of the women had received explicitly a plot from their fathers, uncles or husbands. Many women also entered into land leases or various institutional arrangements with land owners. However, the percentage of women who participated in the project with the sole objective of helping their husbands accelerate the diversification into rubber on the family farm remains difficult to evaluate.

In 2011, the proportion of women who officially joined the project rose to 30 %. This was, in part, due to the project's efforts towards poverty eradication and the deliberate priority it accorded to women. Indeed, it has been established that women tend to spend a large part of their income on education and health of their children. Even otherwise, women were generally very interested in rubber plantation projects. They generally increased the value derived from the plantations in the initial years by interplanting food crops. For another part, one must acknowledge that many farmers responded to GREL's 'priority to women' policy by using the name of their wifes to register.

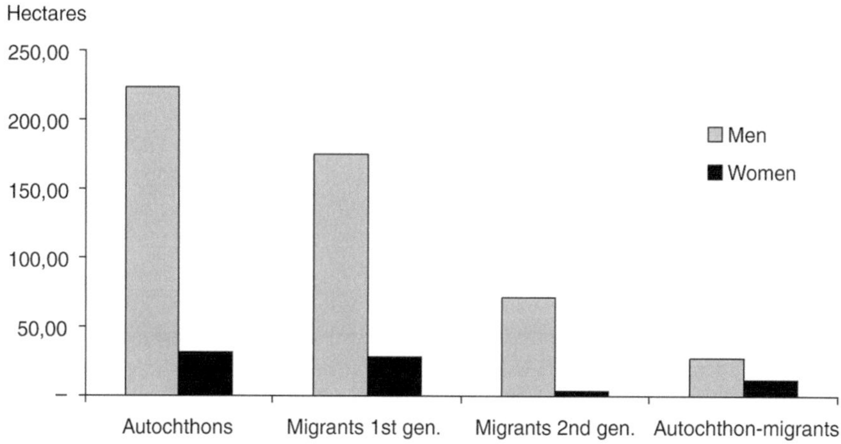

Fig. 8.6 Distribution of rubber farms between men and women among village farmers. (*Source* survey by authors, 2014)

In 2014, we undertook a new survey with 133 village farmers including 27 women. We identified a few successful women who are the real managers and beneficiaries of their rubber farms. For instance in the the Central region, a woman who had inherited some land from her father and who had some capital from a small poultry unit managed to obtain 8 ha under an *abunu* contrat to plant rubber between 2008 and 2012. We roughly estimate that somewhat less than half the number of women are the real beneficiaries of their rubber farms with the majority being just fronts for their husbands. And even if we assume that women remain in control of the GREL registered farms, their economic weight would remain relatively low (Fig. 8.6).

As far as young people are concerned, 16.6 % of the outgrowers were between the ages of 18 and 35 years. This is a respectable percentage when compared to other countries or sectors, but it nevertheless reflects the difficulties young people face in obtaining land. Some of them help their parents and expect to receive a share of plantation at a later date. Others need to pool their savings to work out a land deal. In order to do so, some have to work for a few years in the charcoal or palm wine sectors or undertake any other remunerative activity such as masonry or carpentry, or even work in the education field.

8.5 Rubber, Tropical Forests and Food Security

When cocoa and coconut regions diversify into rubber, where is rubber planted? As in Côte d'Ivoire, rubber is the perennial crop which has the least impact on tropical forests. Only one third of the areas planted with rubber before 2006 were on cleared forest lands (primary or secondary forests that were 30 years or older), in comparison to 90 % for cocoa and 60 % for coconut. Another third of the surface areas planted with rubber were after the felling of aging or dying cocoa trees or oil palm trees and dying coconut trees. Finally, the remaining one third was planted on inherited fallow land which was part of shifting food crop cultivation.

The more recent survey among smallholders in 2014 showed an increasing trend of the replacement of cocoa, oil palm and coconut by rubber (50 % of the rubber plots recorded in 2013) with forest and fallow/food crop systems accounting for 25 % each (Fig. 8.7).

In the initial years, the impact of rubber on food production remained positive as bananas and cassava were intercropped with young rubber plants. From the third year, such intercropping was reduced and the subsequent increase in the surface area devoted to rubber could lead to reduced food production and consequently lowered food security. Though most farmers recognized the risks associated with any reduction in the surface area for food crops, they did not think it would lead to food insecurity since they believed they could rely on the market for any shortfall. Rice could be purchased using the income from rubber.

In reality, almost all the rubber farmers have not been self-sufficient for the past several years and have to buy some rice every year. 'Self-sufficiency' is already partly relying on the market. Many farmers have retained one or two plots that were reserved for food crops and 90 % of village farmers have something between 0.5 and 5 ha of fallow land. However, this reservoir is likely to shrink rapidly under the pressure of rubber. The farmers' strategy is not only to increase revenues. It is also a land-ownership strategy, using the rubber tree as a marker to secure land rights and reduce the risk of conflicts.

Moreover, following the Ivorian model (Chap. 1), the recent advent of the upper middle class investors (UMCIs) in the Ghanaian rubber sector will inevitably put some pressure on food security in the coming years. When a UMCI manages to

Fig. 8.7 Crop that preceded rubber for village smallholders. (*Source* survey by authors, 2014)

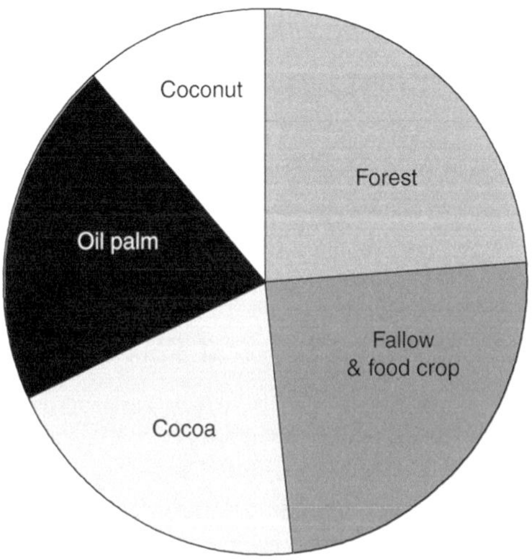

obtain a bloc of 50 ha from a chief, he always finds some villagers who grow some cocoa or oil palm and food crops on that block of land. He pays some additional compensation to them to vacate the land. This approach may reduce risks of conflict in the short term but it also raises some questions for the future, including in terms of food security.

8.6 The Arrival of Upper Middle Class Investors (UMCIs)

While the project funded by AFD was aimed at helping smallholders, rubber revenues became so attractive that, just like in Côte d'Ivoire, it inevitably attracted lawyers, bankers, military and custom officers, politicians, traders, and even a few GREL officials. This category of farmers, which we call upper middle class investors (Chap. 1), started growing rubber in 2005 for the most part. They buy 20–50 ha of land, sometimes even more, from the traditional chiefs, in principle for a duration of 50 years. They have enough capital to plant rubber outside the 'outgrower scheme', paying with cash for planting material and all other inputs obtained from GREL. As most of them do not know anything about farming in general and rubber in particular, they benefit from the technical advice provided by GREL staff. Although figures provided by GREL show that UMCIs are still far behind villagers in terms of global cultivated areas of rubber, they expand quickly (Fig. 8.8). Most of them view rubber cultivation as a pension scheme for when they will retire from their main jobs of lawyer, movie maker, politician, etc.

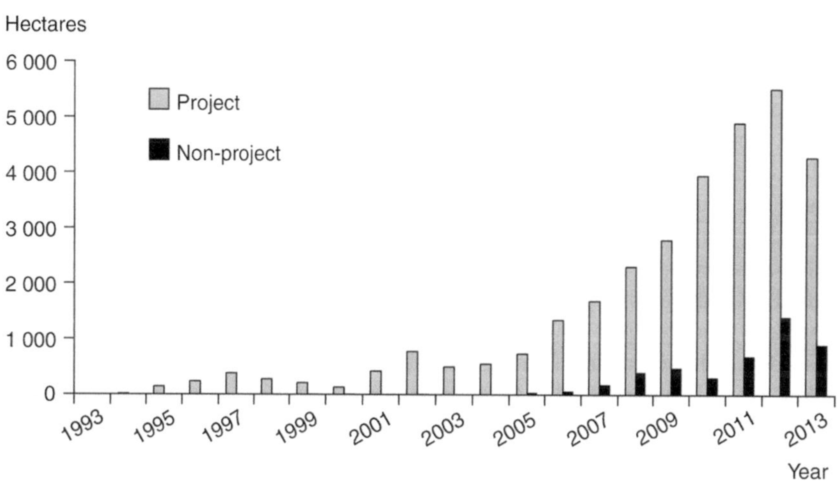

Fig. 8.8 Annual planting by self-financing farmers. (*Sources* GREL records)

Chapter 9
Extensive Fish Farming, a Complementary Diversification of Plantation Economies

Marc Oswald

Perennial crops cultivated in the humid tropics have often been introduced into existing slash-and-burn agriculture systems that are practiced in still-forested areas. This is particularly the case of cocoa and coffee in West Africa. Family plantation economies that developed in this sub-region were capable of providing, for a long time, enough tubers, cereals and other sources of carbohydrates to meet the needs of a growing rural population. Even though there were chronic seasonal shortfalls and even though crises of a more severe nature were not unknown, carbohydrate production remained in surplus for a very long time. However, the situation has been very different with respect to proteins, especially those of animal origin; their structural scarcity has resulted in severe nutritional deficiencies (WFP 2009). Based on observations and on occasional field data, these forest agricultural systems have been observed to be acutely deficient in animal proteins for a long time now. Vague memories refer to a time when game was abundant, during a period of large scale deforestation accompanied by a burst of hunting which quickly depleted animal stocks that were doomed to extinction.

Yet in these regions, hunting and river fishing have always been regarded as the means of meeting basic protein requirements. Thus, the forest tribes of western Côte d'Ivoire have been sometimes described as hunters who are unsuited to agriculture —even though there is evidence that, on the contrary, they are genuine farmers (Léonard and Vimard 2005). In 1985, Dozon (1985) reminded us that collective hunting, traditionally important in some Bété communities before the advent of the plantation economy, provided only a secondary source of food. The earliest descriptions of trade on the northern limits of the forest note a relative abundance of smoked fish from the Niger Delta (Oswald 1997).

In 1957, Brasseur (cited by Fargeot, 1994), was amazed at the rapid disappearance of many large game animals in Kpelle inhabited areas (Guinea)—especially elephants since 1954. Does this then lead us to conclude that all animal

M. Oswald (✉)
Istom, 32 Bd du prt, 95094 Cergy pontoise, Cedex, France
e-mail: m.oswald@istom.net; m.oswald@apdra.org

M. Oswald
Apdra pisciculture paysanne, 24 rue Émile Zola, 91300 Massy, France

F. Ruf and G. Schroth (eds.), *Economics and Ecology of Diversification*,
DOI 10.1007/978-94-017-7294-5_9

species that were a likely source of protein were systematically consumed in these regions (small rodents, fish, reptiles, amphibians, bats and insects)?

There was a similar situation in Ghana where studies undertaken as far back as in the 1930s showed that farming communities depended on dried sea fish (Beckett 1944; Ruf 2007).

Food requirements were thus met through a mainly external supply of protein in the form of fish. This longstanding dependence has only become more pronounced. With an increasing population density stemming from the plantation economy, the quantity of common resources is in decline.

In western Côte d'Ivoire, as in the forests of Guinea, it is mainly women who do the fishing. In fact the women do so with enthusiasm, most often together in small groups, using small traditional spoon-nets. This activity, though socially important, is inadequate; the catch is small and insignificant. However, this contribution is appreciated and considered valuable.

As expected, there have been countless attempts to establish livestock farming in these areas, but these activities were not at all integrated with the farmers' other activities. Furthermore, there are several deterrents to livestock farming: numerous animal diseases, high parasitism and difficulties in maintaining pastures. Brou (2005) and Ruf (2010) have described recent, localized attempts to compensate for aging cocoa through livestock farming, mainly of sheep. There was a renewed interest in pig rearing in Guinea and Côte d'Ivoire, supported by the availability of large volumes of palm-kernel oil cake. Fishing, however, remains the primary source of animal protein.

Grosse (2009) observed that in similar forest areas of central Cameroon, more that 90 % of the animal protein consumed came from fish. In these plantation economy regions, the main food item that is purchased by families is fish, far ahead of cereals and other foods. In this situation, it seemed logically inevitable for fish farming to develop, even without specific policies for the sector. Indeed, numerous attempts were made to develop fish farming in different contexts with different objectives (food self-sufficiency, production of bagrid catfish, shrimp for export, etc.). Unfortunately, these actions were not—and are still not—guided by the desire to meet the needs of rural communities in these regions. Projects propose intensive fish farming systems, giving as a pretext a higher profitability. Subsidies have often been offered to overcome investment constraints and enable a quick start to production, but the realities of fish production in rural conditions were not taken into account. It is not surprising then that most of these attempts ended in failure.

However, when the government, funding entities and non-governmental organizations come together to propose actions designed to meet the expectations of farmers, results are much more encouraging and original, assuming, of course, that enough time is given to these projects to arrive to fruition. Two institutional associations have managed to create such an environment around two projects in Côte d'Ivoire and Guinea: Central-Western Fish Farming Project (French abbreviation: PPCO) and Fish Farming Project in the Forest Region of Guinea (French abbreviation: PPGF).[1] These experiments quickly came to be considered national benchmarks on either side of the border. PPCO's former area of intervention in Côte d'Ivoire is today the country's primary fish farming area (Assi Kaudjhis 2005). Both these fish farming projects took decisions at variance with prevalent norms: extensive models, significant investment implemented with labour, and farmers' capital.

9.1 Methodology

This chapter analyzes the benefits of diversification into fish farming for plantation economies in order to satisfy the strong domestic demand for fish and to fulfil the desire for fish farming. It compares the results of the development of this activity in two villages in different contexts: Gbotoÿe (Guinea), 8 km north of N'Zérékoré, and Luénoufla, 35 km north-east of Daloa (Côte d'Ivoire). These villages benefited from interventions by, respectively, PPGF and PPCO.

We first discuss the regional agricultural situations at the time fish farming was introduced as well as the environment specific to fish farming. The approaches adopted to promote fish farming by the two projects are compared. Based on project documents, we discuss the methods used to disseminate the innovation. The economic performances of farmers who chose to farm fish are analyzed. The consequences of these developments identified outside the farms are summarized.

[1]For PPCO, originally conducted from 1992 to 1998 and subsequently extended to 2000 to encompass those farmers missed out initially, the entities involved were AFVP (French Association of Volunteers for Progress) assisted by CCFD (Catholic Committee against Hunger and for Development), the Ministry of Cooperation and the Directorate of Fisheries and Aquaculture. These institutions welcomed the involvement of other actors in the implementation phase: the Central Region, AFD (French Development Agency), APDRA-CI (Association Pisciculture-Rural Development in Tropical Humid Africa, Côte d'Ivoire) and APDRA-France. Research institutions like IDESSA (Savannah Institute) associated with CIRAD (French Agricultural Research Centre for International Development) and CRO (Centre for Oceanographic Research) were also included as partners. However, this project faced stiff resistance from a large section of the Ivorian research establishment which led to the suspension of the Chinese carp program (Assi Kaudjhis 2005). For PPGF, AFVP (in partnership with APDRA-France) helped implement the project on behalf of the National Directorate of Aquaculture and Inland Fisheries (DNAPC), financed by AFD from 2000 until June 2008. Other partners, including CCFD and MAE (the French Ministry of Foreign Affairs) and USAID (United States Agency for International Development) supported this project that spanned several crises (war with Liberia, the 2003 crisis between the government and funding entities, conflicts arising from national political transitions since early 2007).

9.1.1 *Regional Contexts*

Both villages are located in forested areas. The project area in Côte d'Ivoire receives less rainfall than its counterpart in Guinea. Its dry season is much more severe, but it has richer soils: these are lateritic and slightly to moderately desaturated. The two regions also have different agrarian histories.

9.1.1.1 Luénoufla in Côte d'Ivoire

Luénoufla is located in the Gouro people's territory, not far from the edge of the savannah. Coffee cultivation grew rapidly in the 1930s and 1940s to become a major part of local farming systems. A large number of migrants arrived in the late 1950s and were actively involved on the agricultural frontiers in the centre-west. They mainly grew cocoa and soon it became the main crop in the local agriculture.

Few opportunities were available to farmers to invest money saved from their income from cocoa cultivation. Several attempts were made to diversify into livestock rearing (broilers and laying hens, pasture raised sheep, modern pig farms, and fish farming) but they all ended in failure. Farmers were left with little option but to buy new forest lands to plant cocoa. Some farmers, however, successfully diversified into trade and real estate (Chaléard 1993). The rapid growth in population contributed significantly to environmental degradation. Forests regressed, to be replaced by bush, and the spread of bushfires often endangered plantations. Savannahs began gaining in significance, especially in the lowlands where systematic burning promoted *Pennisetum*-dominated savannah which, subsequently, was gradually converted into food crop fields.

The boom period, based mainly on the cocoa cycle, began to show signs of slowing down in the late 1980s. The continued increase in population made it increasingly difficult to meet domestic food requirements. Land rent for annual crops began to increase, resulting in an enhanced interest in lowlands (Oswald 1997; Ruf 1988).

Consumption of frozen fish dominated the market for a long time despite the availability of beef. Traders interviewed in Daloa in 2000 indicated that the quantity of frozen fish[2] distributed in the markets of Luénoufla, with a population of 12,000 inhabitants, was well in excess of 100 tonnes/year (Trellu et al. 2002).

[2]In Côte d'Ivoire, the frozen fish from Daloa arrives on market days, partially thawed. What is not immediately sold is smoked for resale at a later date.

9.1.1.2 Gbotoÿe in Guinea

Gbotoÿe village is located in an area inhabited by the Kpelle people. Its population density exceeds 100 people/km^2 (Guillaume and Cissé 2000). Agroforests have come up on hillsides in this coffee-dominated region.

The growth of the coffee industry was significantly constrained under the rule of President Sékou Touré. It was not until the late 1980s that this crop was revived with the support of promotional activities (Delarue 2007a). Other tree crops, in particular kola and oil palm, retained an important role in this agroforestry system. The presence of cocoa is marginal here as the soil and climate are not favourable. Some cocoa trees can, however, be found in small plantations around the wetlands.

Upland rice, the main food crop and source of starch, is cultivated on fallow lands. Just like in other humid tropical areas, the optimal fallow period is around 7 years. This fallow period could be maintained until the 1990s as the population density was low. The situation in neighbouring countries at the end of the 1990s forced many Guineans to return to their country, thus greatly increasing pressure on the land and natural resources. With the expansion of plantations and population growth in the most densely settled areas, rice cultivation on 7-year-old fallow lands was no longer possible (Guillaume and Cissé 2000; Wey and Guillaume 2001). Although lowlands were cultivated to overcome this scarcity of land, the daily wage offered for cultivating rice here was much lower. On the whole, rural communities found it increasingly difficult to meet domestic food requirements.

Fish Consumption

While official statistics may indicate a consumption of 1 kg/person/year of fish, the reality is very different. Surveys conducted over a year, between 2001 and 2002, in rural markets showed that consumption exceeded 10 kg/person/year. More than half of this consisted of frozen and imported fish.[3] Humanitarian NGOs working with refugees from Liberia and Sierra Leone also conducted nutritional surveys on people in and around camps. The results clearly indicated that protein deficiency was the most severe problem amongst all the nutritional deficiencies found here. It is known to lead to several ailments in children under 5 years of age, which can result in growth disorders and even death.

Henderson (2001) estimated the number of working people in Gbotoÿe to be 2000. This active labour force was less than half of the total population of the village. The village's annual fish consumption can then be estimated to be more than 40 tonnes.

---

[3] On this topic, refer to PPGF activity reports of 2001 and 2002. From 2003 to date, the only companies to have invested in N'Zérékoré were a saw mill, a bank and a large warehouse for storage and marketing of frozen fish.

In Guinea, project support was even formalized via agreements between the project, the fish farmers' group and individuals who wished to be part of a group for receiving training in setting up fish farming in their village. In exchange for a commitment by the project of a certain production level (0.6–1 tonne/ha/year depending on the type of fish reservoir built) and product quality (fish size), farmers promised to adhere to guidelines, carry out the development with their own funds, share their knowledge and establish an organization that could provide guidance and support on fish farming. The facilitator conducted fish farming classes for groups of candidates for a week every 2 months. He was responsible for assessing the state of fish farming within the group, mainly through site visits and individual interviews. He hosted a workshop on the first evening for the group of motivated farmers to explain immediate priorities. He then conducted collective training, provided individual support and organized informal exchanges.[7] Over time, the services that farmers could reliably provide were transferred to the core group of farmers. The facilitator then took on the role of advisor and evaluator. There was also the idea of creating, later on, a professional organization able to represent all the fish farmers' groups.

Instructions given to the groups of farmers as part of the support extended to them were only offered as advice. When these instructions were not followed, reasons were sought: was the training imparted not properly assimilated, or did personal reasons lead to a different interpretation—or even the questioning—of the suitability of the techniques proposed. Fish farming practices that were performed routinely and the manner in which farmers modified them over time were always regarded by operators as decisive criteria for assessing the suitability of the proposed techniques.

The technical model has been described in detail (APDRA France 2002a, b). In short, it is an extensive fish farming model that enhances the natural production of fish in small catchment reservoirs with the help of a polyculture associating *Oreochromis niloticus* (tilapia), *Heterotis niloticus* and catfish. *Hemichromis fasciatus* is also used as a predator to control tilapia propagation. Sex determination is done for tilapia in order to grow large fish that some village markets demand.[8] The catchment dam ponds are adjacent to service ponds that ensure stocking needs, including pre-fattened sex-determined allevins. The base yield amounts to 600 kg/ha/year. It can easily be increased to 1 tonne/ha/year if the permanent flow of water in the pond can be controlled with proper management. Building a

---

[7]Assessment of extension methods carried out by Borderon (1999) of Agence Française de Développement showed the advantages of this type of organization compared to more conventional interventions in the agricultural sector.

[8]There are many bibliographic references extolling the economic merits for an African fish farmer of producing small fish. In the context of the environments described, these considerations do not reflect the social effects on the development of fish farming: production of large fish allows the processing of much more fish at the same price. This helps foster a spirit of cooperation between producers, resulting in the lowering of the cost of certain production factors (fry and fingerlings, mutual help for fishing, use of fishing gear, such as nets and cages).

diversion ditch (or a bypass canal) is often found to be useful. The yield can be boosted further by the addition of organic fertilizer. However, experience shows that few fertilizer by-products are available in villages with a primarily plantation economy. This situation is easily explained since livestock rearing is not integrated into these agrarian systems, and for carrying the crops from the fields or plantations, the almost exclusive form of transport is on people's heads.

The reasons for farmers opting to develop small catchment reservoirs on the upper borders of lowlands are, a posteriori, quite simple:

- the amount of earthwork in relation to the productive surface created, that is to say, the surface area of the water in which the fish will develop, is minimized, as is, therefore, the investment required;
- the investment can be progressive. Upon completion of a dike that blocks the water flow, the fish farmer obtains a productive area which will progressively increase as and when the dike is built up;
- the existing vegetation on the productive surface perishes as the area gets flooded, making way for aquatic plants to grow over time. This greatly alleviates the work of preparing and maintaining the productive area. This benefit satisfies the apprehension of the farmers who typically wonder, 'How can we reclaim and maintain the lowlands?'

All dam ponds and service ponds were equipped with a monk to help drain water and harvest the fish (even without the use of a net). Such devices also help empty out the production area (for sufficiently long periods to dry the bottom of the pond) and facilitate maintenance of the dam (clearing aquatic plants, for example) or the water holding area (removal of harmful plants).

9.2 Results and Discussion

9.2.1 Chronological Markers in the Development of Fish Farming

9.2.1.1 Luénoufla

Three farmers had already established fish farming activities prior to the implementation of the project in Luénoufla:

- a rich Gouro resident, who had used bulldozers available in the village on several occasions, had built a dike and a monk with help from the fish farming section of the Water and Forests department (which was then part of the UNDP-FAO project). He eventually managed to create a 1-ha reservoir. The investment put into the project, which was started in 1984, allowed him to obtain a fish yield every 2 years starting in 1990. Related structures (storage

Table 9.1 Development of fish farmers in Luénoufla from 1997 to 2002 (Trellu et al. 2002)

Description of fish farmers	Before 1997	1997	1998	1999	2000	2002
Fish farmers in production	3	3	9	22	28	46
Area used (ha)		2.85	7	14.6	16.9	
Estimated production (t)		1.1	8.5	15	25.6	
Duration of action of the project		Continued presence of the project				

ponds) also helped him dispense with aid from the nurseries of the Water and Forests department;

- two simultaneous initiatives by Burkinabé farmers were also implemented in 1991. These farmers also had access to a bulldozer. Their initiatives had two aims: to obtain a yield of catfish at the end of the dry season, and to have a water reserve.

After working with PPCO from 1993 to 1995, the project decided to open up the area in 1996. The ensuing developments, summarized by Trellu et al. in 2002, are described in Table 9.1.

In the course of the ensuing project intervention, the new fish farming facilities proposed to farmers allowed accurate control over the water: the ability to empty the reservoir or maintain the water surface at a desired level. These facilities were quickly perceived and used as farmland in their own right. Nearly 25 % of the area was used to grow paddy as water was easily available because of the monk and the water column created by the dike. Some crops were grown during the rainy season, including rice which was grown on the borders or downstream, and was replaced by maize on the more elevated slopes. In the dry season, women used the dike water to irrigate vegetable crops grown on the borders of the ponds. This activity consumed a lot of water: at the end of the dry season in 2002, two-thirds of the water in the dam was used up to irrigate the adjoining vegetable gardens! Fish from these fish farms is currently the main livestock reared in this area, and is present in every market. With respect to the social conventions that allowed services to come up around fish farming, it must also be noted that a large portion of the catch is fried. It is supplied to the local restaurant market, thereby adding value through this sector. In terms of investment, the use of bulldozers, which was favoured in the early 1990s, was quickly abandoned. Teams of efficient men and women specializing in earthwork soon emerged. Simple standards were used to quantify the volume of dikes to backfill, and channel and spillway levels to be built. A general acceptance of these standards by people who were involved clearly helped shape this development.

After 2002, the Luénoufla region found itself near the buffer zone during the worst period of the Ivorian crisis. Fragmented information[9] showed that, regardless

[9]Periodic information from APDRA-CI, training reports and internal notes of APDRA-F.

of their ethnicity, all the fish farmers who had established themselves during the project continued and expanded their production. While fish farmers who received support over several years through the project successfully consolidated their fish farming activities, many other farmers attempted to diversify into fish farming during the crisis. Interviews conducted in 2008 put this figure at more than 60. However, the quality of the facilities created by farmers who started late portends difficult times for fish farming.

9.2.1.2 Gbotoÿe

In this area too, there was a spontaneous enthusiasm for fish farming when the project was launched. A prospective fish farmer supported by the regional rural planning department has implemented a little pond. Following a request from PPGF, and after much hesitation, two farmers who were already involved in the project to revive coffee cultivation (RC2) accepted the responsibility of coordinating a group of local candidates starting in 2001.

In 2004, there were 12 candidates who owned lowlands and who wished to undertake fish farming. In addition to fish production, all these farmers also maintained a flooded rice paddy in their dam pond. Rice production was estimated at over 8 tonnes in 2004, more than twice the harvest in these lowlands without the installation of state-sponsored irrigated rice plots. The establishment of rice cycles in the pond areas in 2002 and 2003 seemed to have triggered a strong local movement (Table 9.2).

Table 9.2 Development of fish farmers in Gbotoÿe (*sources* PPGF activity reports: 2001, 2002, 2003, 2004 and 2006)

Year	Before 2002	2002	2003	2004	2006
Fish farmers in production	0	3	4	9	13[a]
Area used (ha)		1.35	2.5		>3.4[b]
Estimated production (t)	0	0.04	0.47	1.99	
Duration of action of the project	Continued presence of the project from 2001 to 2004 and since 2006				

[a]This number does not include 8 new fish farmers who were in the process of building their first ponds. In 2006, the large number of candidates in the village, more than a dozen who wished to take up the activity, and in neighbouring villages showed that the significant investment was no longer perceived as a constraint by the farmers, but rather as an opportunity to improve the situation of their farms

[b]This area was measured in the dry season—when it dropped to less than 30 % of the maximum area, in the rainy season—and therefore did not correspond to the area of water when ponds are filled up

9.2.2 *Role of Fish Farming in Farming Systems*

9.2.2.1 Performance of Fish Farming Systems

Interviews with fish farmers showed that fish farming is a profitable and very popular activity in both areas. A comparatively detailed study conducted in 1998 (Coulibaly and Oswald 1999) on a sample of farmers in a neighbouring region (the south-east of Gagnoa) had already highlighted the performance of fish farming activities. Those from Luénoufla also corroborate these observations.

9.2.2.2 First Lessons from the Survey Conducted in 1998 in Côte D'Ivoire

The farming systems of a sample group of farmers (those already involved in fish farming and those evincing interest in it) were analyzed in order to understand the role of fish farming in the farms (Coulibaly and Oswald 1999). This sample included large farms specialized in cocoa cultivation, and whose gross margin per unit area had been calculated (Table 9.4). The surface area of fallow land was included when rice was grown in slash-and-burn plots. The results showed that the gross margin per hectare for fish farming was significantly higher than that for cocoa. This margin could further increase by 30 % when the farmer grew flooded rice inside his dam pond. In fact, a rice-fish association is capable of providing the farmer with a gross margin per unit area almost double that obtained from cocoa.

The overall homogeneity of the value drawn from a day of labour on cocoa plantations highlights the farmers' professionalism. They manage to extract value from their labour force uniformly, regardless of the state of their plantations (Tables 9.3).

As for fish farming (Table 9.5), several farms were coming into production while others had not yet done so. It was thus difficult to determine the value of the working day in a fish farm under normal circumstances. For fish farms that had entered production, the average value of a workday was 16.5 €/day against 5.75 € for cocoa. The exceptional value for farm no. 4 was due to the skilful management of a very large pond. When this farmer was excluded from calculations, the average at 7.2 €/day was slightly above the figure for cocoa. This figure would have been much higher if the sampling included only those farms operating for a sufficiently long period of time.

These figures reveal the ability of this type of fish farming to lead to a real intensification of agriculture per unit area, while, at the same time, improving labour productivity. During the setting up phase, the requirement of labour, and consequently of investment, is high. After that, however, not much work is required to maintain and manage the fish farm. In other words, once the reservoirs are in place, farmers revert to the previous, pre-investment, level of labour requirement. In any case, all work related to fish farms can be done in addition to work required for cocoa cultivation, thus increasing further the remuneration of the working day.

Table 9.3 Economic characteristics of the sample

Farm	Total workers (MWU)	Family workers (MWU)	Capital for cultivation (FCFA)	Cultivated area (ha)	Total gross proceeds (FCFA)	Share of home consumption in total gross proceeds (%)	Labour productivity (FCFA/MWU)
1	10.7	7.5	112,553	33.8	2,444,838	22	203,040
2	5.4	4.1	283,245	12.0	1,777,775	18	263,870
3	8.6	3.7	100,828	45.0	3,252,363	10	337,657
4	21.6	13.6	653,370	64.0	6,051,950	17	224,664
5	9.0	8.7	145,965	16.5	1,795,175	11	128,109
6	10.0	6.5	153,725	26.8	2,521,125	12	201,555
7	7.4	4.1	85,398	18.0	2,254,363	19	275,249
8	6.6	4.2	264,443	17.8	3,054,723	12	406,712
9	10.8	4.5	90,378	69.0	4,620,463	20	406,154
10	22.4	13.3	583,238	100.0	17,123,963	07	696,521
11	11.1	7.4	150,468	55.0	4,368,663	07	338,890
12	7.6	6.1	259,723	23.5	5,599,688	04	653,086
13	8.2	5.6	423,528	40.3	2,701,045	09	219,342
14	9.0	4.5	305,975	22.0	1,462,200	18	111,213
15	8.5	7.0	85,000	18.0	1,935,950	08	204,807
Average	10.5	6.7	246,526	37.4	4,064,285	13.0 %	331,391
Amount in €			376.40		6,205.00		505.90

Table 9.4 Gross margins per hectare of different crops in FCFA

Farm	Fish farming only	Contribution of rice to fish farming (%)	Cocoa	Rain-fed rice on plains and unmanaged lowlands
1	134,667	34	94,344	16,800
2	625,000	–	173,600	15,833
3	218,000	32	122,829	–
4	629,375	21	158,956	10,208
5	246,667	–	247,600	–
6	200,000	–	113,021	6,462
7	314,200	–	201,429	8400
8	232,500	–	388,733	13,938
9	–	–	225,927	9,000
10	224,500	–	422,683	11,523
11	20,833	–	85,744	–
12		–	264,981	18,600
13		–	127,368	7,978
14		–	159,714	5,922
15		–	120,843	11,600
Average	287,380	29	193,869	11,355
Std. dev.	196,364	7	101,801	4,150
Average (€/ha)	438.70		296.00	17.30

Table 9.5 Valuation of the work day in fish farming and a cocoa plantation

Farm	Value drawn from a day's labour exclusively on fish farming	Value drawn from a day's labour on cocoa cultivation
1	2,971	2,769
2	6,375	4,014
3	11,474	3,197
4	53,000	4,731
5	1,510	3,533
6	1,020	2,267
7	n.a.	3,464
8	n.a.	4,952
9	n.a.	4,202
10	9,163	6,186
11	1,020	2,488
12	n.a.	4,801
13	n.a.	3,006
14	n.a.	3,263
15	n.a.	3.196
Average	10,817	3,736
Average (€)	16.50	5.74

9.2.2.3 Review of the Guinea Forest Situation

This study (Barthes 2007), more recent than the one undertaken in Côte d'Ivoire, shows results that are not very different. It was conducted using similar methods on a sample of Guinean farmers, including a number of fish farmers from Gbotoÿe. As in the case of Côte d'Ivoire, not all the fish farmers in this sample (Fig. 9.1) had started producing fish.

The estimated average gross margin was 658 €/year, of which nearly 326 € was obtained from rice cultivation. It should be noted that the performance of fish farming here seems slightly weaker than that obtained in Côte d'Ivoire. This was perhaps due to smaller pond areas and smaller quantities of fish sold on the market, although selling prices were slightly higher. Environmental factors also contributed to this difference: the soils were highly desaturated in Guinea and cooler temperatures could lead to lower yields.

As part of its efforts to compare these systems' performances with the main local cash crop (coffee), the study also analyzed the economic performance of coffee cultivation. However, the fluctuating price of coffee made the task of comparing its effectiveness to that of fish farming very difficult. Two price situations were combined with two different production systems ('wild coffee'[10] and RC2[11]) to

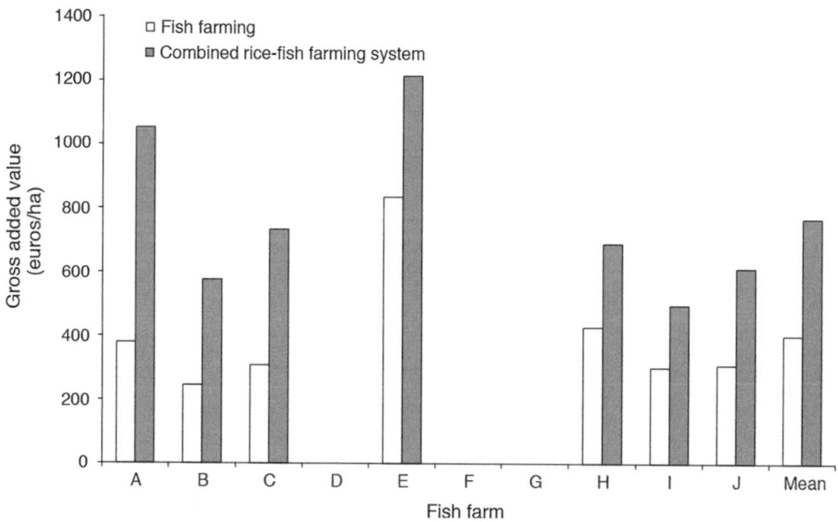

Fig. 9.1 Gross margin per hectare of the fish farming system alone and of the combined rice-fish farming system (Barthes 2007)

[10]Wild coffee ('*sauvageon*' in French) is the local term to describe the coffee planted with young trees taken from old plantations.

[11]RC2: variety introduced by a programme to revive coffee cultivation.

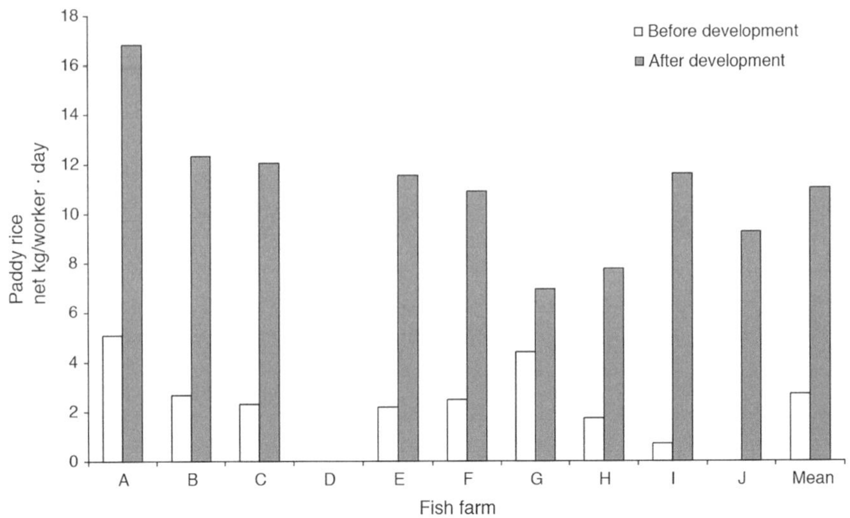

Fig. 9.2 Comparing the value of one working day in a rice field before and after fish-farming development work

obtain the following results: the value of 1 ha of a 'wild coffee' system was always less than that of the rice-fish system (less than 225 €/ha, even in the best price situation). In a situation with an unfavourable price, like in 2004, even the best plots that grew RC2 could not bring in that much revenue. In contrast, when the price was favourable, like in January 2007 (farm-gate prices per kilogram of berries of 0.48 €/kg), the plot of coffee offered a greater margin, subject to a production of more than one tonne per hectare.

A comparison with rain-fed rice, which was in the throes of a crisis in the region, needs no explanation: 149 €/ha considering only the cultivated area. This takes the gross margin per hectare down to less than 30 € if land that is left fallow for at least 5 years is taken into account. The comparison is also clear when we evaluated lowland rice. Farmers, especially those from Gbotoÿe, preferred a traditional arrangement without bunds. The gross margin for this system was only 90 €/year.

Sharp fluctuations in the price of rice and the Guinean currency make it more difficult to estimate the value of the labour put into the rice-fish farming system, especially since most of the fish farming activities account for only a small part of the farmer's working day. On the other hand, the work done in a rice field before and after the creation of a dike for fish farming can be compared based on estimates of the quantity of rice produced in a working day (Fig. 9.2).

The spectacular gains in productivity explain the active involvement in the construction of large dikes. The average pre-development value of the work day was 2.7 kg of paddy/day (excluding seeds). It rose dramatically to 11 kg/day after development.

In Guinea, fish farming addresses the constraints of intensification of surface areas, while adding value to the working day. An added benefit was the potential intensification of agricultural activities without increasing competition in any way. In fact, fish farming alleviates pressure on farmers to grow rice for home consumption (Simon and Benhamou 2009).

9.2.3 *Place of Fish Farming in Earning Revenue*

The share of the total income brought in by fish farming is variable. If the techniques for managing and producing fish are mastered, the limiting factor then becomes access to lowlands that can be developed. This gives rise to several land-use strategies.

9.2.3.1 Significant Income

In any case, fish farming provides a good income to farms, and it is sometimes the most remunerative activity. Most often, however, it is found in the second or third place behind the main plantation crops, but generally ahead of food crops.

9.2.3.2 Marketing, Food Self-sufficiency and Nutritional Balance

Fish is primarily produced as a marketable product. However, the by-products of the catch (unsalable fish or ones considered as pests for fish farming) are consumed within the farm household. It can also be used to pay casual workers hired for the more labour-intensive fishing operations. While it is difficult to arrive at an authoritative ratio, qualitative surveys show that nearly 30 % of the fish caught is finally home consumed. The farms that now have fish farms have recorded an increase in their fish consumption over the years. Some farmers mentioned that their fish consumption has doubled, compared to when they exclusively purchased fish from the market. We can certainly conclude that this has had a positive effect on the nutritional balance of their families. Many farmers involved with fish farming say they no longer buy fish. Mini-ponds have appeared at almost all fish farms to serve as 'pantries' so that farmer families have an easy and continuous supply of fish.

9.2.3.3 Flexibility of Income

Fish farmers in the Luénoufla and Gbotoÿe regions appreciate in particular the flexibility afforded by fish farming. Remember that once the dike walls are built, fish farming no longer requires any extra farm labour. The fish-farming system offers the farmer the flexibility of harvesting the fish when he wants to sell it, allowing him to schedule it when the agricultural calendar permits. Although the value of a workday in a functioning fish farming system is most often greater than that for other crops, the number of days of labour required for this activity remains generally low.

9.2.4 *Unexpected Scale of Investments*

All visitors to the Luénoufla and Gbotoÿe villages who have seen the surrounding lowlands with its fish farms have marvelled at the extent of manually constructed earthwork. The investment is on the order of what would be required to develop primary forests.

How did farmers carry out development work of this magnitude? How much did it cost?

9.2.4.1 What Value to Attribute to Fish Farming Facilities?

The monetary expenditure involved in constructing fish farming facilities is relatively easy to quantify. However, any comprehensive evaluation of the cost of these facilities runs up against two common difficulties inherent to family agriculture. The first is to assess the amount of labour invested in developing the fish farming facilities (especially family labour). The second problem is to estimate the value of this labour, both casual labour and family labour, since it is very difficult to establish the opportunity cost.

For the remainder of this chapter, we shall work with two estimates:

- the first, which overvalues the labour, considers that all labour was paid for (including family labour);
- the second, which undervalues it, considers only the actual expenditure.

A dike in Côte d'Ivoire—of a height of 2 m (from the drain) and 3 m wide, and enclosing 60 m of regularly sloped lowlands—corresponds to a soil volume of 300 m^3. This represents about 60–90 man-days of labour. Only the lean periods in the agricultural calendar can accommodate such investments in rural areas. The cost of a hired labourer (on contract, who ensures the completion of the embankment) varies depending on the hardship involved in the work. It is often 230 € for a derivating pond and 460 € for a dam. A 0.3–0.4 ha facility, with a service pond and a bypass canal, can involve a cost of 850 €/dam (Coulibally et al. 1999) and around 2100 €/ha. The minimum cost for a dam, without earthwork, was 46 €/dam.

In Guinea, the volatile currency makes comparisons more difficult. Barthes (2007) estimates that development work costs 1000 €/ha, by considering that family labour can be valued at market rates. Family labour, on an average, accounted for around 20 % of the total investment cost. The PPGF report (2003) estimates the investment made, taking into account the opportunity cost of family labour, to be, on an average, close to 750 €/ha.

Under specific circumstances, taking the farmer's declared opportunity cost into account triples the investment cost. Thus, a Guinean farmer who declares a direct cost of 180 € knows that the amount he spent to build the dam was actually closer to 405 €. This opportunity cost includes the amount not earned because of a decrease in the area cleared for rice and the reduction in coffee-related work. The cost also, and especially so, includes the loss of earnings from the use of rolling capital that would have allowed him to extract palm oil and cultivate peanuts and rice.

Farmers on both sides of the border establish fish farms mainly by relying on family labour to minimize financial risks. Large farmers in Côte d'Ivoire convert labour contracts from an annual payment basis to the *abusan* system. This is a simplified form of sharecropping, where a third of the harvest is given to sharecroppers as payment, in addition to any other agreement. Under the terms of the *abusan* contract, the sharecropper is required to work an extra day per week for the owner outside of the plot. Here too the goal is to minimize the financial risk posed by the construction of the dam.

Consequently, fish farmers who had learnt the fish farming techniques no longer dipped into their savings for investments. A virtuous process then developed which democratized fish farming to some extent. The presence of the initial fish farms, in fact, helped the poorest farmers gain easier access to certain services (Grosse and Oswald 2010), usually in the form of labour exchanged for investment support. Large farms were the first to use this type of exchange. This kind of relationship, for example, was actually mentioned by a Gouro fish farmer who worked for his elder brother in exchange for the latter's support in constructing his pond.

Finally, it must be noted that for farmers who construct a pond by dipping into their capital, there are few comparable investment opportunities other than in real estate in the town. Statements from this category of farmers indicated that they hoped for an annual income of about 15–30 % of the capital invested. Apparently, this is easy to achieve with the kind of fish farming offered by these projects.

As in other strategies for minimizing risk, prospective fish farmers do not hesitate to use the least fertile lowlands for growing rice. We thus return to the notion of risk associated with an investment: the loss of income is not so important during the construction of the pond. This strategy also highlights the ecological aspect of diversification: fish farm development helps restore a satisfactory level of yields of lowlands that have been degraded by years of rice cultivation. Finally, and more obviously, dikes are mainly constructed when there is less work in the fields. When the agricultural workload is heavy, the groups working on the construction of ponds temporarily suspend their work and wait for a less busy period.

9.2.4.2 How to Assess the Investment Capacity of Production Systems?

The huge investments made highlight the complexity of evaluating the financing capacity of rural farms. We were able to describe this investment capacity here mainly due to a relationship of trust built over time, where each person (farmer, project manager, hired labourers, service providers) assumed his responsibilities. The strong motivation to introduce fish farming and the possibility of sharing skills and know-how with the villagers over a long period of time are certainly key factors in this exceptional mobilization.

The coordination of local groups was a valuable tool for a true adaptation of technical guidelines to farmers' needs. In a PPCO assessment, farmers stated that they acquired their knowledge most often from colleagues rather than through formal training sessions. Nevertheless, such training activities were important for the dissemination of new knowledge (reason for sex determination, management of stocking density, etc.) (Huet 2001; Halftermeyer 2009).

The success of the fish farming sector in these two villages highlights the appropriation of the fish farming project by one or more groups of farmers who, in addition to the obvious economic contribution of this activity to their farms, also regarded it as a tool to further their ambitions. And the larger their ambitions, the larger the capacity of this innovation to fulfil them and to invoke deep interest in others.

We are witness to a model of fish farming development which is very different from the traditional and basic model recommended by extension programs. The innovation is not insignificant at the farm level: it monopolizes the investment capacity of the farm for several years, while demonstrating its sustainability. The events of Côte d'Ivoire help prove this. The 'participatory' aspect then consists of sharing of risks taken by the farmer in the setting up of the facilities, and then during the period his fish farm reaches its full production.

9.2.5 Dynamics on Either Side of the Border

9.2.5.1 A Project that Overcomes Local Constraints Differently

These examples from the two regions show that fish farming can help address water-related matters of the agricultural sector. In this way, the plantation economy context can encompass issues other than the development of cash crops. Other than traditional projects for the development of irrigated rice—which saw mixed success (Delarue 2007b)–, little was attempted to help farmers develop irrigated agricultural systems. In the specific case of fish farming, nothing was done to allow farmers to manage fish farming activities on their own, within the framework of their farm operations.

In Guinea, the exclusive development of fish farming would hardly have had the same relevance. It is only the association with rice cultivation that is capable of producing more per unit area than both systems individually, while also providing a better value for the working day. This outcome has more than adequately addressed the development constraints. Furthermore, the huge investment can be even better justified if it helps make degraded or barren lands fertile. This development underlines the importance of a holistic appreciation of the impacts of fish farming on farming systems.

The example of vegetable gardening in Luénoufla (Côte d'Ivoire) also offers many lessons: the development of water-based agricultural systems is very useful, over and above the simple purpose of producing fish. Water readily available in the fish farming reservoir quickly becomes available for other activities, such as vegetable gardening here.

9.2.5.2 A Project Integrated into Local Policies

At another level, these successes often lead to strong local commitments made by their promoters. Fish farming is fast becoming a political and strategic issue. Two facts are evidence of this:

- in Guinea in 2005, elections were held for the posts of presidents of districts and of rural development communities. Wherever a member of the group of fish farmers stood for the post, the group would proudly announce that their candidate had won. This showed the extent to which fish farming had become an integral part of local debates. Diallo (2003) also noted, with surprise, that local governmental and traditional authorities were well aware of the growing number of fish farmers and their groups and maintained links with various candidates;
- during the Ivorian political crisis, a Sénoufo group, which was a scapegoat of the incumbent government, chose to nominate the only Muslim Gouro who provided several services to them. The Gouro were considered culturally very close to the power centre of the time.

To construct fish farming facilities, several land-related negotiations have to be entered into. Sometimes control over water helps assert the right over a plot or to negotiate for it. We observed here a widespread consistency in the plantation economies of West Africa: a context where land, transferable or exchangeable, is a positive factor in a dynamic of development. Property boundaries in the lowlands often do not reflect ideal demarcations, neither in terms of topography nor water. With a strong process put in place, many land transactions result. Access to lowlands remains a major factor in the process of establishing a fish farm. When fish farming does not meet the expectations of farmers, the area developed for the purpose is neglected; however, such neglect has little impact on the conduct of transactions that are effected over the short term. It is a different matter when the process is sustained. Thus, the Gouro farmer of Luénoufla was denied access to family lowlands by other family members. They were afraid that he would develop the entire lowlands owned by the family and thus lay claim to them. He was first asked to demarcate individual access to the lowlands for each family. In Guinea, the right to purchase lowlands, contrary to existing traditional law, has become common practice, concurrently with the development of fish farming.

9.3 Conclusion

The implementation of the PPCO (Central Western Fish Culture Project, Côte d'Ivoire) and PPGF (Fish Farming Project in the Forest Region of Guinea) projects shows that fish farming is, or should be, an important method of diversification in plantation economies. Its development depends on farm-specific factors: a major issue, in addition to socio-economic criteria, is access to lowlands which can be used for fish farming.

Although the results are not noticeable in national statistics, especially given the large fish quantities imported for consumption, these fish farms demonstrate two major local successes. On the one hand, a true private sector has embraced this activity. On the other, the development of fish farming in these areas had become a significant and structured process, i.e., a transformation and adaptation of the local human environment (Assi Kaudjhis 2005). This step is significant: Chauveau and Richards (2007) emphasize the relationship between, on the one hand, the crises in Liberia, Côte d'Ivoire, Sierra Leone and, to a lesser extent, in Guinea, and, on the other hand, the meagre opportunities offered to the youth and the lack of successful diversification related to water management.

A negative impact of these projects is the building up of a demand that exceeds the capacity to provide support in terms of space and time. This is the price to pay for local success. This impact should be seen as a stage that has definitely been achieved, and which holds lessons for the future.

For the time being, these examples of fish farm development and innovations perfectly illustrate the overall analysis of this book on the economic and ecological determinants of diversification.

References

Apdra-France (2002a) *Techniques de pisciculture: l'aménagement des étangs piscicoles*. Supplementary text, n° 123, CD du Mémento de l'agronome, Cirad-Gret-MAE, 27 p

Apdra-France (2002b). *Techniques de pisciculture: gestion technico-économiques des étangs*. Supplementary text, n° 122, CD du Mémento de l'agronome, Cirad-Gret-MAE, 24 p

Assi Kauhjis JP (2005) Étude géographique de l'aquaculture en Afrique subsaharienne: cas de la Côte d'Ivoire. Doctoral Thesis. Université Libre de Bruxelles, 353 p

Barthes V (2007) Étude de la rentabilité de l'activité piscicole, juin 2006 à avril 2007. Documents du PPGFAFVP and DNAPC-Apdra, Massy, France 28p

Beckett WH (1944) *Akokoaso: a survey of a Gold Coast village*. London school of economics, Monographs on social anthropology, London, 95 p
Borderon A (1999) Programme de développement de la pisciculture continentale artisanale dans le centre-ouest et le sud-ouest en Côte d'Ivoire. Evaluation Report, AFD, Paris 33 p
Brou YT (2005) *Climat, mutations socio-économiques et paysages en Côte d'Ivoire*. Mémoire de synthèse des activités scientifiques présenté en vue de l'obtention de l'Habilitation à diriger des recherches. Université des Sciences et technologies de Lille, Lille, 204 p
Chaléard (1993) Cocoa planters and land investment in towns in Côte d'Ivoire. In: Ruf F, Siwoputranto PS (eds) Reprise des actes du colloque after four centuries of cocoa, what is the future? Chaléard, Bali
Chauveau JP (2000) Question foncière et construction nationale en Côte d'Ivoire: les enjeux silencieux d'un coup d'État. Politique africaine 78:94–125
Chauveau JP (2006) La réforme de 1998 en Côte d'Ivoire à la lumière de l'histoire des dispositifs de sécurisation des droits coutumiers: une économie politique de la question des transferts de droits entre autochtones et «étrangers» en Côte d'Ivoire forestière». In: Les frontières de la question foncière: enchâssement social des droits et politiques publiques, IRD, Montpellier France, 17–19 May 2006, 31 p
Chauveau JP, Richards P (2007) Foncier, transformation de l'agriculture et conflits en Afrique de l'Ouest: enjeux régionaux soulevés par les cas de la Sierre Leone, du Liberia et de la Côte d'Ivoire—revue historique. In: Transformation du monde rural et Développement durable en Afrique de l'Ouest, n° SAH/D(2007)568, February 2007, Club du Sahel et de l'Afrique de l'Ouest. Issy-les-Moulineaux, France, 69 p. www.oecd.org/dataoecd/44/54/39496045.pdf
Coulibaly M, Oswald M (1999) Place de la pisciculture dans un système agraire forestier de Côte d'Ivoire. In: Atelier national sur les bas-fonds du 13 au 15 janvier, Gagnoa, Côte d'Ivoire. CNRA and Consortium Bas-fonds
Coulibaly D, Bamba V, Oswald M (1999) Les techniques d'aménagement piscicole: une alternative de mise en valeur des bas-fonds». In: Atelier national sur les bas-fonds, Gagnoa, Ivory Coast Côte d'Ivoire. CNRA and Consortium Bas-fonds, 13–15 Jan 1999
Darre JP (1999) La production de connaissances dans les groupes locaux d'agriculteurs. In: Chauveau JP, Cormier-Salem MC, Mollard E (eds) L'innovation en agriculture: question de méthodes et terrains d'observation. Éditions de l'IRD, Paris, p 362
Delarue J (2007a) Mise au point d'une méthode d'évaluation systémique de l'impact des projets de développement agricole sur le revenu des producteurs. Étude de cas en région Kpèlè (République de Guinée), INA-PG, Paris
Delarue J (2007b) Aménagement de bas-fonds en Guinée Forestière. Série note de synthèse ex-Post n° 2 Septembre 2007, AFD Paris, 6 p. www.afd.fr/webdav/site/afd/shared/publications/recherche/evaluations/Notes-synthese/02-Notes-synthese.pdf
Diallo (2003) Étude de l'appropriation des aménagements piscicoles dans le cadre du PPGF. Mémoire Cnearc, Montpellier
Dozon JP (1985) La société bété, Côte d'Ivoire. Orstom-Karthala, Paris
Fargeot C (1994) Monographie régionale de la Guinée forestière, document multigrade. Projet de gestion des ressources naturelles, Direction nationale des forêts et de la faune, ministère de l'Agriculture, de l'Élevage et des Forêts, 168 p. plus annexes
Grosse O (2009) La place du poisson dans la consommation alimentaire des villageois des régions Centre et Ouest du Cameroun. Study Report, Apdra, 25 p
Grosse O, Oswald M (2010) The role of the farmers' group in the fish innovation in an extension project's frame. In: Actes du colloque Innovation et développement durable dans l'agriculture et l'agroalimentaire ISDA, Montpellier, 28 June to 1 July 2010. http://hal.archives-ouvertes.fr/hal-00521220/fr/
Guillaume J, Cisse S (2000) Résultats d'enquêtes au sein du Réseau de paysans expérimentateurs de la région de N'Zérékoré. Premières réflexions sur l'introduction de Pueraria Javanica et Arachis pintoï en parcelles paysannes. Irag, Conakry, Guinée, 57 p

Halftermeyer S (2009) Construire un réseau de producteurs ruraux autour d'une nouvelle production: l'exemple du projet piscicole de Guinée forestière (PPGF). Traverses n° 32, pp 1–43. www.groupe-initiatives.org/IMG/pdf/Traverses_32.pdf

Henderson E (2001) *Diagnostic du système agraire du village de Gbotoÿe*. Mémoire de stage dans le cadre de la césure, INA-PG, 147 p. plus annexes

Huet N (2001) *Description de la transmission du savoir-faire nécessaire à l'élevage du tilapia en zone rurale ivoirienne*. Mémoire de stage, INA-PG, 32 p. plus annexes

Keita S (2002) Historique de la pisciculture en Guinée. Communication présentée à l'Atelier de réflexion sur la pisciculture en Guinée forestière; April 2002 N'Zérékoré, DNPCA, Conakry, 12 p

Léonard E, Vimard P (eds) (2005) Crises et recomposition d'une agriculture pionnière en Côte d'Ivoire, Dynamique démographique et changements économiques dans le bas-sassandra. IRD-Karthala, Paris

Oswald M (1997) Recomposition d'une société au travers de plusieurs crises: la société rurale Bété (Côte d'Ivoire). Doctoral Thesis. INA-PG, 290 p. + annexes

Oswald M, Sanchez F (1995) Les sciences sociales sont-elles utiles pour un projet de développement, un opérateur de développement (le projet piscicole du Centre-Ouest) face au diagnostic accessibles, vol 10. Apad, Marseille, France, pp 57–65

Projet Piscicole de Guinée Forestière (PPGF) (avril 2003) Rapport d'Activité, Année 2002, Document du PPGF, AFVP ou DNPCA—Apdra-Massy, France, 46 p

Ruf F (1988) Stratification sociale en économie de plantation ivoirienne. Doctoral Thesis. Université de Paris X, Nanterre, vol 6, 1393 p

Ruf F (2007) Current Cocoa production and opportunities for re-investment in the rural sector: Côte d'Ivoire, Ghana and Indonesia. Contributed paper presented at the World Cocoa Foundation (WCF) meeting, Amsterdam, WCF, Washington, 23–24 May 2007, 35 p. www.worldcocoafoundation.org/about/documents/FRuf_paper.pdf

Ruf F (2010) Les zones forestières et la lente intégration de l'élevage dans les exploitations de cultures pérennes. Ghana, Côte d'Ivoire et Indonésie. In: Thibaud B, François A (eds) Systèmes de production et durabilité dans les pays du Sud. Karthala, Paris, pp 169–190

Simon D, Benhamou JF (2009) Rice-fish farming in Guinée forestière: outcome of a rural development project. Field actions science reports, vol 2, pp 49–56. www.field-actionssci-rep.net/2/49/2009/

Trellu P, Coulibally D, Mangara M, Oswald M (2002) Description d'une dynamique de développement piscicole en milieu rural ivoirien: cas de Luénoufla. In: Pisciculture en Afrique subsaharienne: comment accélérer la diffusion de l'innovation en milieu paysan? Workshop 20 September 2002. Apdra, Massy, France, 12 p

Wey J, Guillaume J (2001) De l'étude de systèmes de productions à base de riz pluvial à la conception de systèmes d'agriculture durable: mise en œuvre de systèmes d'une démarche de recherche participative. Irag, Cirad, N'Zérékoré, Guinée, p 139

WFP (2009) Enquête nationale de sécurité alimentaire (ENSA): République de Guinée, version préliminaire en cours de finalisation, du 6/11/2009. 72p et annexes. http://ochaonline.un.org/OchaLinkClick.aspx?link=ocha&docId=1132997). www.any-documents.com/ebook/Enqu%C3%AAte+Nationale+de+S%C3%A9curit%C3%A9+Alimentaire+%28ENSA%29. Accessed 8 Nov 2011

Chapter 10
Determinants in the Choice of Perennial Crops in Diversified Production Systems of Rubber Growers in South-Western Cameroon

Bénédicte Chambon and Simon Gobina Mokoko

An important part of the strategy of farmers in Cameroon's South-west region for diversifying their production is to invest in the cultivation of rubber. The development of coffee and cocoa plantations in this region dates from the early 20th century. More recently, in the late 1970s and especially the 1980s, alternative crops like rubber and improved oil palm were introduced (Hirsch 2000). In fact, oil palms grown from unselected planting material were present in the region for a long time, but since they were grown for producing oil and palm wine for local consumption, they were not considered a cash crop. Farms in the region have followed different evolutionary trajectories. However, regardless of the first perennial crop planted, the majority of the rubber farms in the area are characterized by diversified cropping systems (Michels 2001).

What makes a farmer select a particular crop for diversification? Based on the results of a survey of a limited number of farmers regarding their decisions to invest in perennial crops, this chapter presents a qualitative analysis of diversification processes and determinants in farmers' choices.

10.1 Methodology

10.1.1 Study Area

The South-west region is located in the humid forest zone which is characterized by a single rainy season. The ecological conditions of the area are favourable for agriculture. Approximately 60 % of the area is occupied by industrial plantations of

B. Chambon (✉)
UPR Systèmes de Pérennes, Cirad-Persyst, F-34398 Montpellier, France
e-mail: benedicte.chambon@cirad.fr

S. Gobina Mokoko
Irad-Ekona, National Rubber Plantation Programme, Latex Plants Programme, PMB 25, Buea, Cameroon
e-mail: smgobina@hotmail.com

F. Ruf and G. Schroth (eds.), *Economics and Ecology of Diversification*,
DOI 10.1007/978-94-017-7294-5_10

oil palm, rubber and banana (Anon 2002). Plantations of perennial crops also play an important role on family farms. Nevertheless, the farmers' production systems and incomes do not wholly depend on perennial crops. Marketable food crops (plantain and tubers, mainly cassava, cocoyam and yam) are also grown to supply the cities of Buea, Limbe and, in particular, Douala.

In addition to this strong complementarity between export crops and food crops (Temple and Achard 1996), the South-west region differs from other regions of southern Cameroon because of two trends: a widespread diversification of perennial crops in family farms and sustained plantation activities.

Several factors favourable to the development of perennial crops in family farms explain this particular situation:

- Two agro-industrial companies have been growing rubber and oil palm since a long time: Cameroon Development Corporation (CDC), created in the 1940s, and Pamol, since 1962;
- Land was readily available despite the presence of agro-industries and the creation of forest reserves by the government (where agriculture is prohibited);
- The province has attracted a large number of immigrants from the West and North-west regions, as well as from Nigeria. Attracted by land availability and industrial, agricultural, port-related, and petroleum-related activities (Anon 2002), these migrants constitute a low-cost labour reserve. The development of perennial crops is linked to this migration of populations, as they constitute the plantations' main labour force (Ruf 2001);
- It is possible for immigrants to purchase land, provided they have the capital to do so. The land market is, in fact, well developed (Temple and Achard 1996);
- The region is well integrated into national and international markets.

10.1.2 Survey

This study was conducted as part of a project to encourage the revival of smallholder rubber cultivation in Cameroon. Data were collected between 2002 and 2003 in areas where CDC was active in supporting the development of smallholder plantations; the number of rubber farmers here was estimated to be less than 500 at the time. A sample of 22 farmers was identified on the basis of two criteria:

- The farmer had to own at least two different perennial crops, one of which had to be rubber;
- The farmer was responsible for taking decisions to invest in perennial crops.

These farmers were selected from a sample created for a previous study on recent dynamics of rubber plantations. This study had identified evolutionary trajectories of different production systems (Chambon 2002). Semi-structured individual interviews were conducted with the heads of the farms—sometimes along with the farmer's son and wife if they played an important role in managing the farm—in order to identify the determinants in the choice of perennial crops. These interviews were conducted in English in the farmer's house to ensure good quality of the data collected. In each farm, the farmers were asked to retrace the history of the plantation's establishment and that of the family. For each plantation that was established or purchased, the farmers were asked to state the reasons and conditions under which the investment was made.

Given the small size of the sample, data were analyzed manually: reconstructing the trajectory of each farm (family events, activities, income sources, land acquisition, and establishment or purchase of plantations), identification of elements that led to the investment and the determinants in the choice of crops.

10.2 Results and Discussion

10.2.1 Establishment of Diversified Production Systems

Common concerns and several factors created a favourable context for diversification.

10.2.1.1 Public Projects

Several government interventions encouraged the diversification of perennial crops in family farms. Smallholder rubber plantations were developed in areas around CDC's plantations in the South-west region. The project, implemented between 1979 and 1986, and funded by the National Fund for Rural Development (FONADER) provided farmers—whose production system was until then based mainly on cocoa—with an opportunity to diversify their income sources. In addition to its direct impact (beneficiary farmers), the project also served as a strong indirect determinant of diversification: it introduced a new crop in the villages that allowed farmers to maintain and increase incomes.

Subsequently, support and extension agencies, such as the National Programme for Agricultural Extension and Research (PNVRA) or the South-West Development Authority (SOWEDA), encouraged farmers to diversify their crops in the 1990s. They provided information on the benefits of diversification, provided technical information on crops and distributed subsidized planting material. These organizations encouraged in particular the setting up of oil palm plantations using improved planting material.

Development projects, which provided technical information and sometimes even capital, thus played an important role in crop diversification. This phenomenon was also observed in Côte d'Ivoire (Naï Naï et al. 2000).

10.2.1.2 Return of Young People to the Village

In addition, many young people, some of them university graduates, returned to the village after having attempted to make a living in the city. This trend seems to correspond to the combined effect of an increase in prices of perennial crops, especially cocoa following the liberalization of the sector in 1996, and the cumulative effects of the crisis that affected the cities and other regions of Cameroon for several years (Ruf 1999). The return of a son to the village to work with an aging parent seemed to infuse new life in farms where the production system, based on a single perennial crop, had stagnated for several years. For a young man seeking better living standards, it was important to diversify production to increase revenue. Moreover, he wanted to establish his own plantations in order to obtain a greater share of the benefit from the production, even while he maintained and managed his father's plantations. In the existing arrangement, even though the son played a major role in managing his father's farm, the income generated by the plantation went to the father. The father decided how much to pay his son.

10.2.1.3 Imitation Effects

The relationship between farmers in the villages contributed to the spread of the dynamics of diversification of perennial crops. The group effect was important for the management of production systems and farming practices: discussions between friends, sharing of experiences and replication of decisions taken by one farmer in another's farm. The effects of imitation in the dissemination of new crops in the village environment, observed in Indonesia in the 1990s (Pomp and Burger 1995), were detected in the diversification process of perennial crops in Cameroon.

10.2.1.4 Maintaining Incomes in the Face of Price Fluctuations

In all cases, the diversification of perennial crops reflected the motivation of farmers to diversify their sources of income over time in order to make it more secure and to increase it. This motivation mainly resulted from price fluctuations. Diversification could be the result of a decline in price of the single cash crop upon which the family's income depended. The family then sought to sustain its income levels over the medium term. Conversely, an increase in price of a particular crop encouraged farmers to introduce that crop in their production system in order to increase their incomes.

Regardless of whether plantations were viewed as retirement capital or as a heritage to pass on to children, diversification was clearly perceived as a strategy to mitigate risk. Taking into account the fluctuations in commodity prices, farmers no longer thought it wise to depend on a single cash crop when they had the opportunity to grow others.

Farmers in the area originally had no opportunity to cultivate a perennial cash crop other than cocoa. Later, when various external interventions made it possible for them to introduce new crops in their production system, many chose to do so. The presence of a single export crop (rubber) represented a transition phase towards a diversified production system in farms being set up, particularly for those farms where perennial crops were introduced only in the 1990s.

10.2.2 Differentiated Evolutions

Several trajectories of the evolution of production systems were observed in the rubber plantations of the South-west region.

10.2.2.1 From Cocoa to Rubber, Palm and Citrus Fruits

Cocoa was the first cash crop on some farms. These farms then diversified in several ways:

- Rubber;
- First rubber, then oil palm. This was the most frequent evolution observed in the survey sample;
- Rubber and coffee, with the purchase of a coffee plantation. This situation was relatively rare. Since several years, coffee was no longer a sufficiently attractive crop for farmers to establish new plantations. It was significantly less profitable than other cash crops in the area (Chambon et al. 2003). Farmers only retained coffee plantations when the pressure on the land was not high enough to force their replacement with a different crop. They provided an additional income;
- Rubber and citrus fruits. The introduction of citrus fruits as a crop to diversify income was a relatively recent phenomenon, occurring only in the late 1990s. In fact, citrus fruits were previously grown, albeit in small numbers, often intercropped with cocoa, with their production destined for home consumption (Chap. 11).

Although production systems were being diversified, farmers were not forsaking cocoa. In fact, some farmers were even expanding their plantations. The majority, however, were simply content to maintain them and replace dead trees. This allowed them to maintain the level of production at the lowest cost, since they avoided the clearing of new plots. On the whole, farmers grew rubber or improved oil palm which they considered to be the most profitable perennial crops. Old cocoa plantations were, however, not cut down for several reasons:

- It was difficult to abandon this traditional crop;
- Land was not yet a limiting factor for setting up new plantations in the area;
- Cocoa helped diversify income sources.

10.2.2.2 From Rubber to Oil Palm and Cocoa

On some plantations, usually the more recent ones, rubber was the initial cash crop. Diversification resulted in two types of farms: one with rubber and oil palm, the other with cocoa as an additional third crop.

10.2.2.3 Diversification Trajectories and the Age of Farmers

Older farmers (aged 50 years or more) started their plantations with cocoa and later diversified into rubber and oil palm. The younger farmers more often planted rubber first. Some had received a cocoa plantation as inheritance or as a means to get established.

The trend was to increase the number of perennial crops in the production system as the farmer grew older. It was observed among autochthons and migrants who had settled in the village for over 25 years, as opposed to immigrants who had arrived more recently. This trend was partly the result of the influence of successive external interventions, as the oldest farms benefited more from projects and financial incentives.

In general, two main types of trajectories coexisted in south-west Cameroon, mainly depending on when the farms were established:

- Diversification of cocoa farms to other perennial crops, observed in the oldest farms;
- The establishment of a rubber plantation, then a gradual diversification to other perennial crops, including cocoa. These were farms where perennial crops had first been planted during the 1990s.

Other than this establishment criterion, farm characteristics did not appear to have a determining effect on their trajectory of adopting and growing perennial crops. In 2002, production systems in the area were generally based on two or three perennial crops with a variable composition.

10.2.3 Determinants of the Choice of Crops

10.2.3.1 Information, Experience and Imitation

In the absence of extension programmes and external intervention, farmers tended to plant crops for which they had sufficient technical experience. When they wanted to diversify their production system, they chose the crop they had gained experience on by working with other farmers of the village or in a plantation company like CDC or Pamol. As seen above, the imitation effect played a role in the choice of the crop.

10.2.3.2 Key Role of Planting Material

At the time of setting up a plantation, farmers chose the crop for which planting material was readily available. The type of diversification crop thus depended greatly on this factor.

Farmers believed that planting material for cocoa was easy to obtain and did not cost much. Indeed, it sufficed to obtain a few pods from farmers in the village to set up a nursery, or plant the beans directly in the field. They thus sowed unselected local planting material. The improved planting material (hybrids) was primarily bred for higher productivity. However, many farmers who had planted such hybrids later realized that these plants had a lower tolerance to disease. Moreover, growing hybrid cocoa was no longer an attractive proposition after the sector was liberalized and subsidies on inputs were withdrawn (Jagoret 2011). This was certainly the reason why farmers seemed disinclined to use hybrids and even preferred the traditional variety (German cocoa).

In the case of rubber, farmers showed more interest in improved planting material (clones or grafted plants). However, when it was not possible to obtain them, the farmers used seeds or planting material from village plantations or from CDC's plantations. So their plants grew from seedlings. The difference between the production level of grafted plants and seedlings is large: under local conditions, a plantation established from seedlings was seen to produce about 500 kg of dry rubber/ha/year. Under similar conditions, the average production of a plantation established with grafted saplings was 1000 kg/ha/year (Delabarre and Eschbach 2002). It appeared that the majority of the farmers were aware of the benefits of improved planting materials, but they still preferred to set up a plantation with limited production potential rather than not grow rubber at all, even though it meant planting seedlings. This observation was consistent with the results of a survey conducted in early 2002 of farmers who had established plantations in the 1990s. It showed that a significant proportion of the plantations were established with seedlings (Table 10.1). It is possible, however, to plant rubber seedlings and then later graft clonal planting material in the field. This was another reason why plantations were established even when the farmer could not procure grafted saplings. However, as far as plots that were planted with seedlings were concerned, no distinction was made in the survey between those farmers who had the intention of a subsequent grafting in the field and those who had no such intention. If the farmer lacked capital or labour at the appropriate time to undertake it, he could forego grafting in the field. Sometimes grafting was done in a portion of the plot, a fact that helped explain the significant presence of rubber plantations with trees grown from a mix of clonal planting material and seedlings (Table 10.1).

Table 10.1 Planting material used for self-financed rubber plantations

Type of planting material	Plantations (%)
Seedlings	36
Plants from grafts	47
Seedlings and grafted plants	17

Sources surveys by Chambon (2002), on a sample of 53 plots

Farmers believed oil palm to be the only perennial crop in the area for which improved planting material was absolutely necessary to establish a plantation. Consequently, the diversification of production systems with oil palm depended much more, in comparison to other crops, on access to improved seeds or plants. This not only meant that farmers had to have access to a source for planting material, but also the ability to pay for it. On some farms, these two conditions prevented the establishment of oil palm plantations, and instead oriented farmers towards cocoa or rubber as crops for diversification.

10.2.3.3 Market Access

Another important determinant in the selection of one or more crops for diversification was the ease of finding an outlet for the production. Farmers could easily sell their cocoa to rural purchasing agents and get paid in cash at the time of delivery. Rubber, on the other hand, was more difficult to sell in the same area due to a lack of a collection network and the virtual monopoly CDC enjoyed in the marketing of the production (Chambon and Eschbach 2009). This forced farmers to arrange and pay for transporting their produce to processing plants. It also involved a long delay between the delivery of the harvest and the payment to the farmers: 3 months on average, sometimes even four. As a result, farmers could be left without an income for several months, despite the fact that they had delivered their produce (for example, there was no payment in May and June 2002). Such difficulties in the rubber sector resulted in a renewed interest in cocoa cultivation.

These disadvantages of the rubber sector also encouraged small farmers to diversify their production systems by setting up oil palm plantations. Oil palm offered several different outlets for the production and had a more diversified market than the other two crops: a small production was more or less guaranteed an outlet. The oil palm's production could be consumed by the household and friends. Farmers could also sell it in the form of bunches and, more commonly, as oil in the local market. However, as long as rural farmers did not invest in presses and as long as they were located far from bunch-processing factories, the areas planted remained constrained by the available processing capacity.

That is why the proximity of CDC's rubber processing plants and the ease of selling the production to these factories encouraged rural farmers to establish rubber plantations, especially in areas located far away from the oil palm processing factories. Transportation costs thus remained relatively low compared to those for transporting palm bunches.

10.2.3.4 Levels of Investment and Income

Smallholders normally plant those crops that are the most profitable at the time they choose to diversify their production systems. Rubber was considered a high-income crop (Table 10.2). Since 1995, the income generated by a rubber plantation under

Table 10.2 Production costs of and incomes from major perennial crops

Costs	Rubber: clones and hired labour	Cocoa: hired labour	Oil palm: hybrids and hired labour
Establishment costs (€/ha)	563	91	211
Maintenance costs for the entire immature phase (€/ha): 4 years for cocoa and oil palm, and 7 years for rubber	374	223	204
Cost of production in the mature phase (€/ha/year)	209	362	44
Gross income (€/ha/year)	549	534	366

Sources Plaza (2003) for costs; based on data from Plaza for gross income

agricultural conditions prevalent in south-west Cameroon was almost always higher than that for cocoa (Fig. 10.1). Income was calculated for a mature plantation, assuming constant production costs and yields. Only the selling prices of cocoa and rubber varied over time.

It was the same for citrus fruits, especially for oranges: in recent years, farmers, mainly the young ones, began to recognize the economic advantages of cultivating citrus fruits, especially when they perceived a demand on the local market. Income levels, however, were not quantified.

From the farmers' point of view, rubber also has the advantage of relatively low production costs in the productive phase, in comparison to those of cocoa (Table 10.2). In fact, cocoa requires significant expenditure on chemicals to ensure a good yield. Depending on the degree of intensification, the cost of inputs varied from 7,000 to 80,000 FCFA/ha/year, or 11–122 €/ha/year in 2002–2003 (Plaza 2003). Rubber, for its part, does not need any inputs during the mature period, except the stimulating paste and acid used to coagulate the latex in case of rainfall

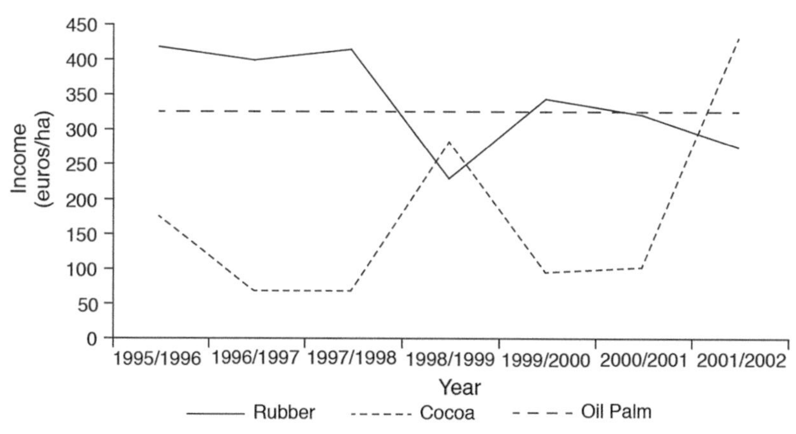

Fig. 10.1 Comparison of incomes from major perennial crops. (*Sources* Surveys conducted by Plaza 2003; CDC and agricultural statistics)

after the plot has started to be tapped. The cost of inputs varied from 1,000 to 27,000 FCFA/ha/year (1.5–41 €). In most cases, however, it did not exceed 5,000 FCFA/ha, or 8 € (Plaza 2003).

On the other hand, the cost of setting up a cocoa plantation remained low because small farmers did not use selected planting material. The investment for establishing a cocoa plantation was much lower than for other perennial crops (Table 10.2). Cultivating cocoa allowed the diversification of income sources at a lower cost.

10.2.3.5 Land-Related Constraints

The location of the land available for plantations determines the choice of the crop. In the case of rubber and oil palm, it is important for the plot to be near the village to limit the risk of theft, especially in oil palm plantations. It is also important for the plot to be accessible to facilitate the offtake of the harvests.

On the other hand, remoter plots can lead some farmers to diversify into cocoa, even if they are not otherwise interested in this crop. The production from cocoa trees is lighter and easier to transport than that of oil palm, or even rubber which is marketed as a coagulum at the end of each month. In the case of cocoa, if a person is assigned to manage the cocoa harvest for a few days, the pods can be split open directly in the field and only the fresh or dried beans need to be transported.

10.2.3.6 Disappearance of Forests and Exhaustion of Forest Rent

Setting up a cocoa plantation on a plot already used previously to grow several food-crop cycles is not an optimal way of growing cocoa. Farmers preferred growing cocoa directly after clearing a forest plot. This pertains to the concept of forest rent and the impact of its exhaustion on farmers' diversification decisions (Ruf 1987). In fact, small farmers choose to plant rubber or oil palm instead of cocoa if the plot was previously used to grow food crops.

10.2.4 Specific Features of Crops

10.2.4.1 Rubber: Regularity of Income and Transmission of Heritage

Historically, rubber was introduced in production systems which were, until then, based on cocoa cultivation. For many cocoa farmers, the main attraction of rubber lies in the regularity of its production, which translates into a steady income throughout the year. Rubber tapping is interrupted only for a month, between February and March, when the trees undergo refoliation. The produce is sold at the end of each month. Consequently, farmers enjoy a regular monthly income which enables them to meet the basic needs of their families.

Rubber is also characterized by a long productive period: if the productive capital is well managed, tapping can take place for more than 30 years, thus assuring a long-term income for the family. Furthermore, if the head of the farm dies, his children are assured of an inheritance. The rubber plantation also guarantees an income to the farmer when he is no longer able to work since, for most farms, tapping work is carried out by hired labour external to the farm and the family.

10.2.4.2 Cocoa: A Quick Return on Investment

Cocoa is both a traditional crop as well as a crop for diversification for farmers who started with rubber, especially young farmers, or those who have recently begun to grow perennial crops. Its main advantage is its rapid entry into production, within three to 4 years in small-farmer conditions. Farmers are thus able to meet the needs of their families faster by growing cocoa than by growing rubber, whose immaturity period lasts at least 7 years under local conditions. It is also regarded as an ideal crop for diversification when rubber farmers are confronted by increased short- or medium-term financial needs.

In the establishment phase, cocoa is traditionally interplanted with food crops such as cassava or plantain. In contrast, extension agencies have long recommended rubber plantations to be kept monospecific. For farmers with limited land access, the association of food crops with perennials is a determinant for choosing crops, making this a possible reason for adopting cocoa cultivation.

In general, farmers who had already grown cocoa were neither interested in reinvesting in this crop nor in expanding their plantations. Several reasons were advanced for this, especially after liberalization, when the price of cocoa began to fluctuate more than those of other crops:

- Cocoa is considered to produce an annual yield (from September to December) and not a monthly one;
- The cost of cocoa production is considered to be higher than that of rubber.

Nevertheless, in the risk-minimizing strategies of the farmers, rubber and cocoa are perceived as two perennial crops with complementary incomes. Ultimately, the role of rubber and cocoa has always been to ensure a monetary income for the family. In the event that the price of one of these two crops declines on the international market, the other can compensate for it and sustain the farm's income.

10.2.4.3 Oil Palm: Diversification of Consumers and Markets

The most frequently observed evolution was the introduction of improved oil palm in a cocoa- or rubber-based production system. The production from oil palm trees is regular throughout the year, with production peaking during the year's first six months. This makes it attractive to farmers whose objective is to obtain a regular income. This can be possible either through crop diversification, or by choosing crops which produce throughout the year.

However, the main benefit of this crop, traditionally grown in the area, is its food value: palm oil and palm kernel oil as well as palm wine. When palms grow too high and it becomes difficult to harvest the bunches, the trees can be cut down and tapped to produce palm wine. This allows farmers to obtain an income even after the tree is felled. The oil and wine produced are consumed by the household, or sold locally when there is a surplus. The rationale for management of this perennial crop is the same as that for annual and perennial food crops. The primary function of oil palms is nutritive. The area on which palms are grown is limited and, on farms smaller than a hectare, it is not uncommon to find plantations with just 20–60 palm trees. The objective is often to fulfil the family's needs and not to sell the produce, although any surplus production is sold. Based on information obtained from specialists in oil palm cultivation in the African village environment, it was estimated that 30 improved palm trees can provide for a family's home consumption (taking into account the traditional African practice of gifting some production to family and friends). These improved palms were planted to replace unimproved palm trees with limited yields. Farmers thus seek to improve labour and land productivity.

Oil palm seems to be a safe crop for farmers. By cultivating this crop, farmers demonstrate their belief that diversification of production systems can help limit risks and ensure an income for the family, even if the price of cash crops (cocoa and rubber) decline and even if CDC or its agents no longer buy their production.

10.3 Conclusion

Over the past 30 years, there have been different evolutionary trajectories of perennial crops in south-west Cameroon. Whether it is a cocoa farm that starts cultivating rubber or a new rubber plantation that ventures into cocoa, neither system ignores the oil palm. However, existing production systems are relatively uniform and are based on two or three perennial crops: rubber, cocoa and oil palm.

For farmers already well integrated into markets, there are many and complex determinants of diversification. They are not only economic in nature, but also ecological and institutional. Although crop specificities are important and real, they are finally only significant in a secondary manner, because ultimately, crop diversification primarily reflects a very simple motivation, that of maintaining—and, if possible, improving—farm incomes.

References

Anon (2002) Document de stratégie de développement du secteur rural. Ministère de l'Agriculture, Yaoundé, Cameroun

Chambon B (2002) Dynamique récente de plantations villageoises d'hévéa dans la province Sud-Ouest du Cameroun. Survey Results, Ekona, Cirad-Irad

Chambon B, Eschbach JM (2009) Commercialisation du caoutchouc naturel des plantations villageoises: un facteur essentiel du développement de l'hévéaculture au Cameroun. Cah Agric 18(4):364–368

Chambon B, Eschbach JM, Plaza C, Gobina S (2003) Diagnostic du secteur hévéicole villageois: modélisation de quelques exploitations agricoles de la province sud-ouest du Cameroun. In: Presentation at the Workshop Olympe, Sept 2003

Delabarre M, Eschbach JM (2002) Les plantes à caoutchouc. In: Mémento de l'Agronome. Ministère des Affaires étrangères, Cirad, Gret, p 1185–1197

Hirsch R (2000) Dynamique récente des plantations individuelles de palmier à huile au Cameroun. OCL 7(2):172–174

Jagoret P (2011) Analyse et évaluation de systèmes agroforestiers complexes sur le long terme: application aux systèmes de culture à base de cacaoyers au Centre-Cameroun. Doctoral thesis. Montpellier SupAgro

Michels T (2001) Petites et moyennes exploitations hévéicoles au Cameroun: systèmes de production, itinéraires techniques et stratégies. Diagnostic technico-économique, diagnostic agronomique et typologie. Ekona, Irad-Cirad-Coopération française

Naï Naï S, Cheyns E, Ruf F (2000) Adoption du palmier à huile en Côte d'Ivoire. OCL 7(2):155–165

Plaza C (2003) Situation de l'hévéaculture villageoise dans une filière en évolution: cas de la région d'Ekona, province du Sud-Ouest, Cameroun. M.Sc. thesis, Cnearc, Montpellier

Pomp M, Burger K (1995) Innovation and imitation: adoption of cocoa by indonesian smallholders. World Dev 23(3):423–431

Ruf F (1987) Éléments pour une théorie sur l'agriculture des régions tropicales humides. I - De la forêt, rente différentielle au cacaoyer, capital travail. Agronomie tropicale 42(3):218–232

Ruf F (1999) Comment et pourquoi la Côte d'Ivoire produit durablement plus d'un million de tonnes de cacao ? Afrique agriculture 268:21–25

Ruf F (2001) Libéralisation et tenaille des prix cacao/intrants. Le cas du sud-ouest du Cameroun. In: Griffon M (ed) Filières agroalimentaires en Afrique. Comment rendre le marché plus efficace? Ministère des Affaires étrangères, Paris, p 269–313

Temple L, Achard R (1996) La gestion de la fertilité dans les systèmes de culture du bananier-plantain dans le sud-ouest du Cameroun. In: Fertilité du milieu et stratégies paysannes sous les tropiques humides, Workshop Proceedings, Montpellier: Cirad-Sar, p 519–526

Chapter 11
Socio-economic Conditions of Horticultural Diversification in Cocoa Production Systems in Southern Cameroon

Ludovic Temple and Jules-René Minkoua Nzié

Two important changes in the economic environment of cocoa farmers in southern Cameroon have taken place over the last 15 years. On one hand, the growth of urban markets has created new market opportunities for food and horticultural crops. On the other, there has been an increased uncertainty of income from cocoa stemming from several factors: aging plantations (Leplaideur 1987); phytosanitary pressures resulting from a shift to monocultures, a price-cost squeeze (prices of cocoa, coffee and inputs); withdrawal of the government from the management of the sector; and an instability in international cocoa markets (Boussard and Gérard 2005).

Such contextual changes have led to the acceleration of diversification strategies at various levels (Malézieux and Moustier 2005): at the plot level, by an intensification of complex agroforestry systems; at the farm level, through economic activities of the household; and at the territorial level, through peri-urban agriculture. However, the increasing demand from cities requires a regular supply of produce which the surplus from home consumption-oriented agriculture—atomized at the spatial and microeconomic levels—is insufficient to fulfil. These new demands require changes in production systems and the functioning of supply chains.

According to some authors (Cour 2004), these changes take place when the level of urbanization (simplified indicator of urban demand) exceeds 50 % of the population. This level has already been reached in Cameroon. The transformations of farming systems (Mazoyer et Rondart 2002) are rarely the result of great revolutions, except in the case of foreign investments by funding entities, for example, the

L. Temple (✉)
UMR Innovation, Cirad-ES, TA C-85/15 (Bât. 15, Bur. 302),
73 rue Jean-François Breton, 34398 Montpellier Cedex 5, France
e-mail: ludovic.temple@cirad.fr

J.-R.M. Nzié
Département d'économie publique, FSEG - Université de Yaoundé II – SOA,
B.P. 1365, 18 OA Yaoundé, Cameroon
e-mail: minkouarene@yahoo.fr

F. Ruf and G. Schroth (eds.), *Economics and Ecology of Diversification*,
DOI 10.1007/978-94-017-7294-5_11

Green Revolution of the 1970s and 1980s. They often result from changes in the farmers' strategies and their attempts to adapt their production systems to take advantage of opportunities. These strategies are guided by objectives of maintenance, improvement or, even, transformation of lifestyles (Yung 1992). Their performance is dependent on external incentives, whether institutional, economic or technical (for example, agricultural research).

Recent studies in Cameroon on adapting cocoa-based production systems to urban market demands (Temple 2007) have helped characterize the process of diversification of agriculture around two axes. The question is how to analyze this evolving diversification today. At the level of the cropping system, the convergence of microeconomic-level diversification choices towards a few dominant species (Aulong et al. 1999) leads to the emergence of pre-specialization zones in certain regions. One consequence is that regional names are widely used to identify local varieties, for example, 'Obala mandarin', 'Makénéné African plum' or 'Munja plantain'. At the level of the production system, this diversification-specialization results in increased labour inputs for production in the case of agroforestry or increased use of inputs in the case of peri-urban horticulture (Moustier and Fall 2004).

However, the determinants or the microeconomic conditions that facilitate these changes remain poorly characterized. This knowledge is, nevertheless, necessary to permit agricultural policies and research efforts help guide or accelerate these changes, while taking into account their impact on sustainable development: the fight against poverty, environmental management, food security, or production of global public goods. We will attempt to increase this knowledge on microeconomic determinants of family farming.

11.1 Data and Methods of Analysis

The survey was conducted in the Central and South-west regions of Cameroon, and included the description of natural conditions, settlements and activities of rural people. In the Centre region, the survey was conducted in Lékié division and Mbam and Kim division—these two divisions happen to be the largest cocoa producers in the Centre region—and the Fako and Mémé divisions in the South-west region.

The Centre region of Cameroon, home to dense, semi-deciduous rainforests and ferralitic soils, is subject to pronounced human activities. The region has an equatorial climate with four seasons, and annual rainfall varies between 1400 and 1500 mm. Lékié has moderately desaturated soils that offer a much higher potential for intensive agriculture by virtue of their fertility. The high population pressure suggests that farmers are aware of the potential of their soils. The Centre region is home to several ethno-linguistic groups (Beti, Bassa, Bafia and Sanaga) originating from a sustained migration of people from Grassfields, eastern Nigeria and Bamiléké areas. Population surveys (Courade 1974) indicate that immigrants make up 85 % of the population, with current population concentrations often the result of forced settlements during the colonial period. The advent of large agribusiness companies and their high demand for labour have also influenced these migration flows.

The South-west province is characterized by abundant rainfall, ranging between 2000 and 3000 mm, and volcanic soils that favour the cultivation of export crops; the yield of cocoa plantations here is about one tonne per hectare. The abundance of rainfall, however, exacerbates phytosanitary problems. Moreover, it is difficult to dry cocoa beans, as the peak harvest season coincides with the period of the most intense rainfall. This greatly impacts the ability to dry the beans under the sun, and forces farmers to use wood stoves, with a consequent increase in production costs.

Data gathered by surveying 234 farmers in the Centre and South-west regions, which are the main cocoa producing regions of Cameroon, revealed a trend towards diversification of cocoa cultivation. The rural households, interviewed by using an open-ended questionnaire, were randomly selected.

In the Centre region, surveys were conducted in Lékié division and Mbam and Kim division (the region's largest cocoa producers). Surveys were carried out in a total of five villages in the Sa'a and Makénéné sub-divisions. In the South-west region, the surveys were conducted in the Fako and Mémé divisions (which play a prominent role in cocoa cultivation). It was shown that cocoa yields recorded in Mémé (25,000 tonnes) were the highest for any division in the country. Three villages with high cocoa production were selected in each division, making a total of six villages in the sub-divisions of Muyuka and Mbonge.

Table 11.1 shows the distribution of farmers by regions, divisions, sub-divisions and villages.

Table 11.1 Distribution of farmers interviewed in the study areas

Regions	Divisions	Sub-divisions	Villages	Number of farmers
Centre	Lékié	Sa'a	Nkolbikak	30
			Ngoksa	16
			Letong	14
	Mbam and Kim	Makénéné	Makénéné-East	43
			Makénéné-Centre	17
			Total	120
South-west	Fako	Muyuka	Ikata	16
			Bafia	23
			Muyengue	25
	Mémé	Mbonge	Small Ekumbe	20
			Big Ekumbe	15
			Kombone	15
			Total	114
Total				234

Source Surveys by Minkoua (2003)

Table 11.2 Cropping systems adopted by farmers in the cocoa farms of the Centre and South-west regions

Cropping system of the farm	Number of farms	Percentage (%)
Cocoa + oil palm + coffee + food crop	1	0.43
Cocoa + oil palm + food crop	6	2.56
Cocoa + coffee + food crop	1	0.43
Cocoa + oil palm	16	6.84
Cocoa + coffee	2	0.85
Cocoa + food crop	84	35.90
Only cocoa	114	48.72
Total	234	100

Source Minkoua (2003)

The statistical study was based on analysis of variance and a means comparison test. Both tests required, to start with, the empirical construction of a diversification indicator. According to Yung (1992), 'In our analysis, diversification is said to take place when a farm varies or expands the range of its products or its clients.' In methodological terms, we propose evaluating the intensity of diversification using an indicator (I_d), which took into account the diversity (n, the number of species) and the number of times each species is adopted by the farmers (f, frequency of occurrence). For the sake of simplicity, we assume that the number of species (n) represents fully the importance consumers attach to diversity: the more the number of producers and species, the greater the well-being of consumers, assuming that all other determinants (preferences, incomes, prices, etc.) influencing this well-being remain constant. Keeping in mind that the production cost of different crops is not the same, and that the price elasticity of the demand for the produce is heterogeneous, these two variables influence the difference in the levels of adoption between farmers. Table 11.2 shows the frequency of occurrence of different crops on cocoa farms.

11.2 Analysis of the Diversification of Cropping Systems in the Centre and South-West Regions of Cameroon

Farms usually employ one of two methods to diversify. First, the farmer can grow a mosaic of monoculture plots, with each plot planted with a different crop. In our case, it means monoculture plots of cocoa, oil palm or coffee. Second, the farmer can decide to mix all the crops, either randomly or in a systematic manner. We can then see, on the one hand, plots where cocoa is the main crop, but in which other crops are also grown, and, on the other, we find food crop plots in which several species are grown; there is no systematic hierarchization of these species (Table 11.2).

We observe a preponderance of cocoa-only cultivation (48.7 % of farms) and the intercropping of cocoa and food crops (35.9 %). In our study areas, these different cropping systems lead to three forms of intensification:

- In the first form, the farmer intensifies farming through labour by planting food crops on ridges or mounds (plantain, cassava). This practice helps restore soil fertility through better recycling of available biomass. It also allows an increase in the density of crops and lets the farmer take advantage of their potential complementarity. It helps stabilize or even increase yields. However, their evaluation is complex because of the intercropping practices adopted and their variability (number of crops, density, etc.). This system replaces shifting cultivation on slash-and-burn forest lands;
- In the second form, the farmer intensifies cocoa plantations by planting other perennial fruit crops (African plum or *Dacryodes edulis*, citrus fruits) or oil palm;
- In some areas (Obala), a third form has been developed through a diversification into tomato cultivation, practiced mainly by young people and women. Cropping patterns range from extensive systems (small plots, shifting cultivation) to pesticide-intensive systems.

The diversification of cropping systems in the study areas reveals a diversity of species grown on the farms. We thus observed changes in the trajectory of cocoa-based cropping systems, exemplified by the introduction of perennial crops (African plum, orange, oil palm and coffee), food crops (plantain and cassava) and vegetables (tomatoes, etc.).

Table 11.2 shows the number of times a given crop was observed in cocoa plots (total per row). We show a maximum of three of the most important species, other than cocoa, which were introduced in the plots. The last column provides, for each crop, its frequency of occurrence in the plots, and thus the probability of its adoption by the farmer. As Table 11.3 shows, plantain and banana, African plum, citrus fruits (orange and mandarin) and oil palm have, in general, the highest presence in the cocoa plots: 27.2, 21.5, 12.3 and 10.6 % respectively, or a total of 71.7 % for these four groups of products. We note that the intensity of diversification is not homogeneous: a more or less random presence of oil palm plants on some farms; a reasoned inclusion of oil palm in farms with cocoa or coffee; or interplanted with these latter crops. The form of this diversification (random placement of more or less isolated palm plants or their planting in a logical grid pattern in cocoa plantations) cannot be accurately distinguished using available data.

Nevertheless, we observe that coffee and cocoa plantations were replaced by food crops (16 %) and oil palm plantations (21 %). This conversion process is also dependent on the crop that preceded the food crop (Fig. 11.1) and the palm plantations (Fig. 11.2), with 31 % of oil palm plantations being established on primary forest land and 26 % on fallow land.

On the whole, we observe a diversification process that take places more or less to the detriment of surface areas devoted to cocoa cultivation.

Table 11.3 Frequency of occurrence of crops in cocoa plots

Type of crop	Scientific name	Crop 1	Crop 2	Crop 3	Total	Frequency
African plum	*Dacryodes edulis*	89	79	22	190	0.2147
Avocado	*Persea americana*	11	24	19	54	0.0610
Sugarcane	*Saccharum robustum*	–	1	–	1	0.0011
Plantain	*Musa*	120	77	44	241	0.2723
Peanut	*Arachis hypogaea*	–	–	7	7	0.0079
Citrus fruit	*Citrus spp.*	42	41	26	109	0.1232
Guava	*Psidium guajava*	–	1	1	2	0.0023
Maize	*Zea mays*	3	3	1	7	0.0079
Cocoyam	*Xanthosoma sagittifolium*	10	24	12	46	0.0520
Oil palm	*Elaeis guineensis*	62	20	12	94	0.1062
Cassava	*Manihot esculenta*	13	15	18	46	0.0520
Coffee	*Coffea spp.*	28	5	–	33	0.0373
Mango	*Mangifera indica*	7	8	6	21	0.0237
Yam	*Dioscorea spp.*	1	–	1	2	0.0023
Bean	*Phaseolus vulgaris*	–	1	–	1	0.0011
Kola	*Cola acuminata*	3	3	9	15	0.0169
Chilli	*Capsicum annuum*	1	–	–	1	0.0011
Pineapple	*Ananas comosus*	–	4	6	10	0.0113
Hazel	*Coula edulis*	1	–	–	1	0.0011
Papaya	*Carica papaya*	–	–	1	1	0.0011
Njansang	*Ricinodendron heudelotii*	1	2	–	3	0.0034
Total		392	308	185	885	1.0000

Source Minkoua (2003)

11.3 Typological Determinants of Diversification in Cocoa Farms

11.3.1 Developing a Diversification Indicator

The diversification indicator proposed here is the number of different crops on the farm, associated with the frequency of occurrence of major new crops (oil palm, African plum, plantain, orange, cassava, etc.) in cocoa plots. The farm is considered here in a holistic sense: it includes cocoa plantations as well as monoculture or polyculture plots juxtaposed with the cocoa plots, plots in which marketable food crops are usually grown. The diversification indicator we propose allows us to create the typology (Table 11.4).

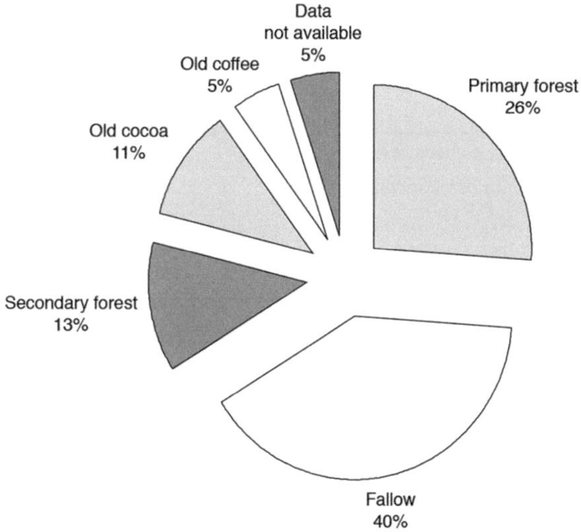

Fig. 11.1 Preceding vegetation on food crop plots. (*Source* Minkoua 2003)

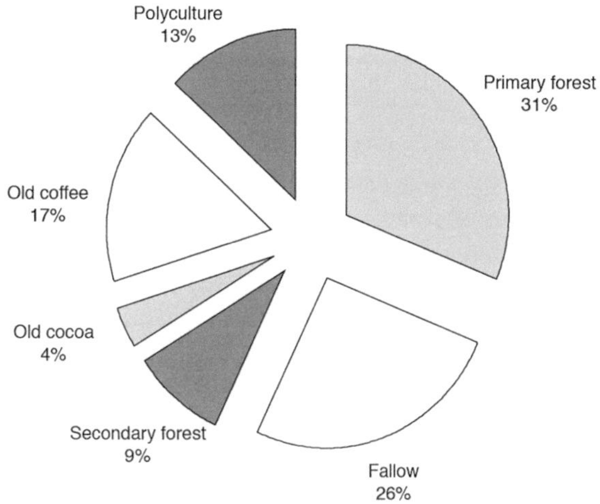

Fig. 11.2 Preceding vegetation on oil palm plots. (*Source* Minkoua 2003)

Table 11.4 Farm typology by diversification indicator

Number of different crops associated to frequencies	Number of farms	Class	Type
0	4	No diversification	0
1–3	79	Low diversification	1
4–5	75	Moderate diversification	2
6–12	76	High diversification	3
Total	234		

Source Minkoua (2003)

Table 11.5 Results of statistical means comparison tests by variable and diversification type

Typology	Size of class	Mean of variables			
		Age	Experience	Farm size (ha)	Size of household (ha)
Low diversification (type 1)	79	47	20	6	8
Moderate diversification (type 2)	75	46	20	5	8
High diversification (type 3)	76	51	21	8	11
Fisher's test		3.0[a]	0.23	4.9[a]	10.4[a]
Z (calculated)	1–2	0.33		1.24	0
	1–3	2.0[c]		2.0	3.71[c]
	2–3	2.5[d]		3.0[d]	4.05[d]

Source Data from the survey by Minkoua (2003)
[a]Overall difference in means
[b]Difference in means between diversification of type 1 and 2
[c]Difference in means between diversification of type 1 and 3
[d]Difference in means between diversification of type 2 and 3

Type-0 households, which were not diversified, are not studied here. We tested different variables that determine the level of diversification of a farm, i.e., low, moderate and high.

Tables 11.5 and 11.6 show the significance tests of the variables. Fisher's test refers to the variance analysis, while the Z statistic reflects the means comparison test. Some variables, such as age of the cocoa plantation, could not be tested due to the difficulty in confirming the reliability of the information collected. In fact, the concept of age of the cocoa plantation varied greatly from farmer to farmer and depended on the intensity of the regeneration process. Its real evaluation involves specific work that is outside the scope of this survey.

The variance analysis indicates that there was a significant difference between the three average ages of farmers in each type, which implies that there is an age difference of 4–5 years between a farmer at the 3rd level of diversification and one at the 1st and 2nd level (which are significantly equal between themselves). We can therefore conclude that the older the farmer, the higher will be his tendency to diversify his cropping system. If we associate diversification with innovation, this result looks very different from those of other studies in which the innovators tend to be the young farmers (Gockowski 2002). However, we can posit that, given the theory of the life cycle (Ruf 1991), there exists a positive correlation between the age of the farmer and that of the cocoa plantation—this variable could not be analyzed, as is explained below. In fact, the intensity of the diversification would then, in part, be related to a change in the evolutionary trajectory of cocoa.

For the farm size, the high Fisher test value (4.96) suggests that the mean areas of the plantations in the three types of diversification are different: the bigger farms are more diversified than the smaller ones. For the household size—which in reality is correlated with farm size,—the result is comparable.

Table 11.6 Results of statistical means comparison tests by variable and diversification type

Typology	Size of class	Mean of variables			
		Cocoa net profit (FCFA/inhabitant)	Savings	Active women	Access to urban markets
Low diversification (type 1)	79	188,462	1.6	2.32	1604
Moderate diversification (type 2)	75	161,981	1.5	2.28	1523
High diversification (type 3)	76	664,388	1.6	2.86	1471
Fisher's test		0.84[a]	1.6	9.5[a]	4.42[a]
Z (calculated)	1–2			0	2.16[b]
	1–3			3.1[c]	3.13[c]
	2–3			3.05[d]	1.09

Source Surveys by Minkoua (2003)
[a]Overall difference in means
[b]Difference in means between diversification of type 1 and 2
[c]Difference in means between diversification of type 1 and 3
[d]Difference in means between diversification of type 2 and 3

Furthermore, if we take into account the specialization of labour in rural Cameroon, which shows that cocoa cultivation is an activity carried out by men (Temple 1997), while food production is looked after by the women (Baumann 1980), the results indicate that diversification increases with the number of active women in the household (wives, and co-wives in the case of polygamy). However, we could not observe many factors that can explain diversification between types 1 and 2, whereas level 3 was observed in large farms run by households with several working women. Market access was estimated by the cost of transporting goods from the village to the major urban markets of Yaoundé and Douala. The result indicates that these markets had a significantly positive effect on the level of diversification of rural farmers in the Centre and South-west regions of Cameroon.

11.3.2 *Farmers' Activities and the Strategic Role of Diversification*

As far as the activities of cocoa farmers in the study areas (Centre and South-west regions of Cameroon) are concerned, our surveys showed that, in general, the owners or heads of cocoa plantations were:

The first concerns the low risk aversion to innovation that older farmers exhibited. We can advance two explanations identified during the survey, even though they contradict with many deterministic conclusions on the subject, but we cannot, however, rank them based on our data.

We can first mention the role of elders in promoting community cohesion: they would like to retain their pioneering role in shaping collective decisions. Since farmers are capitalizing on areas planted with cocoa, they improve their ability as they age to devote financial resources to invest in changes of trajectories. On the other hand, they also reduce their aversion to risk arising from the introduction of new activities.

The second explanation refers to the roles of women in steering the diversification towards food crops; in general, they are the ones who have technical expertise required for cultivating these crops. This reason can also partly be used to buttress the first explanation. Some older farmers who have succeeded in their strategy of capitalizing on cocoa are able to practice polygamy. They thus assign some arable plots to a certain number of women for growing food crops, and use the income from cocoa to meet, at least partly, some of the expenses of their expanded household (education, health, energy, etc.).

The third explanation refers to the life cycles of cocoa plantations which, having depleted the fertility rent, have become less competitive when compared to cultivation on agricultural frontiers. Indeed, such a situation encourages trajectories of adaptation of technical itineraries based on perennial crops. These trajectories result in partial conversions to food or horticultural crops which, in turn, lead to processes of intensification. There are probably other strategies to be discerned in collective or individual actions to adopt innovations in farming practices in order to help manage fertility in the ecosystems or territories under consideration.

These results also highlight the fact that farmers with the lowest diversification indicators were in farms where the sectoral pluriactivity of household members was the highest. A pluriactive farmer who invests in, or maintains, an agricultural activity, generally chooses a specific cultivation and standardizes his cropping system. These results will need to be confirmed in two ways:

- First, through a better characterization of diversification indicators, not only from the structure of cropping systems but from the structure of revenues. This will obviously require more regular and sustained survey mechanisms;
- Second, by increasing the number of variables that may affect diversification, particularly at the level of the mesoeconomic determinants of territories, the functioning of the sectors, and, more generally, institutional incentives.

References

Aulong S, Dury S, Temple L (1999) Dynamique et structure floristique des agroforêts à agrumes au centre du Cameroun. Fruits 55(2):103–114

Baumann E (1980) Monoculture cacaoyère et production vivrière dans la Lekié (Centre-Sud Cameroun). Mémoire de DEA d'Études africaines, Université Bordeaux I, Institut d'études politiques, Centre d'étude d'Afrique noire, année universitaire 1979–1980.

Boussard JM, Gérard F (2005) Libéraliser l'agriculture mondiale: théories modèles et réalités. Montpellier, Cirad

Courade G (1974) Atlas régional Ouest I. Orstom, Yaoundé

Cour JM (2004) Peuplement, urbanisation et transformation de l'agriculture: un cadre d'analyse démo-économique et spatial. Cahiers agricultures 13(1):158–165

Gockowski J, Ndoumbe M (2002) The adoption of intensive monocrop horticulture in southern Cameroon. Agric Econ 30:195–202

Leplaideur A (1987) Stratégies paysannes autour des systèmes cacaoyers et vivriers en zone Centre-sud forestière camerounaise. In: État, développement et paysans, colloquium proceedings Cirad-Mesru, Sept 1987, pp. 74–85

Malézieux E, Moustier P (2005) La diversification dans les agricultures du Sud: à la croisée de logiques d'environnement et de marché: I. Un contexte nouveau. Agricultures 14(3):277–281

Mazoyer M, Roudart L (2002) Histoire des agricultures du monde, du Néolithique à la crise contemporaine, coll. Points Histoire, 2e éd., Seuil

Minkoua JR (2003) Stratégie de diversification agricole en zone cacaoyère dans le Sud-Cameroun. Thesis DEA. Université des sciences économiques. Yaoundé II

Moustier P, Fall AS (2004) Les dynamiques de l'agriculture urbaine: caractérisation et évaluation. In: Développement durable de l'agriculture urbaine en Afrique francophone: enjeux, concepts et méthodes. Montpellier, Cirad

Ruf F (1991) Les crises cacaoyères. La malédiction des âges d'or? Cahiers d'études africaines 121–122, 31(1–2):83–134

Temple L, Fadani A (1997) Cultures d'exportation et cultures vivrières au Cameroun: l'éclairage d'une controverse par une analyse micro-économique. Économie Rurale 239:40–48

Temple L, Minkoua JR, David O (2007) Diversification des exploitations cacaoyères au Cameroun et demande d'innovation. In: Gafsi M, Dugue P, Janin JY, Brossier J (eds) Exploitations agricoles familiales en Afrique de l'Ouest et du Centre. Edition Quae, pp 303–311

Yung JM (1992) Les stratégies des producteurs. In: Bosc PM, Dolle V, Garin P, Yung JM (eds) Le développement agricole au Sahel. Cirad-Sar, Montpellier, France, pp 277–299

Chapter 12
Agroforestry-Based Diversification for Planting Cocoa in the Savannah of Central Cameroon

Patrick Jagoret, Frank Enjalric and Éric Malézieux

Cocoa is a crop most often cultivated after clearing forests. This ensures, at least in the initial years, favourable conditions for good agricultural production: a relatively high level of soil organic matter, nutrient availability (from burning of standing vegetation), limited amount of weeds and reduced pest pressure (Ruf 1987, 1995). In order to establish cocoa plantations on forest land, large trees are either completely or partially cut down. In recent years, farming practices in several countries have evolved towards a requirement of complete clearing (Chap. 2).

However, in Cameroon, especially in its Centre and South regions, forest trees are usually preserved during clearing, both because of their economic value as well as to provide shade for young cocoa trees (Duguma et al. 2001). Native fruit trees, some with medicinal value, are also preserved. Farmers also adopt the practice of interplanting annual crops (maize, groundnut and cocoyam) with perennial crops (cassava and plantain) in the plots of young cocoa trees. At the same time, they also introduce other perennial species (oil palm, orange, African plum, kola, avocado, etc.) which grow in association with cocoa and the originally preserved forest trees (Tchatat 1996; Aulong et al. 1999; Temple et al. 2007; Chap. 2 of this book). These cultivation practices result, in a few years, in a highly diversified cocoa agroforestry system which dominates the cocoa production area of south-central Cameroon (Ruf and Schroth 2004; Zapfack et al. 2002; Bidzanga 2005; Sonwa et al. 2007).

P. Jagoret (✉)
Cirad-Persyst, UMR System, Bâtiment Agro Montpellier SupAgro, 2 Place Viala—Bât. 27, 34 060 Montpellier Cedex 1, France
e-mail: patrick.jagoret@cirad.fr

F. Enjalric
UPR SIA, Cirad-Persyst, Direction Régionale du Cirad - Ampandrianomby, BP 853 101 Antananarivo, Madagascar
e-mail: frank.enjalric@cirad.fr

É. Malézieux
Cirad-Persyst, UPR Hortsys, UPR Fonctionnement Agroécologique et Performances des Systèmes, de Culture Horticoles, TA B-103/PS4 (Bât. PS IV, Bur. 101), 34 398 Montpellier Cedex 5, France
e-mail: eric.malezieux@cirad.fr

F. Ruf and G. Schroth (eds.), *Economics and Ecology of Diversification*,
DOI 10.1007/978-94-017-7294-5_12

This development of cocoa cultivation, whether it is the agroforestry-based system prevalent in south-central Cameroon, or the more monospecific system found in other regions or countries, is very specific to humid tropical forests. Researchers and farmers consider savannahs to be unsuitable for cocoa cultivation because of their climatic constraints. While experts have classified Cameroon's Centre region as having suboptimal conditions for cocoa cultivation (Champaud 1966), farmers have succeeded in growing cocoa there.

In fact, for several decades now, this forest-savannah interface area has witnessed the development of cocoa. This has taken place, not on cleared strips of gallery forests, but well and truly on grassland covered by *Imperata cylindrica* and *Pennisetum purpureum*. The cropping system that is adopted follows a reconstituted agroforestry model. The existence of these multispecific cocoa plantations in the savannah invites us to take a closer look at the potential, characteristics and role of the dynamics of agroforestry-based diversification. The setting up of cocoa plantations on grassland soils is indeed a significant innovation in the history of cocoa cultivation, given that the crop has traditionally been dependent on tropical forests. What is the role of agroforestry-based diversification in this innovation?

12.1 Materials and Methods

12.1.1 Study Area

The study was conducted in the forest-savannah interface area of Cameroon's Centre region, in the villages of Bakoa, Begni, Yorro and Kedia in the Bokito sub-division (latitude 4°30′ north and longitude 11°10′ east, at an altitude ranging from 450 to 500 m) (Fig. 12.1). Most of the farmers are autochthons of Yambassa ethnicity (Jagoret et al. 2006a).

Several constraints render the forest-savannah interface area in central Cameroon a priori a suboptimal area for cocoa cultivation:

- Poor rainfall distribution. This zone has a hot and humid climate, with an annual average temperature of 25 °C. In addition to an average annual rainfall of about 1300 mm, which is considered to be the lower limit for the cocoa ecotype, the forest-savannah interface area of central Cameroon has a dry season of more than 3 months, during which rainfall is less than 70 mm (Champaud 1966). The annual rainfall deficit for cocoa cultivation is thus about 200 mm. When cocoa is grown as a monospecific crop, this deficit causes defoliation of the cocoa trees. When the defoliation is total, a very high mortality usually results (Braudeau 1969; Wood and Lass 1985);
- Poor soil quality. There are two types of dominant soils in the forest-savannah interface area of central Cameroon: ferric acrisols, which are found everywhere in the interfluvial area, and gleysols, which extend on either sides of the talweg (FAO 1998). In both cases, the surface horizon presents 60 % of sand, which is a

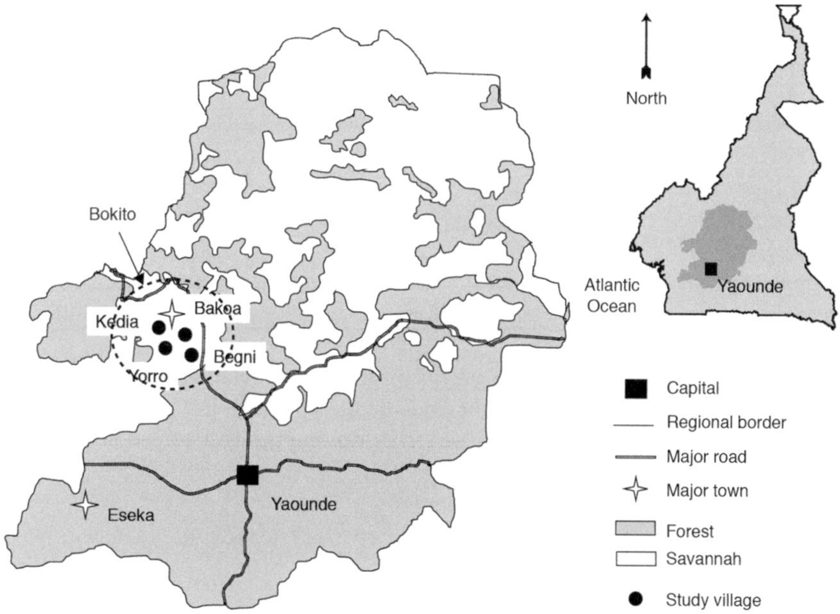

Fig. 12.1 Location of the centre region and the study area

clear handicap for cocoa cultivation. In addition, poor soil quality is not compensated by its levels of soil organic matter (less than 2 %), which can normally improve the soil's texture and its water-holding capacity (Braudeau 1969; Santoir and Bopda 1995);

- Scarcity of forest land. Much of the northern part of central Cameroon was once forested, but has now been subject to heavy clearing by human action. This deforestation has resulted in a forest-savannah mosaic characterized by gallery forests and groves which are normally used for cocoa cultivation (Kuété 1988). Most farmers can no longer benefit from the advantages of cleared forest lands, and forest rent is exhausted;
- The presence of *Imperata cylindrica*. Under human action, several types of savannahs have been developed for cultivation, especially grasslands characterized by *Imperata cylindrica* (Kuété 1988; Santoir and Bopda 1995). The development of these grasslands shows that farmers have been able to gain mastery over this weed. *Imperata cylindrica* is very heliophilous and tends to reappear when the cultivated plots are inadequately maintained. It competes with the crop for water and nutrients (Deuse and Lavabre 1979).

12.1.2 Method

In 2004, 339 cocoa plantations belonging to 282 farmers were randomly selected from lists of members of local cocoa producer organizations. The following information was obtained from each farmer: year of setting up of the cocoa plantation (age of the cocoa plantation in years), type of vegetation that existed before cocoa was planted (savannah or gallery forest), area covered and stated productions for three cropping seasons preceding the survey (2001, 2002 and 2003). The average yield of marketable cocoa for each cocoa plantation was calculated from data on area and production in order to arrive at an agronomic evaluation indicator for the cocoa cultivation system.

Data collected during this survey also helped obtain details about the different types of activities on the cocoa plantation (weeding, pruning, pesticide applications against mirids and black pod disease) and the human and material resources employed (type of labour, number of repetitions of each activity, number of working hours, quantity and cost of inputs).

The economic indicators used to evaluate cocoa systems in the savannah were:

- Gross value added per hectare. This indicator measures the income remaining with the farmer after deducting the cost of inputs used over the farming year;
- Gross value added per hour. This indicator may be interpreted as net labour productivity. It corresponds to the ratio of gross value added divided by the total hours of labour for a crop.

A typology of farmers was arrived at based on the criterion of homogeneity of elements in a class (Volle 1981) in order to compare the economic performance of classes of individuals predominantly having the same characteristics (age of the cocoa plantation, age of the farmer, level of intensification of labour and inputs in the technical itinerary) (Jagoret et al. 2006b).

We also interviewed village chiefs and elders regarding the historical stages of cocoa cultivation in the area.

At the end of the survey of 339 cocoa plantations, field visits revealed that farmers were using one of two strategies to remove *Imperata cylindrica* before planting cocoa in the savannah: the creation of dense shading or the cultivation of an annual crop. This is why, for each of these strategies, we selected five cocoa plantations that were being set up to study the modalities of growing cocoa in the savannah. We did so by conducting site visits, complemented by detailed interviews. These interviews focused on farmer practices and the main reasons behind the diversification of cocoa systems in order to reconstruct their evolution over time.

A change in the content of soil organic matter was observed in 16 cocoa plantations that were between three to 10 years old and which had been established in the savannah on gleysols (6 plots) and acrisols (6 plots), and on acrisols in gallery forests and groves (4 plots). The level of soil organic matter was estimated from the total carbon dosage (Walkley and Black 1934) in composite soil samples taken from depths of 20 and 40 cm.

The agrobiodiversity level of these cocoa systems was estimated from an inventory of forest and fruit species interplanted with cocoa (Messie 2007). These species were recognized on the basis of their local names. Their correlation with the common and scientific names was established with the help of identification guides (Vivien and Faure 1985; Eyog Matig et al. 2006). The agrobiodiversity level was assessed using the Shannon-Weaver index that took into account the proportion of each species present on the plot (Krebs 1985; Frontier and Pichod-Viale 1998).

12.2 Results and Discussion

12.2.1 Cocoa Cultivation in the Savannah, an Activity a Century Old

From our sample of 339 cocoa plantations, 157 were in the savannah and 182 in gallery forests or groves.

Despite the severe constraints that characterize the forest-savannah interface areas of central Cameroon, the distribution of cocoa plantations by age class and by preceding vegetation shows that cocoa cultivation was an old activity (Fig. 12.2). The first cocoa plantations were, in fact, established at the beginning of the 20th century. Interviews with village chiefs and elders helped to distinguish between three phases of development of cocoa cultivation.

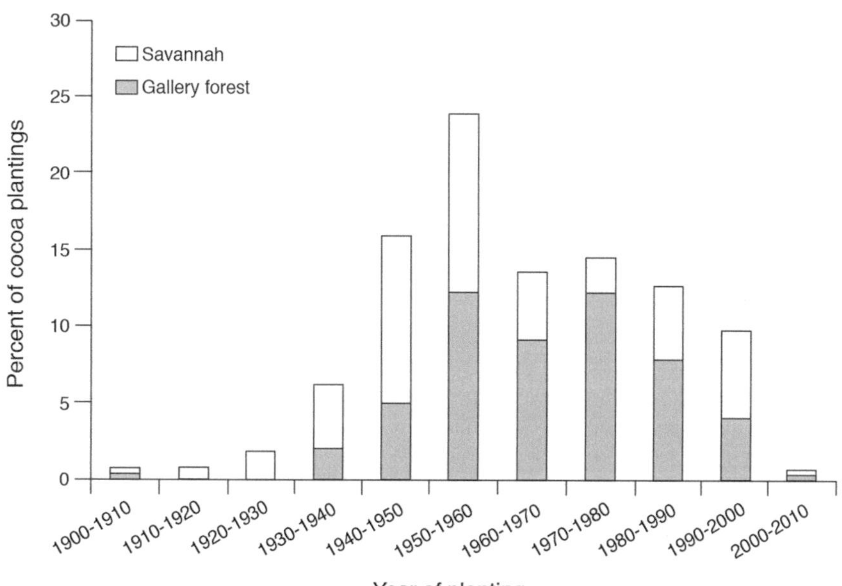

Fig. 12.2 Year-wise distribution of cocoa plantations in the initial establishment period in the Bokito sub-division for both the preceding vegetation types (savannah and gallery forests)

12.2.1.1 1900–1930: Establishment of the First Cocoa Plantations in the Savannah

Some pioneering farmers, convinced of the benefits of cocoa cultivation, smuggled out cocoa beans taken from the plantations of Eséka (Nyong and Kellé division) (Fig. 12.1). At the time, this city was the main centre for paying taxes in the region (Assoumou 1977). Two factors worked against the development of cocoa cultivation in the forest-savannah interface area: the lack of experience of these pioneers and local laws that forbade cultivation activities in gallery forests, which traditionally served as refuges during tribal wars. The first cocoa plantations were thus set up in the savannah. Thus, the social control exerted over gallery forests and groves explained to a large extent the appearance of cocoa cultivation in the savannah during this initial period.

12.2.1.2 1931–1960: Extending Cocoa Cultivation to Gallery Forests and Groves

The French colonial administration encouraged farmers and, in particular, traditional chiefs to grow cocoa on cleared forest lands. After the end of the Second World War, these chiefs lifted the bans that restricted the development of forest areas. They established their own cocoa plantations using cocoa pods distributed by the first German plantations established around Mount Cameroon and in the Kribi region (Michel 1970). They also encouraged their people to follow suit. This period was not only marked by the establishment of cocoa plantations in gallery forests and groves, but also by the onset of conflicts over allocation of forest areas. These areas, which were once collective resources, could now be appropriated by anyone growing cocoa.

12.2.1.3 1961–2000: Resurgence of Cocoa Cultivation in the Savannah

Cocoa cultivation continued to grow after independence, albeit at a smaller scale than in the earlier phase, with the support of agricultural services. The government intervened in the cocoa sector as a result of economic issues that largely resulted from skyrocketing global prices in the 1970s. Starting in the 1980s, the diseases and pests afflicting cocoa were proclaimed to be national scourges, and significant control measures were put in place. In addition to being provided subsidized treatment equipment and phytosanitary products, farmers were given selected planting material and support by the Cocoa Development Corporation (SODECAO) to establish their cocoa plantations (Assoumou-Mba 1981).

Most of the cocoa plantations established during the 1960s and 1970s were set up in gallery forests or groves. This trend reversed in 1980s; there was an increase once again in the number of cocoa plantations set up in the savannah. It became difficult to find fresh forest land, as much of it was already converted to cocoa plantations.

The slowdown experienced by cocoa cultivation starting in 1990 was largely due to the termination of SODECAO's activities following reforms announced by the government in 1989 as part of the structural adjustment programme. Liberalization and volatility in international prices also appeared to repeatedly undermine the profitability of cocoa cultivation in the production basin of south-central Cameroon (Alary et al. 1994; Janin 1999; Alary 2000).

However, despite these risks and economic disruptions, cocoa cultivation still occupied a dominant place on family farms in the forest-savannah interface area in central Cameroon. It represented, on an average, 60 % of their crop rotations, i.e., 2 ha for cocoa compared to 1.6 ha for annual and perennial crops. Cocoa cultivation also remained the farmers' main source of income, accounting for 67 % of their total income, i.e., 495,000 FCFA/year (739 €), whereas the income from other crops was, on an average, 181,000 FCFA/year (276 €) (Jagoret et al. 2006a).

These results were similar to those obtained in the forest areas of central Cameroon where cocoa cultivation accounted, on an average, for 57 % of the crop rotation of farms, and represented 69 % of the farmers' incomes (Jagoret et al. 2006a). They also confirm the values obtained by Weber (1977) and Santoir (1992) who had observed that cocoa represented 60 % of the crop rotation of farms. It was the same for household incomes: a survey conducted in 1954 showed that the sale of cocoa accounted for 70 % of the farmers' incomes (Binet 1956). Thirty years later, the sale of cocoa still covered 50–75 % of the budget of more than 90 % of the households (Leplaideur 1985).

12.2.2 Agroforestry-Based Diversification, an Essential Element of Savannah Cocoa Cultivation

The establishment of cocoa plantations in the savannah initially depends on mastering a key factor: the elimination and control of *Imperata cylindrica*—which is a limiting factor in any savannah area. After that, the success of savannah cocoa plantations depends on controlling two other major factors: shading and soil fertility levels.

Cropping practices used by farmers in the forest-savannah interface area in central Cameroon to establish and manage cocoa plantations in the savannah thus attempt to overcome these three major constraints. Since the reconstitution of an agroforestry system is the key to controlling the system, it becomes both a goal and a process.

Farmers thus developed two major strategies for controlling *Imperata cylindrica*: the establishment of dense shading by the planting of appropriate plant species (S1) or tillage to grow annual crops (S2).

12.2.2.1 Establishment of Dense Shading (S1)

The species most commonly used by farmers to control *Imperata cylindrica* before starting cocoa cultivation in the savannah is the oil palm (*Elaeis guineensis*). Palm seeds were sown through broadcasting in the area identified for the new cocoa plantation. Sometimes, young palm trees are interplanted with mango trees (*Mangifera indica*). The rapid growth and dense shade of these two species eliminates much of *Imperata cylindrica* within 4–5 years.

After this weed is eliminated for the most part, the high density of oil palm trees is gradually reduced. The felled palm trees are used to produce palm wine, thus increasing the farmer's income. At the same time, cocoa is introduced into the cultivation system. This is done either by direct seeding, where two to three pods are planted together in each hole, or by planting seedlings previously raised in nurseries or a germinator. These young plants can be combined with banana trees which add to the shade. The routine maintenance of cocoa trees and the progressive closure of their crown over the next 3–4 years help eliminate *Imperata cylindrica* from the plot.

12.2.2.2 Annual Crops (S2)

Before annual crops can be planted, the soil has to be broken by hand with a hoe. This deep ploughing operation pulls out the roots of *Imperata cylindrica* and exposes them to the sun so that they can be eliminated. Once the annual crop is planted, farmers introduce cocoa into the system, either by direct seeding before weeding (the pods are then sown during the hoeing) or by planting seedlings grown in nurseries or germinators. Young cocoa trees are thus interplanted with a mixture of annual and perennial crops: yam (*Dioscorea* sp.), cocoyam (*Xanthosoma sagittifolium*) and plantain (*Musa spp.*), followed by a succession of maize-groundnut (*Arachis hypogaea-Zea mays*) and pistachio-maize (*Cucumis mani-Zea mays*) associations which are renewed for 3–4 years. During this period, weeding is carried out with a hoe to remove and control any regrowth of *Imperata cylindrica* until the time the cocoa crowns are completely closed.

This technique to remove *Imperata cylindrica* based on tillage and cultivation of annual and perennial crops requires a greater amount of labour in comparison with the planting of dense shade trees. However, it allows farmers to establish their cocoa plantations faster in the savannah, in 5–6 years compared to eight to 9 years for the technique to remove *Imperata cylindrica* using oil palm trees.

12.2.2.3 Managing Interplanted Stands Over Time

Irrespective of the strategy adopted by farmers to control and eliminate *Imperata cylindrica* before establishing cocoa plantations in the savannah, several fruit and forest species are subsequently interplanted with the cocoa trees. The gradual

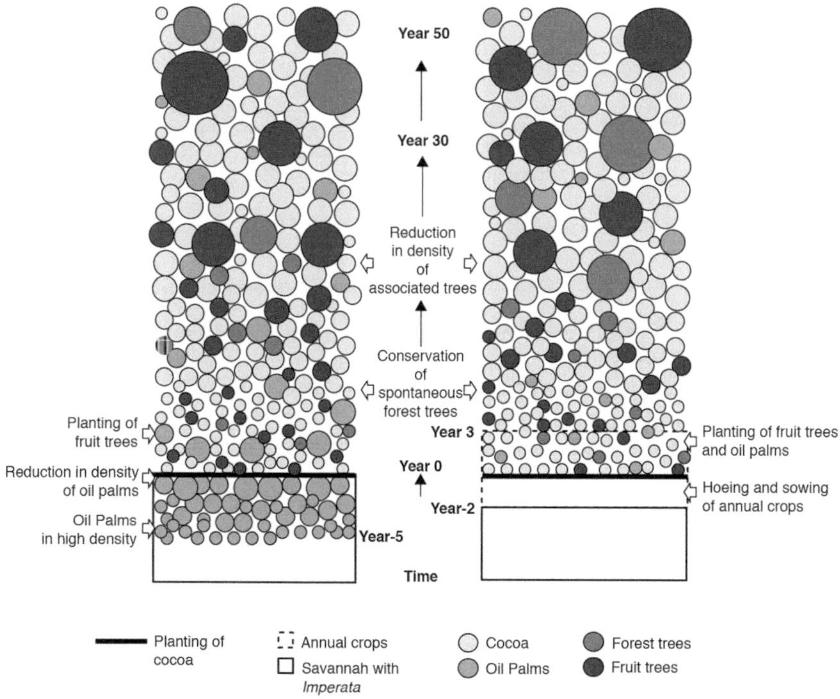

Fig. 12.3 Evolution over time of the savannah cocoa cultivation system based on the two strategies to control *Imperata cylindrica*. *S*1 Establishment of a dense shade; *S*2 Cultivation of annual crops

replacement of the initially planted dense shade trees or annual and perennial crops by fruit or forest species which provide light shading allows farmers to diversify their cocoa cultivation system (Fig. 12.3).

Farmers prefer a high density of fruit and forest trees in young cocoa plantations so that they can create an unbroken and dense ground cover to help control weeds in their plots, as well as to create a suitable shade for cocoa trees. To this end, farmers continue to plant various fruit trees in their cocoa plantations, including orange (*Citrus sinensis*), African plum (*Dacryodes edulis*), avocado (*Persea americana*) and kola (*Cola nitida*). Forest trees are also introduced into cocoa plantations through transplantation, or are preserved when they grow there naturally, like the kapok (*Ceiba pentandra*), iroko (*Milicia excelsa*), African border tree (*Newbouldia laevis*) and flat-crown (*Albizia adianthifolia*).

Subsequently, farmers vary their management practices according to the species but undertake a significant reduction in the density of fruit and forest trees. Farmers try to maintain a canopy cover over their plots which is favourable to cocoa development. They compensate for the growth of the several tree species interplanted with cocoa by regularly removing excess trees, mainly by girdling which causes their gradual demise. By using this technique, they avoid the damage that

Table 12.1 Main purpose of fruit or forest trees that are introduced—or retained—when a cocoa plantation is established in the savannahs of central Cameroon

Objectives	Answers (%) (sample size = 10)	
	Forest species	Fruit species
Home consumption of fruits	0	28.6
Soil fertilization	19	0
Sale of fruits or timber	19	52.4
Shading for cocoa	47.7	19
Utilization of timber	14.3	0

would be caused to cocoa in case these trees are felled. At the same time, the density of oil palm trees remains constant since farmers normally replace the trees that are felled for producing palm wine.

The strategy of reconstituting a diversified cocoa agroforestry cropping system serves several purposes (Table 12.1).

Trees of certain forest species that grow naturally in cocoa plantations are retained mainly to provide a favourable shade for cocoa trees. Some other species have a positive effect on soil fertility (Bidzanga et al. 2009). Farmers may also decide to retain some species which provide valuable timber for construction and for sale. On the other hand, the introduction of fruit species in the system is mainly for home consumption and for selling various products, with shading for cocoa being only a secondary objective.

12.2.3 *Yield of Savannah Cocoa Plantations Equivalent to Those in Gallery Forests*

12.2.3.1 Agronomic Performance

Table 12.2 provides a comparison by age group between the marketable yields of savannah cocoa plantations and those in gallery forests or groves. Regardless of the age class considered, there was no significant difference between these two situations of preceding vegetation. During the first years of cultivation, cocoa plantations on grasslands appeared to be as productive as those grown on forest land. After 20 years, however, not only was the yield of savannah cocoa greater than of cocoa from gallery forests and groves, but it remained relatively stable over time. Furthermore, the yield of marketable cocoa grown on the savannah is similar to that observed by Duguma et al. (2001) for cocoa plantations on cleared forest lands in central and southern Cameroon: between 264 and 500 kg/ha depending on the intensification level of the technical itinerary.

Table 12.2 Evolution in the yield of marketable cocoa in the forest-savannah interface area of central Cameroon for the two different preceding vegetation types (savannah and gallery forest)

Age classes	Average yield (kg of marketable cocoa per hectare)			
	N	Savannah (standard deviation)	N	Gallery forest (standard deviation)
<5 years	2	130 (±19.12)	4	20 (±5.03)
5–10 years	11	128 (±36.85)	6	136 (±23.80)
10–20 years	18	214 (±34.58)	22	223 (±28.11)
20–30 years	10	470 (±40.37)	39	444 (±35.96)
30–40 years	13	447 (±64.34)	38	324 (±21.49)
40–50 years	34	459 (±34.98)	40	383 (±31.12)
50–60 years	40	418 (±32.75)	24	353 (±54.21)
>60 years	29	406 (±37.76)	9	374 (±69.20)

N sample size

Table 12.3 Deriving value from the land for cocoa cultivation in the forest-savannah interface area of central Cameroon for the two preceding vegetation types (savannah and gallery forest)

Farmer class	Gross value added of land (FCFA/hectare)			
	N	Savannah (standard deviation)	N	Gallery forest (standard deviation)
A	33	125,236 (±74,879.51)	28	129,447 (±47,149.92)
B	107	214,910 (±114,180.22)	129	236,440 (±97,920.31)
C	17	386,845 (±136,228.07)	25	399,486 (±143,562.56)

N sample size

12.2.3.2 Economic Performance

The classification developed to group individuals with largely the same characteristics helped identify three categories of farmers: young farmers in the establishment stage (class A), farmers who had adopted an extensive itinerary (class B), and farmers with intensive farming practices (class C).

At the economic level, regardless of the farmers' categories, no significant difference was found between the two preceding vegetation types in terms of the gross value added per hectare in the productive phase (Table 12.3). It was the same for the gross value added per labour hour (Table 12.4).

Although the economic performance of cocoa plantations in gallery forest or groves remained higher than that of cocoa on the grasslands, the difference between the two preceding vegetation types was of the order of 10–15 % in the case of deriving value from the land, and from 3 to 9 % for deriving value from labour.

Table 12.4 Deriving value from labour for cocoa cultivation in the forest-savannah interface area of central Cameroon for the two preceding vegetation types (savannah and gallery forest)

Farmer class	Gross value added of labour (FCFA/h)			
	N	Savannah (standard deviation)	N	Gallery forest (standard deviation)
A	33	2,296 (±792.05)	28	2,600 (±860.39)
B	107	3,628 (±1,299.80)	129	4,223 (±1,430.59)
C	17	2,148 (±740.54)	25	2,496 (±852.01)

N sample size

Table 12.5 Number of species per hectare and the Shannon-Weaver index by type of preceding vegetation and soil type

Soil type	Preceding vegetation	Number of cocoa plantations	Number of species per hectare	Shannon-Weaver index
Acrisol	Forest	6	36	3.3
Acrisol	Savannah	6	77	3.0
Gleysol		4	58	2.9

12.2.3.3 Level of Agrobiodiversity

A total of 62 woody species belonging to 29 botanical families were recorded in the cocoa agroforestry system in the savannah. The average value of the Shannon-Weaver index—greater than 3—reflects the high diversity of this cropping system. This holds irrespective of the type of soil considered (Table 12.5).

The Shannon-Weaver index value of savannah cocoa agroforestry plantations is close to values obtained for cocoa plantations on forest lands in central Cameroon. These latter values were respectively 3.4 in the Lékié division and 3.8 in the Nyong and So'o division (Messie 2007). The level of agrobiodiversity of savannah cocoa was also the same as that obtained by Zapfack et al. in 2002 (4.3) and by Sonwa et al. in 2007 (3.1–3.9) for cocoa grown on forest lands in south-central Cameroon. However, it was greater than that observed for cocoa agroforestry in Nigeria (2.7) (Oke and Odebiyi 2007) and in Ghana (2.6) (Asare and Tetteh 2010).

We thus generally found that the process of a combined planting of woody species and cocoa to establish savannah cocoa agroforestry systems resulted in a high diversity akin to that observed for cocoa systems established on cleared forest lands.

The three dominant species of the savannah cocoa system are oil palm (*Elaeis guineensis*), orange (*Citrus sinensis*) and African plum (*Dacryodes edulis*) (Table 12.6). These fruit species, just like kola (*Cola nitida*) and avocado (*Persea americana*), are usually introduced at the time of establishing cocoa plantations. Their production (palm oil, palm wine, fruits) is meant for household consumption, while the surplus is sold.

The role of these fruit trees in providing shade for cocoa and improving soil fertility is limited. This function is normally provided by forest species, whose

Table 12.6 The main botanical species and families of savannah cocoa cultivation system in central Cameroon's forest-savannah interface area

Species			Family	
Common name	Scientific name	Presence (%)	Name	Presence (%)
Oil palm	*Elaeis guineensis*	17	Arecaceae	17
Orange tree	*Citrus sinensis*	12	Rutaceae	15.3
African plum	*Dacryodes edulis*	11.2	Burseraceae	12.3
Kola tree	*Cola nitida*	6.7	Sterculiaceae	10.7
Avocado	*Persea americana*	5.4	Moraceae	8.8
Mango tree	*Mangifera indica*	3.5	Anacardiaceae	5.9
Iroko	*Milicia excelsa*	3.5	Lauraceae	5.4
African border tree	*Newbouldia laevis*	2.9	Mimosaceae	4.6
Mandarin orange	*Citrus reticulata*	2.9	Bignoniaceae	3.1
Ayous	*Triplochyton scleroxylon*	2.4	Bombacaceae	2.1
Flat-crown	*Albizia adianthifolia*	2.2	Euphorbiaceae	2
Kapok tree	*Ceiba pentandra*	2.1		
Fig tree	*Ficus mucoso*	2		

importance to cocoa grown in the savannah appears to be less than that of fruit species. However, the role of forest species in the farmers' strategy to establish and cultivate cocoa in the savannah is crucial. This is the reason why several forest species are introduced in the cocoa cultivation system or are maintained if they grow there on their own: trees that provide shade to cocoa trees (*Albizia adianthifolia*, *Ceiba pentandra*, *Newbouldia laevis* and *Milicia excelsa*) and those that help improve soil fertility (*Ceiba pentandra* and *Ficus mucoso*).

Some forest species sometimes play a dual role: *Ceiba pentandra* (shading and improving soil fertility), *Newbouldia laevis* (shading and demarcation of plots) or *Milicia excelsa* (shading and timber production). The functions of different fruit and forest tree species in cocoa systems in the savannah thus appear to be complementary at the agronomic and economic levels.

12.2.3.4 Soil Fertility Level

The maintenance or regeneration of soil fertility is a key determinant of the agro-ecological sustainability of a cropping system. Monitoring changes in soil organic matter using a synchronic approach, as an indicator of the fertility and sustainability of cropping systems, helps highlight the performance of the savannah cocoa agroforestry system.

Levels of soil organic matter in cocoa plantations established in gallery forests or groves decrease in the initial years from 3.5 to 2.5 % compared to a control forest

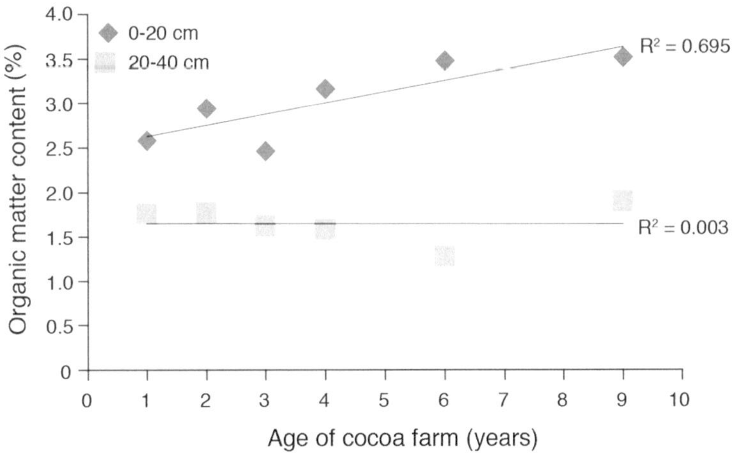

Fig. 12.4 Evolution over time of the level of soil organic matter in cocoa plantations set up on gleysols in the savannah

plot, and then stabilize after 25–35 years of cocoa cultivation (Glatard 2007). On the other hand, for savannah cocoa plantations set up on gleysols, the soil organic matter increases by 2.5–3.5 % over the control plot (uncultivated savannah) 9 years after the establishment of cocoa plantations (Fig. 12.4).

While it is known that soil organic matter of cleared forest land decreases following the establishment of cocoa plantations on it (Snoeck et al. 2010), the maintenance of—or even an increase in—the level of this indicator in the savannah cocoa agroforestry system is therefore a positive development, a result of the farmers' strategy of interplanting different woody tree species with cocoa over time. Such fruit and forest species, interplanted and organized in different strata, help in the process of biomass recycling and resource sharing which ensures a proper functioning of a natural ecosystem.

An increase in the organic matter content in the soils of cocoa plantations, from the low levels typical in grasslands with *Imperata cylindrica* or *Pennisetum*, thus becomes a criterion of sustainability for this relatively old and innovative multi-species cocoa cultivation system (Glatard et al. 2007).

12.3 Conclusion

Agroforestry-based diversification plays a central ecological and economic role in the establishment of cocoa plantations in the savannah. It is a key for the sustainable development of areas previously considered suboptimal for cocoa cultivation. At the agronomic level, the reconstitution of a multispecies cocoa cultivation system in forest lands by farmers in the forest-savannah interface area of central Cameroon helps overcome the main constraints presented by such areas for cocoa cultivation (uneven rainfall distribution, poor soil quality and presence of *Imperata cylindrica*). The spatio-temporal arrangement of several forest and fruit species interplanted with cocoa allows, firstly, the control and elimination of *Imperata cylindrica*. It also helps overcome of the lack of shading during the establishment of cocoa plantations in the savannah. Secondly, the reconstitution of a cocoa agroforestry cultivation system can significantly improve soil fertility in savannah cocoa plantations. Finally, the yield of marketable cocoa from cocoa plantations is similar to that from cocoa plantations in gallery forests and groves. The sustainability of cocoa plantations in the savannah is thus ensured.

At the economic level, the benefits of a cocoa agroforestry system in the savannah are also confirmed. The performance of cocoa agroforestry plantations in the savannah, evaluated in terms of gross margin per hectare and gross margin per labour hour, appears to be similar to those of cocoa agroforestry systems in gallery forests or groves, regardless of the intensification level of the technical itinerary adopted by the farmers. Finally, the positive environmental impact of the savannah cocoa agroforestry system on agrobiodiversity is also confirmed.

References

Alary V (2000) Les cacaoculteurs camerounais face aux risques. Essai de modélisation. Paris, France, L'Harmattan, 236 pp

Alary V, Courade G, Janin P (1994) Permanence et flexibilité des cacaoculteurs béti à l'heure des ajustements. In: Courade G (ed) Le village camerounais à l'heure de l'ajustement. Paris, Karthala, pp 170–183

Asare A, Tetteh DA (2010) The role of complex agroforestry systems in the conservation of forest tree diversity and structure in southeastern Ghana. Agrofor Syst 79:355–368

Assoumou J (1977) L'économie du cacao. Agriculture d'exportation et bataille du développement. Paris, France, Jean-Pierre Delarge, 351 pp.

Assoumou-Mba D (1981) L'encadrement des planteurs de cacao au Cameroun. Cas de la Sodecao. In: 7e conférence internationale sur la recherche cacaoyère. November 1979. Douala, Cameroun. pp 645–648

Aulong S, Dury S, Temple L (1999) Dynamique et structure floristique des agroforêts à agrumes au centre du Cameroun. Fruits 55(2):103–114

Bidzanga NL (2005) Farmers' ecological and agronomic knowledge about The management of multistrata cocoa systems in Southern Cameroon. Ph.D. thesis, University of Wales, Bangor, Grande Bretagne, 258 pp

Bidzanga N, Fotsing B, Agoume V, Birang AM, Onguene AN, Zapfack L (2009) Mycotrophie et connaissances paysannes des essences fertilitaires dans les agroforêts à base de cacaoyers du Sud Cameroun. Cameroon J Exp Biol 5:79–86

Binet J (1956) Budgets familiaux des planteurs de cacao au Cameroun. Paris, Orstom, L'Homme d'Outre-Mer, n 3, 154 pp

Braudeau J (1969) Le cacaoyer. Collection Techniques agricoles et productions tropicales. Paris, Éditions Maisonneuve et Larose, 304 pp

Champaud J (1966) L'économie cacaoyère du Cameroun. Cahiers Orstom, série Sciences humaines 3:105–124

Deuse J, Lavabre EM (1979) Le désherbage des cultures sous les tropiques. Paris, France, Techniques agricoles et productions tropicales, Maisonneuve et Larose, 312 pp

Duguma B, Gockowski J, Bakala J (2001) Smallholder cacao (*Theobroma cacao* Linn.) cultivation in agroforestry systems of West and Central Africa: challenges and opportunities. Agrofor Syst 51:177–188

Eyog Matig O, Ndoye O, Kengue J, Awano A (2006) Les fruitiers forestiers comestibles du Cameroun. Cotonou, Bénin, Ipgri/Saforgen/Irad/Cifor, 204 pp

FAO (1998) World reference base for soil resource. Rome, Italie, Fao, ISRIC, ISSS, 109 pp

Frontier S, Pichot-Viale D (1998) Écosystèmes : structure, fonctionnement et évolution. Paris, Dunod, 447 pp

Glatard F (2007) Analyse de la conduite technique de cacaoyères au Centre Cameroun (village de Kédia): liens avec le milieu (forêt/savane) et les caractéristiques de fertilité des sols. Thesis Engineer in Tropical Agriculture, Montpellier, France, Cnearc, 83 pp

Jagoret P, Couve C, Bouambi E, Menimo T, Domkam I, Nyassé S (2006a) Caractérisation des systèmes de cacaoculture du Centre Cameroun. Yaoundé, Cameroun, Irad-Cirad, 107 pp

Jagoret P, Batomen F, Domkam I, Bouambi E, Menimo T (2006b) Analysis of Cocoa Farming Systems of Centre Cameroon by the Construction of Typologies. 15ème Conférence internationale sur la recherche cacaoyère. San-José, Costa-Rica, 9–14 octobre 2006

Janin P (1999) L'avenir des planteurs camerounais. Karthala, Paris, 246 p

Krebs CJ (1985) Species diversity. In: Krebs CJ (ed) Ecology: the experimental analysis of distribution and abundance. Harper and Row, New York, pp 507–534

Kuété M (1988) L'évolution récente des paysages autour de Bafia. Incidences sur les activités humaines. In: L'homme et la montagne tropicale. Bordeaux, France, Colloquium Proceedings, Société pour l'étude, la protection et l'aménagement de la nature dans les régions intertropicales, pp 95–102

Leplaideur A (1985) Les systèmes agricoles en zone forestière: les paysans du Centre et du Sud Cameroun. Paris, France, Cirad-Irat, 615 pp

Messie AC (2007) Caractérisation de la richesse floristique en fonction de l'âge des agro-forêts à base de cacaoyers du bassin du Centre Cameroun : cas de Bokito, Obala et Ngomedzap. Mémoire du diplôme d'Ingénieur de Conception des Eaux, Forêts et Chasse. Dschang, Cameroun, Faculté d'Agronomie et des Sciences agricoles, Département de Foresterie, 96 pp

Michel M (1970) Les plantations allemandes du mont Cameroun, 1885–1914. Revue française d'Histoire d'Outre-Mer 207:183–213

Oke DO, Odebiyi KA (2007) Traditional cocoa-based agroforestry and forest species conservation in Ondo State, Nigeria. Agric Ecosyst Environ 122:305–311

Ruf F (1987) Éléments pour une théorie sur l'agriculture des régions tropicales humides. I - De la forêt, rente différentielle au cacaoyer, capital travail. Agronomie tropicale 42(3):218–232

Ruf F (1995) Booms et crises du cacao: les vestiges de l'or brun. Paris: Éditions Karthala. Collection Économie et développement, 455 pp

Ruf F, Schroth G (2004) Chocolate forests and monocultures: a historical review of cocoa growing and its conflicting role in tropical deforestation and forest conservation. In: Schroth G, DaFonseca GAB, Harvey CA., Gascon C, Vasconcelos HL, Izac AMN (eds) Agroforestry and biodiversity conservation in tropical landscapes. Washington, Island Press: 107–133

Santoir C (1992) Sous l'empire du cacao. Étude diachronique de deux terroirs camerounais. Éditions Orstom Paris, 191 pp

Santoir C, Bopda A (1995) Atlas régional Sud-Cameroun. Éditions de l'Orstom, Paris, 53 p

Snoeck D, Abolo D, Jagoret P (2010) Temporal changes in VAM fungi in the cocoa agroforestry systems of central Cameroon. Agrofor Syst 78:323–328

Sonwa DJ, Nkongmeneck AB, Weise SF, Tchatat M, Adesina AA, Janssens MJ (2007) Diversity of plants in cocoa agroforests in the humid forest zone of Southern Cameroon. Biodiversity and Conservation, 16:2385–2400

Tchatat M (1996) Les jardins de case agroforestiers des basses-terres humides du Cameroun: étude de cas des zones forestières des provinces du Centre et du Sud. Doctoral thesis. Paris, Université de Paris 6. 145 pp et annexes

Temple L, Minkoua Nzie JR, David O (2007) Diversification des systèmes de cultures dans les exploitations cacaoyères au Cameroun et demande d'innovation technique. In: Gafsi M, Dugué P, Jamin JY, Brossier J (eds) Exploitations agricoles familiales en Afrique de l'Ouest et du Centre. Quae, Inra, pp 303–311

Vivien J, Faure JJ (1985) Arbres des forêts denses d'Afrique centrale: espèces du Cameroun. Paris, Agence de coopération culturelle et technique, 565 pp

Volle M (1981) Analyse de données. 2e édition. Paris, Economica, 319 pp

Walkley A, Black LA (1934) An examination of Degtjareff method for determining soil organic matter and proposed modification of the chromic acid titration method. Soil Sci 37:29–38

Weber J (1977) Structures agraires et évolution des milieux ruraux. Le cas de la région cacaoyère du Centre-Sud Cameroun. Cahiers Orstom, série Sciences humaines 2:113–139

Wood GAR, Lass RA (1985) Cocoa. Longman Scientific and Technical, New York

Zapfack L, Engwald S, Sonke B, Achoundong G, Birang AM (2002) The impact of land conversion on plant biodiversity in the forest zone of Cameroon. Biodiversity and conservation, 2:2047–2061

Chapter 13
Diversifying Central American Coffee Agroforestry Systems via Revenue of Shade Trees

Philippe Vaast, Mario Martínez, Axelle Boulay, Benito Dzib Castillo and Jean-Michel Harmand

Coffee has been the main agricultural crop and source of export earnings in Central America for more than a century. During the 2000s, coffee production provided a livelihood to 300,000 farmer households in this region. Consequently, the last coffee crisis, from 1999 to 2005, led to a severe social crisis, with rural unemployment and poverty reaching alarming levels (Varangis et al. 2003).

Coffee plantations have also large-scale environmental impacts as they cover almost 1 million hectares of the Central American isthmus, which is one of the world's biodiversity hotspots. These coffee plantations are often located in fragile

P. Vaast (✉)
Icraf, United Nations Avenue, Gigiri, 30677 Nairobi 00100 Kenya
e-mail: philippe.vaast@cirad.fr; pvaast@cgiar.org

P. Vaast
UMR Eco&Sols—Écologie Fonctionnelle & Biogéochimie des Sols & Agroécosystèmes—(Montpellier SupAgro–Cirad–Inra–IRD), Cirad, 2 Place Viala (Bat. 12), 34060 Montpellier cedex 2, France

M. Martínez
Carrera Diagonal 16 No. 44-26, Bloques el Mirador Bl C apto 402, Pasto, Nariño, Colombia
e-mail: mmartineas@gmail.com

A. Boulay
Inter-American Development Bank, 1300 New York Avenue, Washington DC, NW 20577, USA
e-mail: axelleb@iadb.org

B.D. Castillo
Instituto Tecnológico de Chiná, Calle 11s/n entre 22 y 28, Colonia Centro, 24520 Campeche, CP, México
e-mail: benitodzib@hotmail.com

J.-M. Harmand
Cirad, UMR Eco&Sols—Écologie Fonctionnelle & Biogéochimie des Sols & Agroécosystèmes—(Montpellier SupAgro–Cirad–Inra–IRD), 2 Place Viala (Bat. 12), Montpellier Cedex 2 34060, France
e-mail: Jean-michel.harmand@cirad.fr

F. Ruf and G. Schroth (eds.), *Economics and Ecology of Diversification*,
DOI 10.1007/978-94-017-7294-5_13

mountainous ecosystems, many of them in watersheds that supply water to urban centres. Around 75 % of the coffee is cultivated under more or less dense shade in this region (Beer et al. 1997; Vaast and Harmand 2002). Therefore these coffee agroforestry systems play a major role in the conservation of biodiversity, especially in highly anthropized landscapes where shade trees contribute to regional forest cover and survival of associated fauna. They help facilitate movement and maintain viability of wildlife populations such as migratory birds along the Mesoamerican Biological Corridor (Somarriba et al. 2004).

However, the 1999–2004 recession contributed to the elimination of thousands of hectares of these agroforestry systems, sometimes up to 20 % of the area. Falling coffee prices encouraged farmers to revert to cultivating annual food crops on steep slopes prone to soil erosion. In the short term, of course, this has compensated for lost income from coffee.

One of the strategies farmers can adopt to reduce their vulnerability to coffee price volatility while preserving natural resources is to conquer new markets and produce certified products. In exchange for a quality product or a certification of organic or eco-friendly farming, the consumer agrees to pay a premium for coffee. Such strategies can contribute to the sustainability of coffee farms in the Americas (Vaast et al. 2005). Environmental awareness is increasing in developed countries and is helping the development of niche markets for sustainable coffee (Ponte 2004; Salazar 2005). However, these markets remain niches and represent less than 10 % of exports of Latin American coffee despite the fact that this inter-American corridor is the world's most advanced region in this respect. In addition, many non-governmental organizations such as Rainforest Alliance, UTZ Certified and Fair Trade, as well as major roasters such as Starbucks and Nespresso, have chosen this region to develop sustainable and socially responsible coffee farming (Salazar 2005).

Do other strategies exist to increase the income of coffee growers? Income diversification through the production and marketing of timber and fuel wood from shade trees can also help improve the living conditions of the coffee growers and their families (Galloway and Beer 1997). Thus, large furniture retail chains in Europe and the United States increasingly label the source of wood produced and used in a sustainable manner, such as the wood with SmartWood certification by the Forest Stewardship Council and the Rainforest Alliance (2004).

Finally, some shade trees can also have a positive impact on the quality of coffee. Producers and cooperatives can therefore sell their coffee on niche quality markets, which are more remunerative and less subject to price fluctuations (Guyot et al. 1996; Muschler 2001; Vaast et al. 2005).

The key question about the future of Central American coffee agroforestry is the same as in Africa: If it is promoted on diverse and diversified markets, can the shade tree become the key to the revival of agroforestry diversification? Unlike their counterparts in West Africa, coffee farmers in Central America have a relatively open access to the wood market. Appropriate strategies of Central American farmers can therefore lead to paths of diversification and improved livelihoods. The central issue of this chapter is therefore to assess the economic significance of income derived from shade trees in coffee agroforestry systems in four regions of Costa Rica and Guatemala, as well as to suggest ways of helping farmers increase incomes in these countries and possibly in other coffee-producing countries on other continents.

13.1 Methodology

From 2003 to 2005, three sets of field surveys were conducted with coffee farmers in Costa Rica and Guatemala with the aim of assessing the income derived from the sale of wood from trees associated with coffee plants.

13.1.1 Three Regions in Costa Rica

In the three regions of Turrialba, Pérez Zeledón and Central Valley, the tree stratum providing shade to coffee usually consists of a leguminous tree (mainly *Erythrina poeppigiana*, sometimes *Inga* sp. or *Gliricidia sepium*) intercropped with a single species of timber wood whenever it is present in coffee plantations (Dzib 2003). The wood of the leguminous species is mainly used as mulch, sometimes as fuel wood. Indeed, the urban and rural households are often equipped with an electric or gas stove.

Our case studies have therefore concentrated on coffee agroforestry systems shaded with timber tree species most often encountered in coffee plantations:

- *Eucalyptus deglupta* in the high-altitude region (about 1200 m) of the Central Valley;
- *Terminalia amazonia* in the low-altitude region (<800 m) of Pérez Zeledón;
- *Cedrela odorata* in the low-altitude region (<800 m) of Turrialba.

Therefore, these are the three forest species mainly planted by the coffee farmers, to which is added *Cordia alliodora* which regenerates naturally and is then managed by the farmers.

Sixty-six coffee farms were surveyed. The volume of saleable wood was estimated for a rotation of 4–25 years depending on the tree species and end use (Dzib 2003; Boulay 2004). The significance of income from timber, as compared to that from coffee, was estimated for a 25-year life cycle of a coffee plantation and a yield of 1000 kg of green coffee per hectare per year. In 2004, this yield corresponded to a gross income of coffee of USD 782/ha/year and a net income of USD 382 (at the purchase price of USD 34 per bag of 46 kg and with a production cost of USD 400/ha/year). The price of timber (sold as standing trees or as logs) and costs of tree planting, maintenance, transportation, sawing and drying were evaluated locally through interviews with various stakeholders of the wood-supply chain.

13.1.2 North Pacific Area of Guatemala

The second study was undertaken in 2005 in the North Pacific area of Guatemala, in the Ocosito watershed where coffee is grown between altitudes of 500 and 1400 m. The potential revenue derived from timber and fuel wood in these coffee agroforestry systems was estimated by assessing the exploitable volume in coffee plantations and according to local market prices for standing trees with a diameter greater than 40 cm. Field surveys were conducted on 36 coffee farms (Martínez 2005). The economic contributions of timber, fuel wood and coffee were determined for the six types of farms found in this watershed. The survey assessed the volume sold over the previous 3 years and production costs during the same period.

13.1.3 Surveys of the Wood Commodity Chains

In both countries and the four regions studied, the key stakeholders of the local timber and fuel wood commodity chains were identified and interviewed through informal discussions (Boulay 2004; Martínez 2005) and the interactions between them analyzed. Revenues from products of these shade trees were evaluated, not only through meetings with coffee farmers but also with local end users (restaurants, artisans, sawmill, etc.).

13.2 Results and Discussion

13.2.1 Public Policy Recognizing Environmental Services

A pioneer in the Central American region, Costa Rica has recognized the importance of preserving its natural resources. It has set up a national system of payment for environmental services to pay owners to preserve private forests (Fonafifo 2007a). These environmental services of national interest provided by forest areas are:

- reduction of greenhouse gases via carbon sequestration;
- protection and regulation of rivers for human uses and hydropower;
- protection of biodiversity for its conservation and sustainable use for commercial and scientific purposes (for example, pharmaceutical);
- scenic landscape beauty for tourism.

Since 2003, agroforestry systems that are used to cultivate perennial crops, including coffee, have been included in the national programme of payment for environmental services. This is also an official recognition of the role of these systems in the preservation of biodiversity, conservation of natural resources (water and soil) and the contribution to scenic beauty in a country which emphasizes ecological tourism. According to a publication by Fonafifo (2007b), to be eligible producers must be part of a development project, an approved cooperative or an indigenous reservation. They must plant 40–250 timber trees or 40–277 fuel wood leguminous trees per hectare of coffee plantation. A list of approved trees has been drawn up. It includes 17 timber species (including the four concerned by this study) and four leguminous species: *Erythrina* spp., *Inga* spp., *Gliricidia sepium* and *Leucaena leucocephala*.

Each producer can receive a minimum subsidy for the planting of 350 trees and maximum for 3500 trees, with a subsidy amount of USD 1.30 per tree in 2007. In return, the farmer agrees to maintain the trees in the coffee plantation during the 5-year contract with the Costa Rican government. However, this type of incentive leads to an increase in the density of trees on plots to a level that is not conducive to coffee production. This fund is financed by a tax on fuel and forest products, donations and international funding (World Bank and European governments) and national and international hydropower companies. The governments of the Central American region have all expressed a keen interest in implementing such schemes for payments for environmental services in their respective countries. But irrespective of government subsidies, timber and fuel wood are acquiring an ever-increasing intrinsic value on the local and international markets.

13.2.2 Costa Rica: Already 20 % of Revenues from Wood

In low-altitude areas (<800 m), which are sub-optimal for growing Arabica coffee, our 2003 and 2004 studies showed that the sale of timber from *Eucalyptus deglupta*, *Cordia alliodora*, *Terminalia amazonia* and *Cedrela odorata* can account for 15–34 % of the total income generated by coffee during the 25 years of its productive life. Therefore, timber income can make up 13–25 % of total farm income (Table 13.1).

Most coffee farmers sell their standing trees to intermediaries who are responsible for their felling, transport and marketing. Timber buyers are mainly local lumber yards, sawmills, furniture makers and construction companies.

Instead of using intermediaries, farmers could benefit more by hiring—on their own or through their cooperative services—contractors to fell and transport their trees and sell the timber themselves directly to wholesale traders, much as they do already for collecting, processing and marketing of coffee. The costs of these different operations were evaluated during the surveys. By eliminating intermediaries and undertaking these operations by themselves, farmers' timber revenue could increase significantly: 30 % of total farm income in the case of *Cedrela odorata*, 42 % for *Cordia alliodora* and up to 51 % for *Terminalia amazonia*.

Despite a lower market value per cubic metre, *Eucalyptus deglupta* can prove to be more interesting than other, more valuable species (Table 13.1). The sale of standing *Eucalyptus* for making poles and boards accounted for 23 and 28 %

Table 13.1 Importance of timber (*Cedrela odorata, Cordia alliodora, Terminalia amazonia* and *Eucalyptus deglupta*) in coffee agroforestry systems in three low-altitude (<800 m) Costa Rican regions

Tree species	Tree density (trees per ha)	Timber production (m^3/ha/year)	Rotation period (years)	Price of timber (USD/m^3)	Actualized revenues[a] (USD/ha/year)	Timber revenues (% of coffee)	Timber revenues (% of total revenues)
Cedrela (St)[b]	75	3	25	117	350	23.4	18.9
Cedrela (Dsl)[c]	75	3	25	169	507	33.7	25.2
Cordia (St)	200	5	25	36	290	19.3	16.2
Cordia (Dsl)	200	5	25	85	500	33.3	25.0
Terminalia (St)	125	5	25	47	233	15.5	13.4
Terminalia (Dsl)	125	5	25	95	475	31.6	24.0
Eucalyptus (St-poles)	120	4	4	39	425	28.3	22.0
Eucalyptus (St-boards)	120	8	8	17	345	23.0	18.7

[a]Based on an inflation rate of 10 % per year as recorded in Costa Rica for the last 10 years;
[b]St: sale of standing trees to intermediaries;
[c]Dsl: direct sale of logs by farmer to wholesaler

respectively of coffee revenues accumulated during the period under consideration, about 20 % of total farm income.

Therefore, these revenues were equal or higher than the sale of standing trees of other species deemed more valuable and of a higher market value. This is due to the fact that *Eucalyptus* grows faster than the other three species: three to five rotations can take place over 25 years. Thus, farmers can receive income three to five times over 25 years instead of waiting 15–25 years to fell and sell slower-growing timber species.

In 2004, revenues from the sale of timber accounted for a maximum of 33 % of income generated by coffee, with *Cedrela odorata* being the tree species most in demand. However, these revenues are likely to increase significantly in the future. Due to growing demand in the region, the price of wood has increased steadily over the past 20 years (Fonafifo 2004). Indeed, the price of *Cedrela odorata* increased 6.3 times and that of *Cordia alliodora* 5.7 times between 1985 and 2005 in Costa Rica.

13.2.3 Guatemala: Considerable Heterogeneity

The Guatemalan study showed contrasting results depending on farm type and altitude (500–1400 m) in the Ocosito watershed. Six main types of coffee farms were identified in this area:

- Traditionally managed small farms, typical of the country (mean area of 26 ha) and at low altitude (<800 m);
- Traditionally managed medium-sized farms (40 ha) at low altitude (<800 m) with high timber extraction;
- Traditionally managed large farms (250 ha) at low altitude (<800 m);
- Intensively managed large farms (120 ha) at low altitude (<800 m);
- Intensively managed large farms (195 ha) at medium-to-high altitude (>1000 m);
- Very small organic farms (2.2 ha) at high altitude (>1300 m).

For the 36 farms surveyed, the mean density of shade trees was 235 trees per hectare, which is higher than that recommended by regional technical institutes (between 75 and 150 trees/ha). In addition, the tree stratum was very diversified with a total of 34 tree species observed, 44 % of timber species, 26 % of fuel wood species (mainly genus *Inga*), 15 % of fruit trees (9 % of genus *Citrus*) and 6 % of other perennial crops (cocoa, macadamia and rubber). The most traditional farms, small and at low altitude and with shade consisting mainly of fuel wood species, were the least profitable due to low coffee productivity and a low level (<10 %) of revenues from wood harvesting (Table 13.2).

It is on the intensively managed large farms at medium-to-high altitudes (>1000 m) that the highest revenues were recorded. These revenues were derived mainly from coffee, due not only to very good yield but also because of the higher prices that the excellent quality of high-altitude coffee commands. Income from the sale of wood was marginal.

Table 13.2 Main characteristics of farms, revenues from coffee and from wood of agroforestry systems in the Ocosito watershed, North Pacific area of Guatemala

Farm type	Farm size (ha)	Mean altitude (m)	Tree density (trees/ha)	Coffee production (kg/ha)	Coffee revenues (% of total revenues)	Fuel wood revenues (% of total revenues)	Timber revenues (% of total revenues)
Trad-small[a]	26	775	259	526	90	3	6
Trad-medium[b]	41	750	237	644	23	25	52
Trad-large[c]	253	700	219	493	74	7	13
Int-large[d]	123	800	204	874	99.6	0.4	0
Int-Alt[e]	195	1200	205	736	87	8	3
Bio[f]	2.2	1400	273	440	69	5	0

[a]Trad-small: traditionally managed small farms at low altitude (<800 m)
[b]Trad-medium: traditionally managed medium-sized farms at low altitude (<800 m)
[c]Trad-large: traditionally managed large farms at low altitude (<800 m)
[d]Int-large: intensively managed large farms at low altitude (<800 m)
[e]Int-Alt: intensively managed large farms at high altitude (>1000 m)
[f]Bio: small organic farms at high altitude (>1300 m)

In contrast, medium-sized farms obtained a predominance of their revenues from wood. Indeed, the sustainable exploitation of timber, originating from natural regeneration, and fuel wood generated up to 76 % of annual incomes of medium-sized coffee farms at low altitudes. These farms have a high density of shade trees with a mixture of leguminous and timber species. The current rate of timber extraction on these farms is generally sustainable, with 1.1 m^3 of wood extracted per hectare per year, much lower than the growth rate, which is estimated at 3 m^3 per hectare per year. However, there is recent trend of increasing extraction. This is primarily due to the desire to offset the loss of revenue from coffee caused by the price collapse since the 2000s. Farmers are also trying to take advantage of the growing demand from local wood commodity chains. Thus, on intensively managed large farms, fuel wood is being extracted at a higher rate (14 m^3/ha/year) than that of growth of forest species (10 m^3/ha/year). This could prove detrimental by leading to the depletion of wood resources and forest cover.

Finally, for small farms practicing organic farming, the sale of shade trees represented less than 5 % of the farm income, even when accounting for fuel wood used for drying coffee and household consumption.

13.2.4 In Both Countries, Poorly Organized Wood Commodity Chains

In both Costa Rica and Guatemala, the timber and fuel wood commodity chains remain poorly organized. There is no cooperation or coordination between the various stakeholders and regulatory and support systems are non-existent. Therefore, there are great opportunities for improvement. In both the countries, local markets for wood are significant and reflect a growing demand, especially for high-value timber. These markets seem open to the introduction of new timber species such as the relatively unknown *Terminalia amazonia*.

In Guatemala, unlike Costa Rica, urban and rural households are still rarely equipped with electric or gas stoves and hence the demand for *Inga* fuel wood is growing rapidly.

13.2.5 Subsidies and Markets

As described above, the availability of government subsidies to farmers who plant trees is part of a pioneering Costa Rican public policy. It is of particular interest because it recognizes the role of the farmer as a provider of environmental services.

Large furniture retail chains in Europe and the United States have increased their procurement of timber grown and harvested in a sustainable and socially responsible manner. This is the case of wood stamped SmartWood by the Forest Stewardship Council in partnership with the Rainforest Alliance (2004). The demand for sustainable-certified timber is a new market opportunity for farms and cooperatives to explore, especially for those that already produce certified coffee.

13.3 Conclusion

The presence of shade trees contributes significantly to the economic sustainability of coffee farms in Central America, mainly through the diversification of revenues: either through government subsidies and/or through the sale of forest products. The production and sale of timber or fuel wood is an almost inevitable choice for coffee farmers in Central American regions at low or medium altitudes, where coffee plants are usually heavily shaded to buffer unfavourable conditions. With a local demand for wood that has been increasing constantly over the past 20 years in the region, this trend is expected to increase to the point that wood could become the primary source of income for 'coffee producers'.

In addition to a diversification of revenues, the presence of shade trees on coffee farms has a positive impact on coffee quality as well as beneficial environmental effects. Indeed, wood extraction from coffee plantations reduces the exploitation of forest reserves and tree fallows. There is no need to venture outside the coffee plantation for fuel wood or to fell valuable species to address pressing financial needs. In such cases, pressure on preserved areas that are critical for biodiversity conservation tends to decrease.

Subsidies for landscape-concerned farmers (currently being tested in Costa Rica) and certification of wood are two avenues to explore and develop. But for now, the farmers and cooperatives clearly need training in wood supply chains and their management in order to obtain higher revenues from their trees by marketing products directly rather than relying on selling standing trees to intermediaries.

Acknowledgements The authors thank ANACAFE (Association of coffee producers of Guatemala), ICAFE (Coffee institute of Costa Rica), coffee farmers, coffee and wood traders, and other stakeholders of these commodity chains in both these countries for their valuable assistance during the surveys. The authors also thank the European Commission (project CASCA, ICA4-CT-2001-10071) and the Science and Cultural Cooperation Centre of the French Embassy in Costa Rica for their financial support of this study.

References

Beer J, Muschler RG, Kass D, Somarriba E (1997) Shade management in coffee and cacao plantations. Agrofor Syst 38:139–164

Boulay A (2004) Production de bois d'œuvre issu des systèmes agroforestiers caféiers au Costa Rica. Thesis, Montpellier, Engref-Cirad 118 pp

Dzib B (2003) *Manejo, secuestro de carbono e ingresos de tres especies forestales de sombra en cafetales de tres regiones contrastantes de Costa Rica.* Master Thesis, Catie, Turrialba, Costa Rica, 114 pp

Fonafifo (2004) *La madera en Costa Rica, situación actual y perspectivas.* 7 pp

Fonafifo (2007a) *Environnemental services-concept.* www.fonafifo.com/paginas_english/environmental_services/sa_concepto.htm

Fonafifo (2007b) *Manual de pago por servicios ambientales e incentivos forestales.* www.fonafifo.com/text_files/servicios_ambientales/Manuales/MPPSA_2007.pdf, 15 pp

Galloway G, Beer J (1997) *Oportunidades para fomentar la silvicultura en cafetales de America Central.* Serie Técnica Informe Técnico, n. 285. Catie

Guyot B, Manez JC, Perriot JJ, Giron J, Villain L (1996) Influence de l'altitude et de l'ombrage sur la qualité des cafés arabica. Plantation recherche développement 3:272–280

Martínez M (2005) *Contribución económica del componente forestal en diferentes tipos de fincas cafetaleras en la bocacosta pacifica de Guatemala.* Master Thesis, Catie, Turrialba, Costa Rica. 123 pp

Muschler R (2001) Shade improves coffee quality in a sub-optimal coffee zone of Costa Rica. Agrofor Syst 51:131–139

Ponte S (2004) *Standards and sustainability in the coffee sector: a global value chain approach.* International institute for sustainable development (IISD) and UNCTAD, 49 pp

Rainforest Alliance (2004) *Sustainable forestry, Smartwood.* www.rainforestalliance.org/programs/forestry/index.html

Salazar M (2005) *Análisis de rentabilidad financiera del programa Cafe practices de Starbucks en diferentes tipologías de productores cafeteros de altura en Costa Rica.* Master Thesis, Catie, Turrialba, Costa Rica, 138 pp

Somarriba E, Harvey CA, Samper M, Anthony F, Gonzalez J, Staver C, Rice RA (2004) Biodiversity conservation in neotropical coffee (*Coffea arabica*) plantations. In: Schroth G, Da Fonseca GAB, Harvey CA, Gascon C, Vasconcelos HL, Izac AMN (eds) Agroforestry and biodiversity conservation in tropical landscapes. Island Press, Washington, DC, pp 198–226

Vaast P, Harmand JM (2002) Importance des systèmes agroforestiers dans la production de café en Amérique centrale et au Mexique. *Plantations, recherche, développement*, pp 35–43

Vaast P, Van Kanten R, Siles P, Dzib B, Frank N, Harmand J, Genard M (2005) Shade: a key factor for coffee sustainability and quality. In: Proceedings of the 20th international congress on coffee research (ASIC) Bangalore, India. pp 887–896

Varangis P, Siegel P, Giovannuci D, Lewin B (2003) Dealing with coffee crisis in Central America: Impacts and strategies. World Bank policy research working paper, n 2993. World Bank, Washington DC

Chapter 14
Coconut- and Cocoa-Based Agroforestry Systems in Vanuatu: A Diversification Strategy in Tune with the Farmers' Life Cycle

Laurène Feintrenie, Frank Enjalric and Jean Ollivier

Malo is an island located in the centre of the Vanuatu archipelago, south of the island of Espiritu Santo, between 15.6° and 15.8° latitudes south and meridians 167° and 168° east. This small island has a maximum length of 18 km and an area of 185 km^2. Its proximity to Espiritu Santo and, more importantly, to Luganville, the second largest urban centre in Vanuatu after its capital, Port Vila, has helped boost the economic development of Malo. This island was discovered in 1767 by Bougainville. The first contacts with the Europeans led to an outbreak of epidemics. To worsen matters, locals were conscripted for forced labour, resulting in a sharp population decline, from about 3000 people in 1890 to less than 500 people within two decades (Allen 2001). The Presbyterian mission, established in 1887 at Avunatari, helped bring local people together and turned this village into a focal point for social and economic activities for the island's western part.

Before it became a cash crop, the coconut was long a part of Vanuatu's agrarian system as a food crop in its own right. It is still habitually used to prepare traditional dishes and coconut water is consumed as a fresh beverage. It also forms part of the staple diet of pigs and poultry in family farms. The coconut tree was thus present on Malo in agroforestry systems adjoining houses, as well as in forests and forest fallows on the island (in rotation with vegetable gardens) well before it was cultivated as a cash crop.

L. Feintrenie (✉)
CIRAD-UR B&SEF, Biens et Services des Ecosystèmes Forestiers tropicaux, Département Environnements et Sociétés, Campus International de Baillarguet, TA C-105/D, 34398 Montpellier Cedex 5, France
e-mail: laurene.feintrenie@cirad.fr

F. Enjalric
Cirad-Persyst, UPR SIA, Direction Régionale du Cirad - Ampandrianomby, BP 853 101 Antananarivo, Madagascar
e-mail: frank.enjalric@cirad.fr

J. Ollivier
Cirad-Persyst, UPR Systèmes de pérennes, Avenue Agropolis, TA B-34/02 (Bât. 2, Bur. 110), 34398 Montpellier Cedex 5, France
e-mail: jean.ollivier@cirad.fr

F. Ruf and G. Schroth (eds.), *Economics and Ecology of Diversification*, DOI 10.1007/978-94-017-7294-5_14

The industrial cultivation of coconut as a cash crop was started in the country by European settlers in the early 20th century. They set up commercial plantations on the coastal strip on the eastern part of Malo between 1887 and 1920. These settlers offered employment on their plantations that not only attracted island residents but also led to an influx of migrants from neighbouring islands (Allen 2000). These migrants used the money they saved from the salaries received from the plantations to buy land in the centre of Malo where they planted coconut trees starting in the 1930s (Lamanda 2005). Under the urging of missionaries settled in Avunatari village and following the example of European settlers, local farmers also developed mono-specific coconut plantations starting in the 1930s (Labouisse 2004; Lamanda 2005) on Malo's western coastline. However, this cropping system was soon modified by farmers who adapted it to their needs by intercropping coconut trees with fruit and timber trees as well as palm for artisanal use (Labouisse 2004). They sometimes set up agropastoral systems in mono-specific plantations on coastal fringes involving herds of cattle, sheep and goats. Thus, in 1969, there were a total of 2450 ha of coconut plantations, accounting for 12 % of the island's area. These plantations were divided between 1740 ha of colonial commercial plantations and 710 ha of family farmers' smallholdings (Lamanda 2005).

Other cash crops also developed in Malo. Cocoa was introduced in 1955 by the European farmers. It has been cultivated as a monoculture or in agroforestry plantations in villages since the early 1970s (Labouisse 2004). In 2000, vanilla made its appearance as the new crop for diversification (Feintrenie et al. 2010). These two new cash crops were gradually combined with coconut plantations; complex agroforestry systems were slowly created, especially systems based on coconut and cocoa (Lamanda 2005).

The land tenure pattern in use before the establishment of European plantations was based on a use right related to the development of the land. The land ownership was held by a group as village landholding. Usage rights could be transmitted through inheritance in a patrilineal system, with women having no land rights. Land began to be privatised and started acquiring a market value following the development of coconut plantations by European settlers and the influx of workers from neighbouring islands. As on several other islands of Vanuatu, the creation of a land market led to frequent conflicts between natives and Europeans. These conflicts resulted in the expulsion of Europeans and in the declaration of independence in 1980. Vanuatuans remained locked in disputes amongst themselves over sharing rights over the land and plantations left behind by the European settlers. Following independence, the new Constitution recognised traditional land laws as the only land rights (Declaration of Independence 1980). Judgments determined customary rights by retracing the history of the land in question over generations to determine who first developed it and bequeathed the right of use to his descendants.

In 2005, agroforestry systems represented an essential component of several traditional farming systems in the humid tropics, particularly on Malo Island. These cropping systems are complex and associate several perennial and annual or multiannual food crops. The main perennial crops are coconut (*Cocos nucifera*), cocoa (*Theobroma cacao*) and vanilla (*Vanilla planifolia*). They are intercropped with several fruit trees including breadfruit (*Artocarpus altilis*), the cut nut (*Barringtonia procera*), citrus or nut trees and food crops including cocoyam (*Xanthosoma sagittifolium*), yam (mainly *Dioscorea nummularia*), taro (*Colocasia esculenta*), the Kanak cabbage (*Abelmoschus manihot*) and sweet potato (*Ipomoea batatas*). These cropping systems have multiple ecological, economic and social benefits. They protect the soil with their permanent cover and help maintain a large agrobiodiversity. From an economic point of view, they allow a diversification of production by combining food and cash crops, thus securing income while increasing the value of the land. Their social contributions are also significant since these systems help establish settlements, boost food security and also increase labour productivity.

But do coconut- and cocoa-based agroforestry systems meet the needs of families on Malo? A socio-economic study conducted in 2005 in three of the island's villages offers answers to this question based on the analysis of the economic performances of these systems and their integration into the family life cycle.

14.1 Methodology

The crop cycle in the agroforestry system was reconstructed with the help of a survey conducted in 2005 on 30 family farms located in three villages of the island (Feintrenie 2006). These surveys focused on details related to agroforestry systems as well as to farmers' families, including aspects like family composition, age and activities of each member, involvement in work related to the farm and to the agroforestry system in particular, contribution to family income and external support (funding or labour), if any. The survey was also designed to evaluate the performance of agroforestry systems in terms of production, working hours and income. The survey thus helped identify the farmers' strategies, their agricultural practices, economic performance and the decision-making rules they follow.

The composition in useful species (based on criteria defined by farmers) of an agroforestry system during its production cycle was determined from surveys of a sample of 15 plots. An additional survey was also conducted with the same sample of 30 families. These families described the floristic composition of each of their agroforestry system's plots, as well as the production of various species. The sample plots were chosen to form chronosequences that represented various phases in the development of the agroforestry system. In this synchronic approach, plots within a same chronosequence belong to the same agroecological environment and have a similar cultivation history and agricultural practices. A comparison of floristic surveys of these plots with their owners' oral descriptions showed that these descriptions were reliable (Feintrenie 2006).

The description of agroforestry systems, in terms of the composition of useful species and of the production based on the age of the main crop (coconut), led to the definition of agroforestry system models. The economic performance of various models was analyzed using Olympe, a software application to model a farm's economic activities (CIRAD et al. 2007; Deheuvels and Penot 2007; Feintrenie et al. 2011).

The analysis of economic performance is based on indicators for value derived from land and from labour, on the labour calendar and the maximum area that can be cultivated by an adult.

14.2 Results

In 2005, the island's cropping system was based on a combination of perennial cash crops (mainly coconut, cocoa and vanilla) and food crops. Cash crops are either cultivated as a monoculture or in an agroforestry system. In an agroforestry system, coconuts, cocoa and vanilla are associated in a time-space relationship among themselves and with food crops and woody plants for use by the family. Food crops are grouped under two types of cropping systems: the Melanesian system of garden food crops based on tubers, in alternation with a long tree fallow period, and agroforestry systems in which they constitute the main production in the early years. Very often coconut-based agroforestry systems represent the predominant cultivation activity in terms of area (more than 50 % of cultivated land), labour time and income source. These systems provide food and raw materials for handicrafts and artisanal use (including material necessary for constructing houses, canoes, etc.). They are frequently associated with cattle farming and thus can be called agropastoral systems. They play a central role in the organization of farms on Malo.

Farms in the study area are small, with the extent of agricultural areas most often ranging from 5 to 10 ha. Production systems are almost always extensive. Agriculture is not mechanized and agricultural tools are limited to the crowbar, the digging stick and machete. The only form of a permanent workforce is family labour.

14.2.1 The Dynamics of Land Cultivation Driven by Farmer Objectives

The main objectives of farmers of Malo are to ensure:

- adequate production of food to feed the family, of raw material for house construction and maintenance, as well as of material for handicrafts;
- sufficient income to pay for the children's education, clothes, medical expenses, etc.;
- creation of land capital for male heirs and setting aside a dowry for their daughters-in-law;

- a social identity through the recognition of the quality of their work, as well as wealth generated from the area under cultivation, the number of pigs and cattle offered on special occasions.

Strategies to achieve these objectives are based on land acquisition. Land is the most important resource on the island. It plays a strong role in shaping the identity of the individual, as well as of the group he belongs to. The farmer seeks to capitalize on the land and guarantee his right to it by cultivating it. Seen from this perspective, planting the land with perennials is to assert ownership and be able to transmit it eventually on to his son. However, if a farmer wants to develop a large area he must first clear it, a task that is long and tedious. The superposition of cash crops and food crops in the same area helps draw the maximum benefit from the area and reduces clearing and maintenance work.

The process of establishing agroforestry systems is progressive. Each year, farmers clear a small forest area (about 625 m^2 corresponding to an average-sized garden on Malo) to establish a food garden intercropped with coconuts and cocoa. This garden gradually develops into a multi-layered agroforestry system. The density of planted coconut trees is high (600 germinated nuts/ha) in order to achieve a density of mature coconut trees of about 170–180 trees/ha, which is higher than the conventional recommendation of 140–160 trees/ha.

The density of cocoa is about 400 trees/ha, well below prevalent standards. The first year of cultivation is devoted to yam and beans. Cocoyam, Kanak cabbage and banana are grown in the second year; to this is added manioc and papaya in the third and fourth years. Coconut or cocoa trees continue to grow in the meantime, but they have yet to achieve their full production. In the fourth year, the undergrowth is cleared and the garden is abandoned as the shade of the perennials has increased and no longer allows cultivation in the understory.

This plantation system is maintained continuously in the farms. The area cultivated as garden thus remains more or less constant and equal in the first, second and third years (Feintrenie et al. 2010). This progressive and continuous cultivation process combines well with the strategy of land acquisition. At the end of the fourth year of cultivating the garden, the agroforestry system has become an extensive cultivation that requires little labour and will last more than 80 years. The extent of perennial plantations increases steadily as more forest area is cleared.

A small area (often 1 ha) of forest is nevertheless preserved close to the home for food garden rotations. These lands initially meant for gardens are increasingly being used to plant crops that are highly labour-intensive or are valuable, like vanilla. Then the rotation system is made more complex by involving a combination of short and long fallow periods. The short fallow period begins at the conclusion of gardening activities, 4 years after the land is first cleared, and usually lasts no more than 2 or 3 years. This short rotation is associated with a long rotation: a new site is selected after three to four cycles of the short rotation (18–24 years). This site is most often located away from the home due to scarcity of available land, with nearby plots having been planted with cash crops such as coconut and cocoa

(Feintrenie et al. 2010). Thus, a farmer may cultivate two or three different sites over his lifetime.

The process of a progressive implementation of the agroforestry system has to be matched with the farm's life cycle. Usually, a farmer settles down at the age of 20 and starts a household. The farm begins to grow in area and in value. When the farmer attains 40 years of age, his daughters get married while his sons become independent from the family. This results in a reduction in the labour force on the paternal farm. During this same period, one of the sons, usually the eldest, begins to assume the responsibility of taking decisions regarding the farm and gradually replace his father as the head of the farm. However, the son who takes over the farm assumes the responsibility of looking after the parents.

Progressive cultivation entails clearing activities that are labour-intensive. This phase of expansion of land area occurs when the farmer is young and is in the process of settling down (Fig. 14.1).

The cultivated area has increased over the years when the maximum amount of family labour was available to the farm.

This strategy, based on capitalization of land and perennial plantations, seems to relegate food gardens to a secondary role. This is not actually the case, since the garden is an integral part of the agroforestry system and is the source of all the food produced in the first 4 years. In a scenario of an annual clearing of land and expansion of plots, the usefulness of a food garden can span 20 years for a plot of 1 ha. A young farmer usually cultivates 2 or 3 ha of perennial crops (within and outside the agroforestry system) at the same time. The age of the plantations depends mainly on his heritage and investments. An area of 3 ha of perennial plantations can be achieved quite rapidly by a couple depending on the farm's initial planted capital and the speed of expansion chosen by the farmer. The conservation of forest areas meant for cultivation systems that alternate gardens and fallow periods also forms part of the strategy of the farmers of Malo.

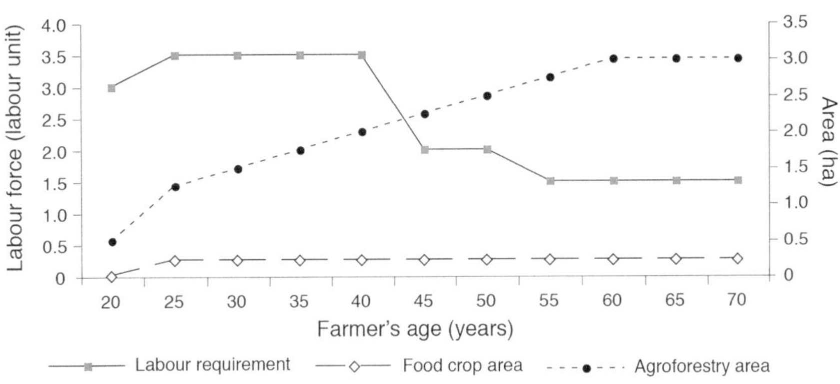

Fig. 14.1 The relationship between the dynamics of cultivating an agroforestry plot and the farmer's age and labour force

The farm's production system is based on a combination of perennial cash crops (mainly coconut and cocoa, and more recently vanilla) and food crops. Agroforestry systems often occupy half of the land planted with cash crops and are often the first systems established by the young farmer. The role of land occupation is added to those of subsistence and income generation; agroforestry systems thus contribute to increasing the farmer's capital.

14.2.2 *Production and Performance: Production Phases in Terms of the Components of the Agroforestry System*

Coconut trees are cultivated for 80 years; this crop is thus trans-generational. Cocoa trees are grown for 40 years, which means that two cycles of cocoa take place during one coconut cycle. Plantations suffer losses over time: 23 % of the coconut trees perish after 50 years and 36 % after 70 years. The proportion of cocoa trees which die is 30 % after 7 years and 35 % after 30 years. These dead plants create openings in the canopy that are then used to cultivate cocoyam. This practice is followed on Malo.

During the phase of establishing an agroforestry system on the plot, each year about 1/16th of the first-year gardens are set up on cleared land. In each of the subsequent 3 years, a similar size of garden is added on newly cleared land. Most of the labour required is for clearing of the new areas and maintenance of food crops, since the perennial plants are yet to become productive. We thus have an initial production phase which mainly pertains to food crops. These crops provide an income to the farmer while the perennial plants are still unproductive at the beginning of the agroforestry system cycle.

Subsequently, the perennial crops gradually take over as and when cocoa and coconut trees develop. Food crops are stopped from the 20th year when the perennial plants take over fully. This step corresponds to the peak production level of the perennial species and a decrease in labour availability in the farm due to the departure of the children and the aging of the farmer.

It was possible to model and analyse this establishment process using the Olympe software. A model was defined for a 1 ha plot that was cleared over 16 years (i.e. 625 m^2 annually). The plot therefore consisted of 16 segments, with one segment being planted each year: first with food crops and then with perennial crops. In this way, the 1 ha plot is fully under cultivation after 16 years. Each segment continues its cultivation cycle during the 80 year lifespan of the coconut trees. The agroforestry system models thus continue for 96 years. The establishment and production phase therefore spans 20 years in the case of food crops, and nearly a 100 years for the production of perennial crops.

To evaluate economic indicators on the basis of production recorded or observed in 1 ha of an agroforestry system, all the products—including those for home consumption—were assigned a monetary value (local market prices for food crops, farm gate prices for cash crops). Thus, the value extracted from one cultivated hectare under a coconut-cocoa agroforestry system was evaluated on the basis of the time elapsed since the start of the plantation (Figs. 14.2 and 14.3).

The vertical axis represents the annual productivity of the land in €/ha, the horizontal axis is the number of years after clearing the first segment of the plot. The 'agroforestry system' curve indicates the results of the cropping system as a whole. The other curves describe land productivity for each component of the cropping system: dry copra, fresh cocoa beans and food crops production. The first

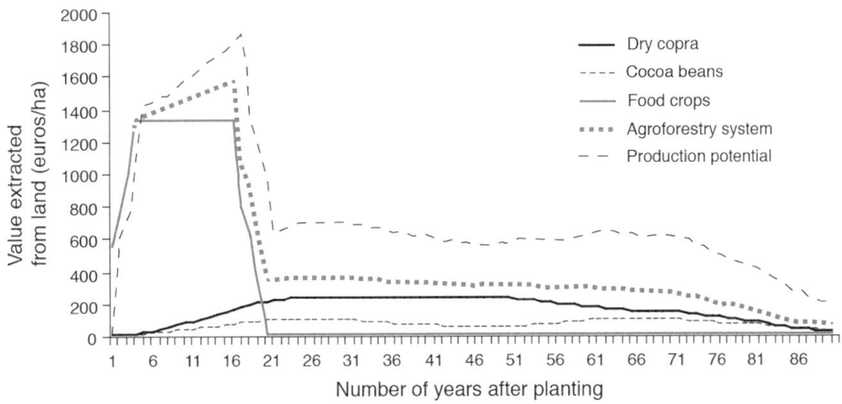

Fig. 14.2 Simulation of the dynamics of income from 1 ha of agroforestry land as a function of the age of the agroforestry system: value extracted from 1 ha of land

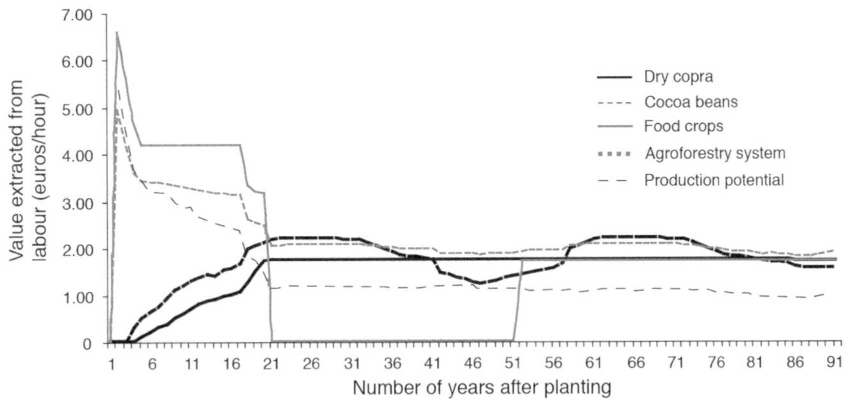

Fig. 14.3 Simulation of the dynamics of income from 1 ha of agroforestry land as a function of the age of the agroforestry system: value extracted from each labour day

20 years show a high productivity, due to the start of production of food crops in different segments of the plot and the beginning of production by the perennial crops.

Simulating the gradual establishment of an agroforestry plot shows a land productivity that corresponds to the sum of productivities of each associated crop. There is a staggering of income corresponding to successive plots becoming productive according to their age. We see that such agroforestry systems offer a per-hectare productivity that is greater than or equal to 1400 €/year for 15 years. Perennial crops start producing from the fifth year onwards and replace food crops as income providers, generating 600 €/ha for the subsequent 50 years.

14.2.3 Role of Diversification in Increasing Land and Labour Productivity in the Coconut-Cocoa Agroforestry System

Agroforestry systems can significantly increase land productivity through the additive effects of different productions on the same surface area. The contributions of the different crops, however, vary naturally over time and depend on their respective production cycles. Land productivity is relatively high during the first 20 years as the production cycles of food and perennial crops are both in play. It is highest at the time when the farmer has to deal with his major monetary expenditures, i.e., in the first 20 years when he has to look after his children and pay for their education.

The farmer's cost of living goes down significantly after this initial 20 year phase, coinciding with a decline in land productivity, and finally stabilises over time. Crop associations and the staggering of their cultivation allow him to plant perennial crops. The yield from perennial crops ensures regular income over the long term, especially when less labour force will be available due to his age and the departure of his children. In fact, between 40 and 60 years, not only is there a reduction in the available labour force with the departure of children (girls get married and boys set up their own farms), but there is also a decrease in activities due to aging or the transfer of responsibilities to the eldest son.

There is thus a strong correlation between the potential economic performance of agroforestry systems and the requirements of farm households. The system is driven by food crops during the first 20 years, before the perennial crops (cocoa and coconut) take over the burden of production. After 70 years, the production of perennial crops starts declining. This is partially offset by the re-cultivation of food crops, cocoyam in this case, in gaps created by fallen trees and palms.

Studies of labour hours helped assess labour force requirements of the various components of agroforestry systems and the amount of value a working day can generate. The latter are different for coconut and cocoa cultivation due to the difference in workload associated with these crops. Labour required for maintaining (mainly for ensuring its height and for sanitary harvesting) and harvesting cocoa are greater than that required for coconut, and this is not offset by the selling price of fresh beans (based on data for 2005). Work time invested in coconut gives better returns than for cocoa when the latter is valued through the sale of fresh beans.

At the beginning of a cycle, it is the food crops that ensure a high labour productivity. This productivity is reduced since labour is required to plant perennials. The system's labour productivity presents a level that is the average of the respective productivities of coconut and cocoa. When the cultivation of food crops is stopped, the potential productivity decreases and stabilises around 1.10 €/h, which works out to an average monthly salary of 20,000 Vatus (160 €), despite the fact that it only requires a small portion of the available time (about 20 %). Although the transition between two cocoa cycles reduces land productivity due to a break in cocoa production, it increases labour productivity. In fact, the workload decreases considerably in the absence of maintenance and harvesting work for cocoa. When farmers do not carry out such cocoa maintenance work, which is usually the case, the value derived from one labour day for this crop, and the entire agroforestry system, increases to about 2 €/h. This result is quite unexpected and highlights the effectiveness of the farmers' practices and choices. It shows that the time that could be spent on pest control and the maintenance of the height of the cocoa tree adds little value in the current context of marketing of the crop as fresh beans.

This method of managing perennial crops allows farmers to adapt to price fluctuations and labour availability. Priority is given to efficient labour practices as compared to a normalized and more productive technical itinerary. The performances of agroforestry systems are thus relatively high compared to local opportunity costs; they point to a level of labour remuneration consistent with the local economic environment.

14.2.4 Diversified and Productive Systems Adapted to Low Labour Availability

The monthly labour requirements were estimated based on a survey of a 1 ha plot cleared at the rate of one-sixteenth part every year. Most of the labour in the first 20 years pertains to the newly cleared area and the gardens, since the perennial plants are yet to become productive (no yield, and small height of the cocoa trees). Figures 14.4 and 14.5 show these labour requirements. The period from August to October, corresponding to the harvest of yams and other food crops, requires more labour than the rest of the year. However, the peak labour requirement (in September) is only 92 h for the entire month, or roughly 3 h a day for one person. Considering that a man can work 208 h a month (working 8 h per day, and 26 days per month), a single person could maintain an area of 2.25 ha (208/92 = 2.25 ha) of a coconut-cocoa agroforestry system during the establishment phase.

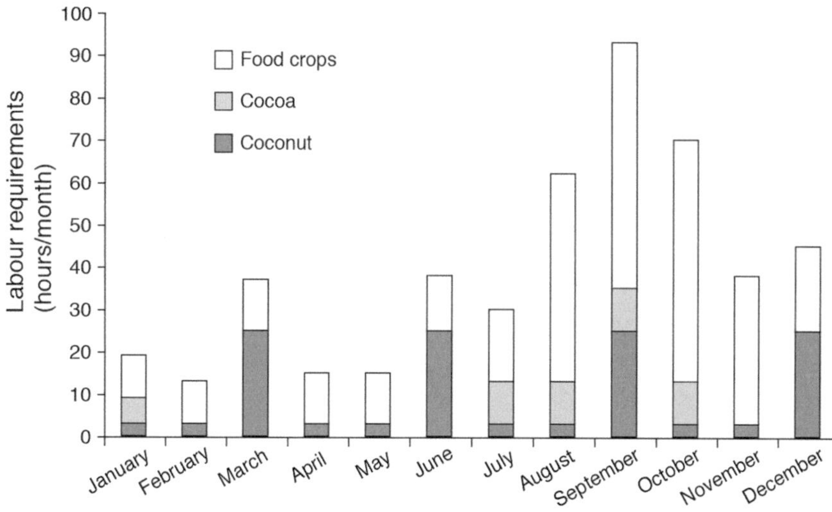

Fig. 14.4 Change in monthly labour requirements for 1 ha of agroforestry land: a 7 year-old food crops-cocoa-coconut plot

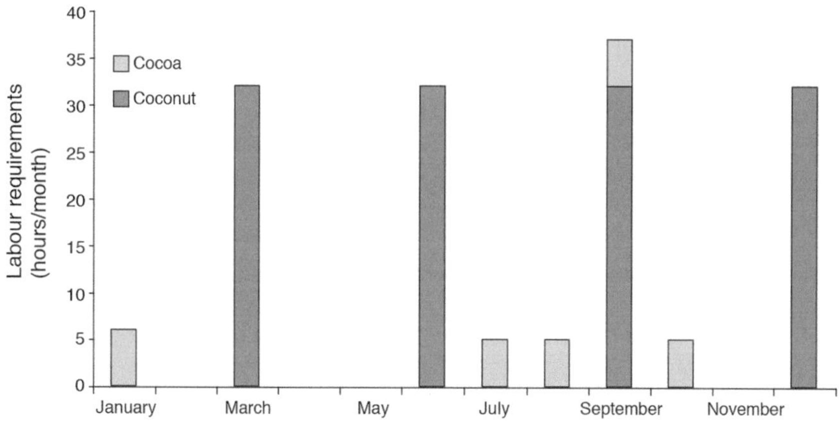

Fig. 14.5 Change in monthly labour times for a 1 ha agroforestry plot: a 30 year-old cocoa-coconut plot

This result illustrates the harmony between practices implemented in village plantations (not very demanding in terms of time) and the time required to maintain social and customary networks.

Farming operations change with the age of the system. There is no work required to maintain food crops and for slash-and-burn activities during the transitory productive phase. Figures 14.4 and 14.5 show the basic labour requirements of a coconut-cocoa agroforestry system. During the transitory phase, labour is required for harvesting and shelling nuts every 3 months, and for harvesting cocoa pods between July and October. Cocoa trees are sometimes pruned in January. Labour requirements peak in September. If coconut is harvested at the same time as cocoa pods, the demand reaches 37 h, i.e., a little less than 2 h per day per hectare. Harvesting is generally done with self-help groups consisting of about ten people. This then requires a half-day's labour for 1 ha, to which must be added the time required for travel, transport, drying of copra, sorting of beans, etc.

A mature agroforestry plot requires less labour than during its establishment. Although copra harvesting is the most labour-intensive cultivation activity, it does not impose a fixed date on the cultivation calendar, thus giving the farmer the flexibility to take care of other activities.

14.3 Conclusions

The study describes, at the farm level, the farmers' production methods, their objectives and farming strategies. Farms in the study area are small (usually a cultivation area of 5–10 ha) and the production systems are almost always extensive. The only form of a permanent workforce is family labour.

The objectives of farmers on Malo Island are: food self-sufficiency, adequate income to take care of the family's needs (medical expenses, clothing, children's education, etc.) and the creation of land capital that can be bequeathed to their sons. This results in a land-use strategy where the farmer seeks to capitalize on the land and safeguard his right on the land by cultivating it. For this purpose, planting perennials on the land asserts ownership and ensures that he builds up an inheritance to pass on to his sons.

Modelling has allowed us to demonstrate the economic potential of agroforestry systems and assess the contribution of various components to the overall production. The benefits of this cropping system are not limited to cash crops or food products. Other plants (palms, trees or grasses) contribute to the wealth of the system.

The harmony between the production cycles of a coconut agroforestry system and those of the farmer's household has to be stressed. The production of an agroforestry system and its labour requirements are in tune with the farms' requirements and means of production. Cultivation of cash crops and food crops in the same space allows the extraction of the maximum possible value from the land surface, while also reducing the work of clearing. This agroforestry system enables the farmer to achieve his objectives, while allowing him time for other activities (remunerative or not) that are essential to the social bonding of the village.

Farmers clear an increasing forest area in a context of constant population growth and the dynamics of cultivating perennial cash crops. Consequently, the land required for long fallow periods between two food garden cycles decreases sharply. This leads to an increased risk of shortened fallows, with a resultant increase in land use and sometimes even the risk of soil degradation. It thus seems important to prevent the risk of environmental degradation resulting from uncontrolled intensification. These agroforestry systems, established spontaneously by the local inhabitants, offer alternatives to forest clearing by developing gardens in which food crops are intercropped with perennial plants.

References

Allen MG (2000) Subsistence or cash cropping? Food security on Malo Island, Vanuatu. Food security for Papua New Guinea. Aciar Proceedings n 99, pp 100–111

Allen MG (2001) Change and continuity; land use and agriculture on Malo Island, Vanuatu. Master Thesis. Australian National University, Canberra, 201 pp

Cirad, Inra, IRD, IAMM (2007) Olympe, simulateur technico-économique des exploitations agricoles. www.olympe-project.net. Accessed 10 Jan 2010

Deheuvels O, Penot É (2007) Modélisation économique des exploitations agricoles: modélisation, simulation et aide à la décision avec le logiciel Olympe. l'Harmattan, Paris, 184 pp

Feintrenie L (2006) Contribution à l'évaluation de systèmes de culture agroforestiers à base de cocotiers sur l'île de Malo (Vanuatu). Thesis of Engineer in Tropical Agriculture, Cnearc-IRC/Supagro. Montpellier, Cnearc-Cirad, 110 pp

Feintrenie L, Ollivier J, Enjalric F (2010) How to take advantage of a new crop? The experience of Melanesian smallholders. Agrofor Syst 79(2):145–155

Feintrenie L, Ollivier J, Enjalric F (2011) L'utilisation d'Olympe dans l'étude des systèmes agroforestiers complexes à base de cocotiers au Vanuatu, île de Malo. In: Penot É (ed) Modélisation économiques des exploitations agricoles avec le logiciel Olympe, vol 2. De l'analyse à l'action. Paris, Quae

Labouisse JP (2004) Systèmes agraires et économie du cocotier au Vanuatu: historique et perspectives. Journal de la société des océanistes 118:11–33

Lamanda N (2005) Caractérisation et évaluation agroécologhique de systèmes de culture agroforestiers: une démarche appliquée aux systèmes de culture à base de cocotiers (Cocos nucifera L.) sur l'île de Malo, Vanuatu. Doctoral Thesis, INA-PG, 200 pp

Chapter 15
The Place of Cocoa and Coconut Cultivation in Family Plantations in Peninsular Malaysia

Pierre Dupraz and Murielle Morisson

The Malaysian cocoa cultivation cycle was one of the fastest and shortest in history. Starting with a yield of less than 1,000 tonnes in the early 1960s, Malaysian cocoa output vaulted to a level of 200,000 to 220,000 tonnes between 1987 and 1993, before collapsing to a plateau of 30,000 tonnes in the 2000s. Large company- and individual-owned plantations played a major role in this process. Nevertheless, the participation of family farms in the cocoa cycle, and what cocoa cultivation represented for these family farms at one time is very instructive. The areas planted with cocoa in smallholdings increased from 8,000 to 35,000 ha between the 1970s and the 1990s. Cocoa became the diversification crop for the oldest systems mainly because of its skyrocketing price in the international markets in the early 1970s. A steep decline in cocoa prices 20 years later, in the early 1990s, led to a second stage of diversification when farmers progressively abandoned cocoa.

This chapter examines a scenario in which cocoa cultivation became financially attractive, prior to being abandoned. We find here the guiding hypothesis of this book: diversification mainly stems from ecological changes, the circumstances of the unfolding of the cocoa cycle, with the start of aging of cocoa orchards and the difficulties of their replanting. The study looks at the 1970–1990 period, when existing systems were transformed by the adoption of cocoa. The enthusiasm for cocoa cultivation, starting in the 1970s, was accompanied by a great diversity of agricultural situations. Cocoa, as tree-capital, was a source of income and contributed to social transformation in Malaysia in different ways in different regions.

Malaysia has undergone dramatic changes following its independence in 1957. We will present an overview of these social and political transformations and discuss how they, in conjunction with prices, influenced farmers to adopt cocoa cultivation.

P. Dupraz (✉)
INRA, UMR 1302 Smart, 35 000, Rennes, France
e-mail: pierre.dupraz@rennes.inra.fr

P. Dupraz
Agrocampus Ouest, UMR 1302 Smart, 35 000, Rennes, France

M. Morisson
18 Avenue Pasteur, 78 580 Maule, France
e-mail: mmorisson@gmail.com

F. Ruf and G. Schroth (eds.), *Economics and Ecology of Diversification*,
DOI 10.1007/978-94-017-7294-5_15

This chapter examines diversification in three villages in three different regions of peninsular Malaysia, starting with Pamah Kulat in Pahang state. This diversification is representative of an extension of cocoa cultivation on cleared forest lands, on the hills and mountains in the centre of the peninsula. This cocoa trend was similar, at least initially, to that seen in other cocoa-producing countries, especially in West Africa.

We must remember, however, that the pioneer cocoa cultivation in Malaysia is secondary to a specific cultivation system common in this country: the intercropping of cocoa in the understory of coconut trees, especially along the peninsula's western coast. The history of human settlements and agriculture in this region dates back over a 100 years. Using the typology of the farmers in the two villages of Sungai Batang and Parit Haji Abdul Salam, we will attempt to address the issues of cocoa cultivation in this part of Malaysia.

15.1 Methodology

The aim of this study was to understand the economic role of cocoa cultivation in small Malaysian plantations. In Malaysia, small plantations are defined as those farming less than 40 ha. They are usually family farms, as opposed to large commercial plantations whose structure and management were inherited from the colonial period.

A preliminary phase of the study involved retracing the history and financial details of the introduction of cocoa in Malaysia, including:

- the role of the macroeconomic context and prices;
- the role of policies for agricultural development and poverty eradication;
- the role of public and private research, development and marketing organizations.

This phase was based on a review of the literature and interactions with various researchers, development officials, village heads and traders of agricultural products ranging from collectors to exporters. This analysis allowed us to ascertain the initial characteristics of the Malaysian cocoa boom of the 1980s. In comparison to Africa, the soil and climate conditions, expertise and quality of the beans here were worse. However, its effective and efficient collection and marketing systems—largely controlled by private operators—explain Malaysian competitiveness to a large extent. In fact, Malaysian cocoa farmers received nearly 90 % of the international cocoa price, while their Cameroonian counterparts obtained less than 50 % during the same period.

The main part of the study focuses on the economic behaviour of smallholders and how they formed part of the local economy, the labour market and, in particular, the agricultural land market. The analytical framework is based on a farmer life-cycle hypothesis. According to this hypothesis, in our context, the farmer anticipates and prioritizes his consumption needs over his entire adult life. He considers income opportunities related to his work as an employee, and then as a farmer once he inherits or purchases land. Finally, he factors in the efforts to save money necessary to accumulate capital that will ensure sufficient income in his old age. The relevance of this hypothesis, however, became only gradually apparent during the initial surveys. In fact, the villagers and farmers did not have any system for retirement pensions, nor any long-term bank savings or credit facilities. Furthermore, the oldest amongst them attached great importance to the maintenance of a young labour force in the village. Thus, agricultural assets such as land, trees and livestock played a crucial role, not only in terms of production but also as assets to pass on to the next generation.

This chapter's first part is based also on a systematic survey of about 60 family farmers. This survey included some landless rural families from two villages on the west coast: Sungai Batang in Perak state and Parit Haji Abdul Salam in Johor state. In addition, some villagers employed by the non-agricultural public and private sectors were also interviewed in an attempt to understand how access to non-agricultural income and an institutional retirement system could upset the economic performance and the economy of the entire village. This is the case of Parit Haji Abdul Salam which, unlike Sungai Batang, is located close to an industrial development zone. The sampling, with no claims to representativeness, focused on describing the diversity of professional and family situations. In accordance with the life-cycle hypothesis, the survey focused on describing household size and composition, monetary savings and debt, use of family labour, revenue structure, and composition of capital and its evolution over time.

The second part explores the economic performance of different agricultural products. For each family surveyed, the goal was to reconstruct the family budget in terms of money and work time, as well as to analyze investments, costs, work time and the income from each crop. Fishing and home consumption of food products (shrimp, cockles, fruits, roots and tubers) cannot be ignored for low-income households, but these were not addressed in detail due to lack of time. Interviews lasted between 1.5 and 2.5 h. In Sungai Batang, a second survey was conducted on a dozen individuals to correct identified accounting inconsistencies.

15.2 Results and Discussion

15.2.1 Cocoa in a Rapidly Developing Area: The Pahang Region

15.2.1.1 Clearing of the Forest

Cocoa cultivation is characterized by the planting of cocoa on cleared forest land and in various crop associations. The historical development of farming systems has changed the conditions under which cocoa is cultivated and its role in cropping systems. The Pahang region is characterized by hilly terrain and significant forest cover. The potential of arable land is high: of a total of 23 million hectares of land, some 0.75 million hectares are suitable for agriculture, of which only 0.2 million hectares are currently in use. The district of Raub has good infrastructure despite the fact that it has little industrial development. All agricultural activities are carried out in small village farms. Rice fields are found in the lowlands, while the slopes are covered with rubber trees. These rice fields are generally managed by women with production destined mainly for home consumption. Rice is processed in artisanal village processing plants. Despite government efforts to help farmers obtain two crops instead of one a year, rice cultivation has been losing steam since 1965.

Farmers believe that large scale clearing of forest land has had an adverse impact on water availability in some plots. In addition, young rural people do not want to work in the paddy fields. In 2012, only one rice processing plant remained functional, and this too operated only twice a week. Paddy fields located around homes were converted to orchards by the end of the 1960s; this was made possible largely due to the distribution of fruit tree saplings by the government. Just like rice, rubber cultivation too faced a labour shortage, with land owners finding it difficult to get workers to tap rubber trees. In fact, casual labourers had several alternatives available to them: either work in industries in the city, find employment in a Felda plantation[1] (which would require them to leave the village), or even acquire land in the village itself.

To summarize, even before the advent of cocoa cultivation, farmers were cultivating rice but only sufficient to help meet their own home consumption requirements. However, they were already diversifying their crops to overcome problems of unpredictable labour availability for rubber cultivation.

15.2.1.2 Cocoa, an Additional Opportunity

In the early 1980s, cocoa was introduced on the government's initiative by the Malaysia Agricultural Research and Development Institute (Mardi). The political machinery within the villages then helped farmers undertake cocoa cultivation. Several economic factors made this possible.

A Rapid but Controlled Colonization of the Land

New areas, especially former government reserves, were opened up through a controlled redistribution by the State. This increase in arable land surface was, however, at the expense of the forest. In addition, the low price of land, about 50 times cheaper than on the west coast, allowed families to accumulate landholdings.

[1]Felda Global Ventures is a large industrial group which manages more than 450,000 ha of plantation estates in Malaysia.

This opportunity to acquire land also attracted many young people from the west coast, where there was not enough land to go around.

An Agriculture Oriented Towards Diversification and Intercropping

Diversification helped farmers reduce risks related to fruit price fluctuations in the market, and the lack of regular labour to tap rubber. Cocoa thus seemed to be an added opportunity for diversification, rather than the principal crop. Since cocoa production depended mainly on family labour, the success of cocoa was assured even in a scenario of limited availability of casual labour.

Cocoa was initially intercropped with banana and fruit trees. Banana trees provided shade for young cocoa plants even as they ensured a steady income, while the farmer awaited the entry into production of cocoa from the second year onwards. Banana trees were removed after 4 years, when cocoa trees became tall. *Gliricidia* then provided shade, until such a time when fruit trees began providing it.

Forest Rent Favourable to Cocoa Production

Plantations established on cleared forest plots enjoyed several advantages. All the benefits derived from forest rent described in West Africa (Ruf 1987) could be found here. Requirement for inputs were lower than for land that was already cultivated. Cocoa trees growing in the ecological conditions prevalent in Pahang produced a harvest at the end of 2 years. Cocoa thus evinced greater interest than did rubber. Yields for cocoa grown in this region were much higher (3–4 times) than on the west coast of Malaysia, with cocoa planted in the understory of coconut trees.

A Highly Subsidized and Therefore Increasingly Attractive Region

In order to assert control over new lands made available by the State for agriculture, sectoral agricultural development agencies tried to one-up each other by offering subsidised inputs and technical support. In this way, they hoped to get farmers to grow the crop over which they had responsibility. In the Raub area, for example, this mainly resulted in a competition over the development of new lands and abandoned rice fields between the Rubber Industry Smallholders Development Authority (Risda), wanting to extend rubber plantations, and Mardi, which wanted to increase the areas under cocoa cultivation.

In this region, home consumption was high and production costs were low. Farming households were engaged in increasing their savings and endeavoured to acquire new forest plots that they could clear. It was also obvious that the pace of economic and demographic growth depended on the rate at which the government released forest lands for agriculture.

15.2.2 Smallholders on the West Coast

The western coast of peninsular Malaysia is a flat area with alluvial soils. The vegetation consists of perennial crops such as rubber, oil palm and coconut. Coconuts were cultivated mainly smallholders of Javanese origin in the two villages studied. The British colonial administration had encouraged their immigration to this region in the early 20th century. They demarcated out village lands and assigned plots to the immigrants. In this way, the British hoped to develop a dynamic rural paddy cultivation system that could feed workers of the tin mine and the large commercial plantations estates that were in full expansion.

However, problems of water availability and markets turned farmers away from growing rice. Coconut plantations emerged as a compromise between the desire of farmers to become part of profitable agricultural sectors (for example, rubber) and the government's wish to restrict them to growing food. In fact, in Malaysia, from the time of colonization to the present, the status of land was linked to the main crop grown on it. Consequently, the choice of the basic crop depended on the government. Farmers thus sought to avoid any competition with large farms, which mainly grew rubber.

15.2.3 Social Aspects of the Introduction of Cocoa

The technical modalities of the introduction of cocoa cultivation, as well as the role of this crop in the coconut-based farms, varied considerably. They were part of an evolution of family farms, and were subject to demographic and macroeconomic constraints. This diversity of agronomic and social factors as they pertained to cocoa cultivation can be understood through a typology of the farmers' productive behaviours.

Smallholder coconut plantations are categorized based on a demographic determinism dating back to colonial times. In the early 20th century, the colonial British government distributed land among farmers, large plantations and the State. This land distribution was not subjected to any review following the country's independence. Indonesian migrants quickly populated the west coast and cleared the entire area reserved for ethnic Malay farmers. A population explosion in this limited land area resulted in a decrease in the per-capita arable surface. The size of family farms decreased as a result of egalitarian rules of land inheritance. Due to the rural exodus, this trend is now being challenged. The main macroeconomic determinants are summarized in Box 15.1.

Box 15.1. Major social and economic events that explain crop diversification in Malaysia in the 1960–1990 period

1960–1970

- GDP growth of 6 %.
- Diversification of major rubber plantations to oil palm, which became a major competitor of copra.

- First downstream industrialization in the agricultural sectors.
- 1969: Race riots. This social crisis led to the implementation of the New Economic Policy (NEP) based on a programme for the eradication of poverty and social restructuring that favoured ethnic Malays. This resulted in numerous subsidy programmes for smallholdings and an improvement of rural infrastructure.

1970–1980

- GDP growth of 7.8 %.
- Fall in the purchasing power of copra.
- The two main driving principles of the NEP encouraged a professional and geographical exodus from old rural areas, mainly from coconut plantations.
- Setting up of labour-intensive industries (textiles and electronic components).
- Felda accelerates its policy of setting up smallholdings within its schemes (1976–1977) in the context of very high cocoa prices: MYR 8/kg of dry beans, i.e., USD 2.6/kg (1 Malaysian Ringgit (MYR) = 0.37 USD).
- Massive expansion of cocoa plantations.

1980–1990

- 1985–1986. The country recorded a negative economic growth as a result of a concomitant decrease in the price of most Malaysian export products.
- Continued decline of cocoa prices.
- November 1989: MYR 2/kg of dry beans, i.e., USD 0.7/kg.
- January 1990: MYR 1.80/kg of dry beans, i.e., USD 0.6/kg.

The role played by coconut plantations in the social organization of villages in the early 1970s was still relevant in 1990 in one of two villages studied despite, or rather because of, the introduction of cocoa cultivation. Rural coconut plantations were the main reason for an exchange of labour between families and farms. It provided an income to villagers belonging to different social classes:

- Landowners who sold coconuts or copra to traders;
- Daily wage or casual labourers in the agricultural sector who offered their services during coconut harvests. This strenuous work can only be undertaken by villagers who are between 20 and 50 years old;
- Chinese or Malay traders who collected the nuts and copra and sold them to the outside world.

This relationship of the coconut between this social role and the one it fulfils in family farms can be elicited by using the life-cycle concept. It allows village farmers to be divided into four groups based on their resources, family constraints

Table 15.1 Coconut plantations in the village life cycle and social organization in Malaysia

Age of the head of the family	Level of fixed expenses (food and schooling)	Income sources	Flow of work	Flow of coconut plantation asset	Purpose of monetary savings
Old: type I	Low	Profits obtained from ownership of the coconut plantation	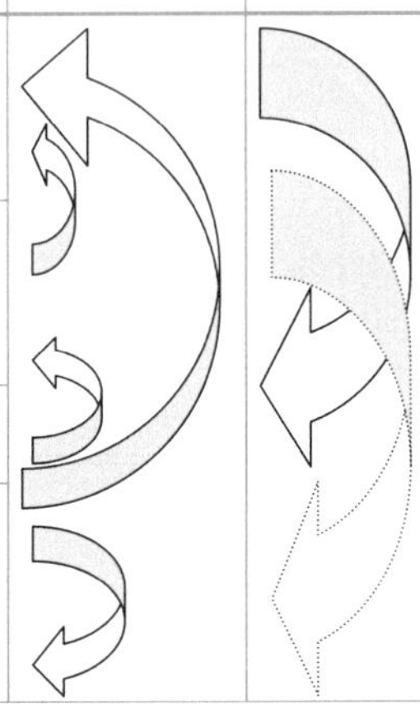		Pilgrimage
Middle-aged: type II	High	Profits from coconut plantation and casual labour			Coconut plantation
Young: type III	Low	Casual labour			Coconut plantation
Entrepreneur 'not part of the life cycle': type IV	Variable	Profits from several activities			Various profitable investments

and interdependencies. These groups constituted the analytical framework which helped explain the different productive behaviours presented in Table 15.1.

Modigliani (1980) defines, in a neoclassical framework, the concept of an individual's life cycle in its simplest form. This approach is based on an intertemporal utility function ordering the various 'consumptions' by the individual during his lifetime according to his preferences. The reference individual, for our purposes, is the head of the family. Given his intertemporal preferences, the head of the family adjusts savings and indebtedness according to his income opportunities. These latter depend on his capacity for work and capital. This should allow him to tailor his resources in the most efficient manner to the desired consumption level at each phase of his life cycle.

The Modigliani (1966) life cycle is illustrated in a very simplified manner in Fig. 15.1. An individual is termed as an economic agent when his income from his work reaches a level Y and enables him to meet his needs C. This income Y(T) is constant over a period T up to the age N and his consumption C(T) is constant until his death at age L, which is greater than N. There is no uncertainty about or any transmission of his wealth. In order to achieve his goal, the individual must put money away (savings) during his active working life (T between 0 and N) to maintain his consumption level during his inactive period by drawing on his accumulated wealth (decapitalization). His wealth A(T), which was initially zero, increases until the age N when it peaks, and then returns to 0 at age L. This pattern can be adapted to different consumption rates and income opportunities.

To take the reality of Malaysian family life into account, we have to understand that this desired consumption level varies over an individual's life. It is defined here

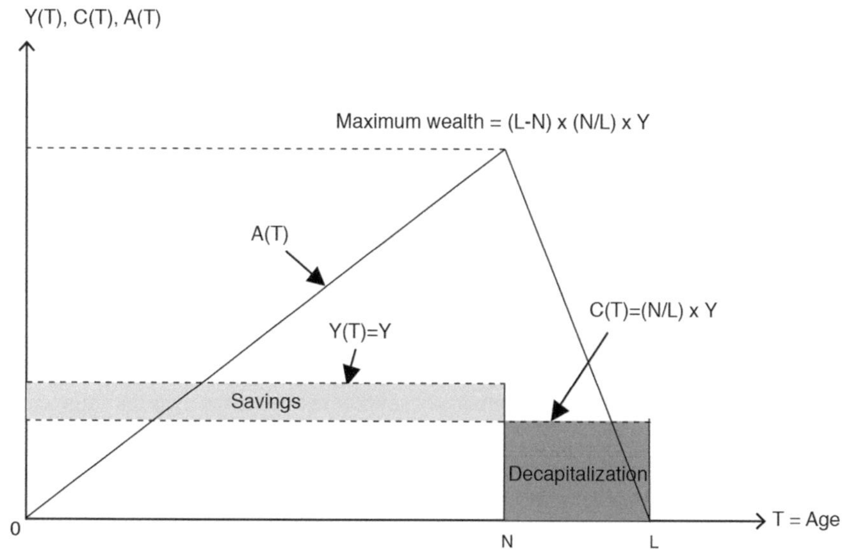

Fig. 15.1 The farmer life-cycle hypothesis and the requirement for capital (*source* Modigliani 1966)

by the stage of development of the family. This concept is adapted from Chayanov (1966). When the basic unit is the nuclear family, it is subject to an internal biological determinism. This determinism is expressed by the number of children—which initially increases and then decreases with the age of the family head—who he must feed and educate. Such food and educational needs are called incompressible needs; the head of the family can predict these needs over the long term. If the nuclear family is also the unit of production, the amount and nature of work done by the head of the family also depends on his age. Similarly, they constitute an internal biological determinism which raises issues of the subsistence of elderly people, whose ability to work shrinks faster than their incompressible needs.

We can thus categorize villagers into different groups, depending on their current life-cycle phase. This categorization of farmers in relation to age necessitates a prior differentiation between farms. In fact, it is important to distinguish between farms whose main source of income and accumulation is (or was) based on family labour and larger farms where labour, on the whole, is in the form of salaried employment. The life-cycle approach cannot be used in the latter case. These farmers are then considered entrepreneurs and are classified as type IV (Table 15.1).

15.2.4 The Life Cycle of the Coconut Farmer

The life cycle of the farmer is characterized by the changing needs of his family, his capacity for work and his wealth in relation to his age. As Table 15.1 shows, the life cycle is divided into three phases (I, II and III). If we group rural families, at a given time, on the basis of the life phase of the head of the family, we get a representation of the social organization in the village based on coconut cultivation. Exchanges of land, money, work and products find their usefulness in the light of the main objectives of the head of the family at each phase of his life cycle, and in accordance with the perception he has of his needs during his lifetime.

The incompressible needs of a young head of a family (phase III) are low and he has not yet bought or inherited coconut plantations. In these circumstances, his only source of income is the wage he receives as a casual worker during harvest and post-harvest periods in coconut plantations owned by older (phase I) or wealthier (phase IV) farmers. A major portion of this income can then be saved with the goal of eventually buying a coconut plantation.

Profits arising from the ownership of a coconut plantation as well as his wages will allow a young head of a family to meet his projected increase in needs. This will mainly concern expenses related to food and children's schooling in the next phase of his life cycle (middle age, phase II). In addition, the current coconut plantation, as well as any that could be acquired later (purchase, inheritance) will constitute the capital that will generate profits and provide him with his only source of income when he becomes too old (phase I) to do the kind of work on his coconut plantation(s).

Middle-aged heads of families (phase II) will continue to save, as far as possible, to develop their retirement capital: the coconut plantation.

For the elderly heads of families (phase I) with no longer any dependents, the income from their capital of the coconut plantation could generate a surplus that would allow them to make a pilgrimage to Mecca. In the worst case, they could dissolve their capital by selling the coconut plantation and spend their old age in the care of their children. It was noticed that village plantations did not exceed 5 ha, while those of entrepreneurs (phase IV) could be as big as 20 ha.

15.2.5 The Predominance of the Life Cycle in the Social Structure

The coconut harvest is sold to traders. This can be in the form of nuts immediately after they are plucked, or in the form of copra. For harvesting or post-harvesting work, casual workers are paid either in kind (a portion of the harvest, usually half of it, is given to the workers) or calculated on the basis of a fixed rate per nut for each type of work. Finally, agreements for short-term loans link coconut plantation owners to traders. Coconut harvests, once about every 2 months, help repay loans.

Chinese or Malay traders frequently own land. They constitute the biggest percentage of villagers who we classify as entrepreneurs. On the one hand, the size and diversity of the activities in their plantations is such that they have to rely primarily on paid labour; on the other, they have sufficient capital which can be rapidly moved and deployed to take advantage of the most profitable investment opportunities. Their farms are generally larger than those of farmers. However, two factors have prevented land concentration:

- The establishment of Malaysian reserves, a legacy of the British times, with a prohibition on the sale of certain lands to the Chinese;
- The richest Malay Muslims often became polygamists. Consequently, a coconut plantation established during the life of the head of the family was divided among his many children, in accordance with traditional rules of transmission.

In consequence, these entrepreneurs do not play a major role in the social control of land. Coconut plantations therefore represent well the fundamental connection between the life cycle of the farmers and the social organization of the village.

15.2.6 The Role of Cocoa Cultivation in the Restoration of the Social Fabric

It is important to distinguish between the role of cocoa cultivation in the village economy and various external factors that led to its introduction. The interpretation proposed here considers the introduction of cocoa (between 1970 and 1980) as a method of re-establishing a specific social functioning of villages, which had been based on coconut plantations. In the late 1960s, the conditions for establishing social equilibrium, solely through coconut plantations, were no longer being met. They were:

- the purchasing power of copra had to increase in order to compensate for the decrease in farm sizes;
- the new generation had to ensure availability of labour required to harvest coconuts;
- there had to be some security in supply and marketing.

This last point did not seem to be a problem, but renewal of the labour force implied that coconut plantations must offer the same benefits as other economic sectors, in terms of wages, access to credit and pension systems. This was a particularly important consideration for young men who did not yet own land and therefore constituted the most mobile labour force.

However, the low price of copra between 1965 and 1975 led to a failure to meet the growing needs of the entire population dependent on coconut cultivation. Faced with this situation, young people took up jobs in other sectors of the booming economy. This exodus of professionals threatened to unravel the social organisation described above.

Before explaining how the various categories of villagers reacted to this crisis in the coconut economy, we should present the factors that led one of the villages to adopt cocoa extensively in 1975 and thus restore stability, in terms of the aforementioned scenario, to the shaky social organisation in the village. In the other village, where the economic importance of coconut was waning gradually, a very small number of farmers adopted cocoa cultivation before 1985.

Some factors were common to both villages: price of the products, production factors and the national government's agricultural policy. On the one hand, the price of cocoa applied to everyone—a few geographic-based variations aside—and the particularly high prices in 1976–1977 helped launch the growth of cocoa cultivation in Malaysia. On the other hand, the 1969 race riots highlighted the marginalization of Malaysian rural society within a growth model retained from colonial times. The government had to, henceforth, take into account this section of the population whose demographic importance continued growing. Small coconut farmers, who represented one of the poorest social groups in Malaysia, soon benefitted from numerous projects to improve their means of production.

From 1972–1973, the research and development entities which were set up for this purpose advocated planting cocoa in the understory of coconut trees as the solution that best met the needs of farmers. They did so on the basis of the following qualities:

- cocoa could start producing quickly, from the 3rd or 4th year;
- cocoa could be harvested every week so it would bring in a regular income;
- cocoa farming required much more work than coconut plantations, and was considered suitable to keep family labour engaged (considered to be under-employed) and check the exodus from rural areas;
- under the aegis of development policies, cocoa saplings and sufficient inputs could be provided to farmers for 3 years prior to the crop's entry into production;
- the socio-economic data at the local level varied widely from village to village.

Only a road separates Sungai Batang village from large commercial plantations which had been cultivating cocoa in the understory of coconut trees since the 1960s. This proximity to commercial plantations not only helped organize the labour and produce markets, but also played a significant role in disseminating techniques and information about cultivating cocoa under coconut trees and other know-how pertaining to cocoa cultivation. The rural labour market was based mainly on coconut-harvesting work. The remuneration for casual labour kept pace with that of workers in large plantations near the village, which was negotiated between labour unions and company management at the national level. Indeed, remuneration in the village tended to be higher to prevent an exodus of village labour to the large plantations. Due to a lack of availability of labour and difficulties faced by owners to recruit casual labourers, labour wages varied little, unlike in large plantations, and barely tracked the fluctuations in copra prices. The lack of industries near the village forced villagers to move far away in search of work, thus affecting the renewal of village labour. This exodus weakened intergenerational relationships and risked depriving elderly farmers of their children's support.

Parit Haji Abdul Salam village is part of the Batu Pahat district, which is an industrial area. The professional exodus, in this case, did not require people to move far away and thus did not lead to an isolation of the elderly.

As far as the opportunities to plant cocoa trees were concerned, the response of these two villages depended largely on local characteristics. We will study, in each village, the behaviour of individuals in different age groups with respect to the opportunities offered by the local environment, and how each of these behaviours is linked to the others.

15.2.7 Cocoa, a Response to the Coconut Crisis in Sungai Batang

15.2.7.1 Establishment of Cocoa Cultivation

High cocoa prices in 1976–1977 resulted in a massive increase in the cultivated area of cocoa in villages. The presence of large plantations and the existence of a marketing network also supported this development. More importantly, cocoa helped meet the immediate need of retaining young workers in the village in order to maintain sufficient labour for coconut plantations.

The introduction of cocoa helped retain young people in the village by establishing cocoa-specific farm tenures. A farmer who owned a coconut plantation and was burdened with high familial needs usually planted saplings himself. He added cocoa to his plantation capital, and accorded it as much importance as coconut. This allowed him to increase his total income from the plantation. The cocoa plantation also helped him diversify his tree capital.

Farmers with little or no requirements allowed their son or sons-in-law to plant cocoa in their coconut plantation. The son grew cocoa and maintained his father's coconut trees in the process. The cocoa harvest belonged to him, while the coconuts still belonged to his father. The cocoa trees then represented a capital for the young farmer. The arrangement between father and son was not bound by time, and was valid even after the death of the father and during the division of assets between his children. The same land thus had two plantations that were managed by two different people.

This unusual mode of tenure of cocoa plantations helped retain the young labour force in the village. In fact, for the young, cocoa was a source of income and a heritage. Besides, it offered the prospect of accumulation which had seemed unlikely until then due to the partitioning of the land.

The key role of coconut plantations in the life cycle was thus restored. Even though this free transfer of the right to use land from the elders to the young amounted to a sacrifice, it was necessary for social cohesion in the village.

15.2.7.2 Heritage Status of Cocoa and Coconut Plantations

The advantage of the concept of inheritance, as distinguished from those of capital and savings, lies in the modalities of acquisition and transmission of assets which are considered as the farmer's heritage. Thus, while the coconut plantation, inseparable as it is from the land on which it grows, is bequeathed to all children equally, with the occasional difference between boys and girls, the right to use the land to grow cocoa is accorded only to children who remain with the father.

We documented an instance of the transmission of a cocoa plantation. The father, who worked on the land since 1976, passed it on to his 20-year-old son. A part of this cocoa plantation was held by the father under a tenant farming model; this too was transmitted. Thus, a cocoa plantation is clearly distinct from a coconut plantation, as much in its functioning as in its inheritance characteristics.

The fairness or, at least, the equality of such sharing amongst the children is thus challenged. Should this cocoa innovation be related to changes that occurred in the renewing of the village population, when only one or two children per family—which has six children on an average—now stay in the village?

15.2.7.3 The Increase of Areas Under Cocoa Cultivation in 1980

Whenever possible, farmers first planted cocoa on their own plots or on those of their parents. Thereafter, their aim was to plant cocoa on other favourable plots. This was facilitated by the emergence of sharecropping.

15.2.7.4 Types of Sharecropping

Sharecropping constitutes an opportunity for a landless farmer or one owning very little land—and with little or no prospects of inheriting a father's plantation–, to plant cocoa. Sharecropping is an arrangement between a villager who wishes to plant cocoa and the coconut plantation owner who has emigrated, has another occupation or is just too old. Normally, the landowner gets one-third of the harvest of fresh beans. This quantity may increase to two-thirds of the crop, especially when the owner himself helps in the planting.

The areas under cocoa thus increased considerably and incomes increased without a proportional increase in labour. It was a matter of taking advantage of high prices and of subsidies available for inputs.

It is mainly the young farmers who enter into sharecropping agreements: their family needs are still low and they have surplus labour time that can be used to plant and maintain young cocoa saplings. Since sharecropping contracts are generally for a duration of 25 years, it is mainly the young farmers who are interested.

In contrast, older farmers have higher incompressible needs. Given that they farm their own cocoa—which are of small size–, they have no other alternatives than to intensify labour and inputs to benefit from high cocoa prices. Inputs for the cocoa plantation are then bought with the revenue generated from cocoa.

Figure 15.2, created using data collected from farmers in 1989, shows the structure of farms as a function of the age of the farmers. In conformity with Fig. 15.1, which summarized Modigliani's life-cycle model, the polynomial function was adjusted to the area of the coconut plantation owned, based on age. It

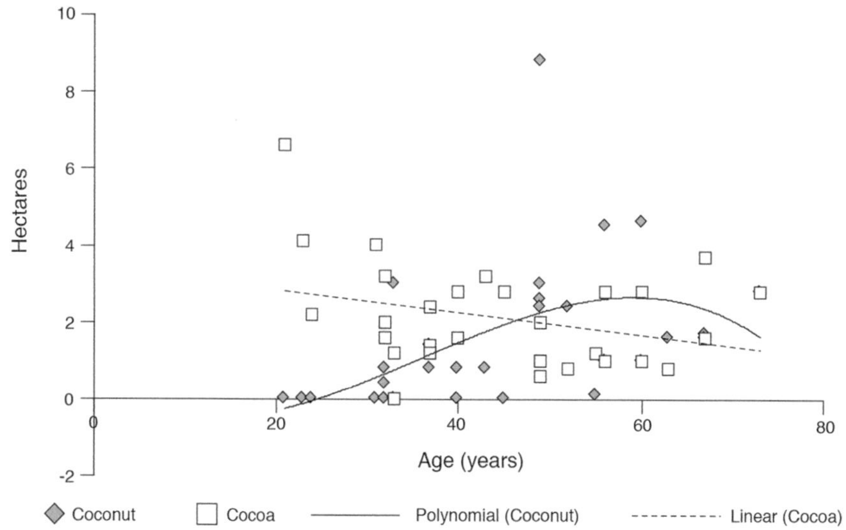

Fig. 15.2 Coconut-farm assets and the area under cocoa: polynomial function adjusted to the area of self-owned coconut farm (*sources* authors' surveys 1989)

clearly shows a tendency of an increase in the coconut plantation asset until about the age of about 60, after which the value of this asset declined.

In contrast, the area under cocoa decreased linearly with age: the additional income from cocoa mainly went to the young people, who divide their time between cocoa and coconut cultivation. Cocoa plantations helped retain the youth in the village and, to some extent, also helped maintain the labour force necessary to maintain the coconut plantations of the elderly.

Figure 15.3, created with data collected from the same farmers, illustrates the diversity of land tenure types that were used for cocoa cultivation (Table 15.2).

15.2.7.5 Technical Performance of the Cocoa System Under Coconut

Yields recorded in village farms were very low compared to those in neighbouring large plantations. In fact, the Bleinheim-Estate plantation recorded a dry beans yield of 1.5 tonnes/ha/year in a non-grafted orchard that was 20 years old. The intensive cultivation method allowed a steady increase in yields despite the age of the cocoa.

A survey published by Shaaban Bin Sahar, Malaysian socio-economist, showed that yields in village farms were between 170 and 320 kg/ha/year of dry beans for orchards ranging in age from 4 to 8 years in the same district. A small number of farmers achieved a yield of 600 kg/ha/year of dry beans.

In 1989, estimates obtained from our surveys confirmed these figures for orchards that were 10–15 years old. Farmers who were interviewed affirmed that the maximum yield was achieved between the 6th and 8th year after planting, after which the yield decreased gradually. These low yields can be explained by the

Table 15.2 Structure of orchards and land tenure types of cocoa plantations in Sungai Batang in 1989

Individuals	Age	Area of self-owned cocoa plantations (1 acre = 0.405 ha)	Area under cocoa plantations (acres)	Owner farmed	Tenant farmed	Sharecropping	Farming for free
1	21	0	16.5				16.5
2	23	0	10.25		4		6.25
3	24	0	5.5		5.5		
5	31	0	10		5		5
6	32	0	5		3		2
7	32	2	4	2			2
8	32	1	8	1		7	
9	33	7.5	3	7.5	1	3.75	5.75
10	33	0	0				
11	37	3.5	3.5	3.5			
12	37	2	3	2	1		
13	37	3.5	6	3.5			2.5
14	40	0	7				7
15	40	2	4				4
16	43	2	8			5	3
18	45	0	7			7	
19	49	22	5	5			
21	49	7.5	1.5	1.5			
22	52	6	2	2			
23	55	0.25	3	0.25			2.75
24	56	2.5	2.5	2.5			
25	56	11.25	7	7			
26	60	2.5	2.5	2.5			

(continued)

Table 15.2 (continued)

Individuals	Age	Area of self-owned cocoa plantations (1 acre = 0.405 ha)	Area under cocoa plantations (acres)	Owner farmed	Tenant farmed	Sharecropping	Farming for free
27	60	11.5	7	7			
29	63	4	2	2			
30	67	4.25	9.25	1.25		8	
31	67	4	4	4			
32	73	7	7	7			
33	49	6.5	2.5	2.5			
34	49	6	2.5	2.5			
Total			150.5	66.5	19.5	30.75	56.75
Average			5	3.3	3.25	6.15	5.15

source authors' surveys 1989

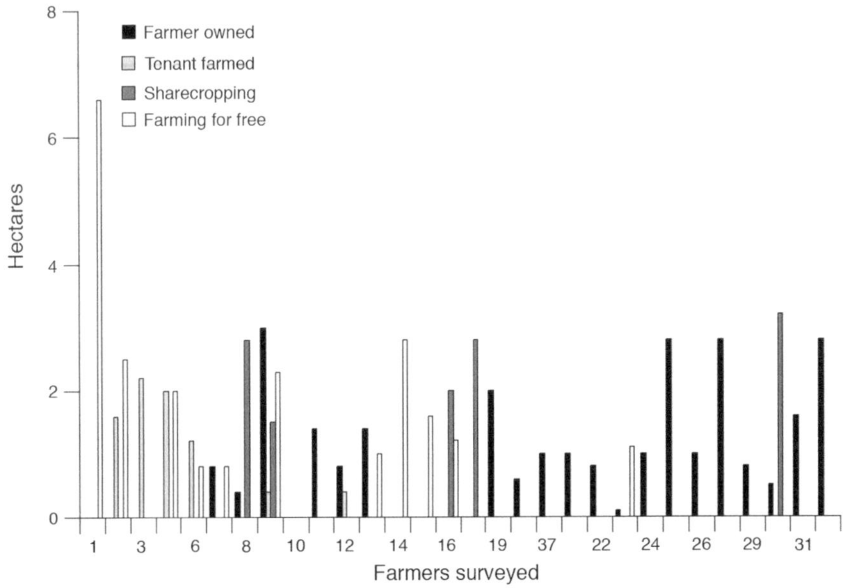

Fig. 15.3 Land tenure types for surface area under cocoa (*sources* authors' surveys 1989)

agronomic characteristics of cocoa grown in the understory of coconut trees in village plantations. Normally, two rows of cocoa are intercropped between two rows of coconut trees, with densities of 750 cocoa and 153 coconut trees per hectare. It should be noted, however, that the Malaysian extension authorities recommended a density of 133 trees/ha for coconut.

The introduction of cocoa also brought in a more widespread use of herbicides that reduced weeding work in plots, which was, in any case, already reduced because the shade of cocoa trees inhibited the growth of grass and weeds. Lime was widely used to counter the effects of flooding on cocoa trees. The use of fertilizer, on the other hand, must be studied more carefully. According to reports published in the 1980s and our surveys conducted in 1989, most farmers stopped using fertilizer on their plantations when subsidies were withdrawn. Their expenditure on inputs was less than MYR 200/ha (USD 70/ha).

Although some entrepreneurial and middle-aged farmers had a relatively small cocoa area measuring about 1 ha each, their family needs were high. They used fertilizer, and sometimes insecticides too. Their expenditure on inputs amounted to about MYR 500/ha (USD 185/ha). Their yields, more than 400 kg/ha/year of dry beans, were high compared to other farmers, but they were still much smaller than those of large plantations.

Researchers and developers suggest that the yield was low because, despite intensification, there was too much shade on the cocoa trees and inadequate drainage in the plantations. These parameters limited the effectiveness of intensified use of inputs and labour.

15.2.7.6 Effects of Cocoa-Coconut Intercropping

The maintenance of cocoa trees cannot be separated from that of coconut trees. It seems that the increase in coconut yields is due to the use of subsidized fertilizer meant for cocoa trees. The harvesting of coconuts, on the other hand, has become more difficult as cocoa trees hamper movement, and care must be taken not to break their branches when coconuts are being plucked. Plantation owner-farmers observed that an intensification of cocoa increased the coconut rent.

Farmers regarded cocoa and coconut as two distinct assets. Research and development centres have suggested that cocoa yields could be improved through intensification, shade management and grafting on adult cocoa trees. They attributed no economic importance to coconut, stating that its primary purpose was to provide shade to cocoa trees. But what, in reality, is the place of the cultivation of cocoa and coconut in the family economy?

15.2.7.7 Cocoa in the Family Economy

The reality of farming families can be very different from the way the research and development entities perceive it to be. According to surveys conducted by Shaaban Bin Sahar, when the procurement price of cocoa was between MYR 4.5 and 5.6/kg of dry beans (USD 1.6–2.1/kg) in 1979–1980, cocoa accounted for roughly 50 % of the household income. In 1989, due to falling prices and stagnating yields, this contribution dropped to less than 25 % for most of the households surveyed. Income from cocoa then became secondary in the family economy.

The organization of the working day too was representative of the lower importance of cocoa as compared to coconut. Four to five hours of uninterrupted work was reserved for harvesting coconut in the morning by farmers of type II and type III. Work related to cocoa was taken up in the afternoon, from 14 to 17 h, between the two prayer times. Moreover, social activities (weddings and political meetings) or religious activities which took place in the afternoon often took priority over cocoa-cultivation activities. Cocoa cultivation was then regarded as an addition to the normal workday and did not necessarily form a part of it.

As far as income flow was concerned, harvesting work for coconuts provided a daily or almost-daily income. It allowed the meeting of daily expenses on food and children's schooling, which made up the family's incompressible needs. In this regard, we showed that the share of casual labour needed to harvest coconuts in the total working time tended to be especially high when irreducible needs were high and the assets of the family head were small (Dupraz 1989).

Income from cocoa came in for only 4–6 months in a year and it was also subject to climatic variations stemming from quasi-annual floods and to price fluctuations. This income was then added to the daily wages, and sometimes even substituted it when cocoa prices were at their peak.

This income often allowed farmers to meet new expenses which arose following electrification in 1970. Contrary to the expectation of agricultural research and development institutions, cocoa only had secondary importance in the family economy of farmers. It represented a diversification of the tree capital whose main role was to restore the economic and social role of coconut cultivation in the village. There is thus a disconnect between research efforts to improve cocoa yields and the perception of farmers.

15.2.8 Increasing Disconnect Between Farmers and Research and Development Institutions

Government entities were initially created to develop coconut-associated cultivation systems with a goal of eradicating rural poverty. We take a look at how well they have reoriented themselves around cocoa cultivation in terms of production. In the laboratory and field station, the technical and economic performance of cocoa relegates coconut to second place. The breeding programme for cocoa through grafting on adult trees of selected clones, set up by the Malaysian Agricultural Research and Development Institute (Mardi) in 1985, is an illustration of this development. It was expected that this method would lead to the breeding of high-yield clones with larger-sized cocoa beans.

Mardi set up pilot facilities to supply cocoa grafts, mainly in Sungai Batang. An extension agent was appointed for a 2 year period to distribute the grafts and the associated technical package: fertilization, felling of old trees and optimizing shade by removing some coconut trees. The non-compliance with fertilization recommendations affected the development of the clones and the fruits failed to mature. Furthermore, shading had to be maintained within bounds for fertilization to be effective. The farmer received free all the necessary inputs only when all the trees are grafted. This strict condition partly affected the development and dissemination of grafting. In fact, the survey suggested that the results of this endeavour by Mardi were, on the whole, unsatisfactory. Few farmers (<6 %) grafted all the trees on their plots. The identified causes of failure were mainly moisture or drought which ruined the graft and, consequently, delayed fruiting (waiting period >2 years).

In fact, the technical package could not be fully applied by farmers who usually did not use fertilizer routinely for their cocoa (extensive farming practice), or who did not own the coconut trees. Moreover, the not-so-promising results of the first grafts quickly discouraged farmers.

Thus, the existence of distinct property laws for cocoa and coconut plantations and the extensive farming practice for cocoa were factors Malaysian research did not take into account. Its output-oriented perspective was contrary to the farmers' plantation management practices.

It also appeared that farmers, for the most part, were not inclined to specialize in cocoa cultivation. They did not believe that intensification would lead to the promised increase in yield. The move towards diversification of agricultural activities appeared to be more pronounced in villages.

15.2.9 The Diversification of Activities: A Response to a Decline in Cocoa Prices

The most important diversification activities consisted of the initial processing of beans, livestock rearing, cultivation of bananas and various fishing activities. Their essential feature was not to disturb the usefulness of cocoa as a tree capital, which was true for most farms. These activities resulted from opportunities that the environment provided and became more attractive as cocoa prices declined. Type II and III farmers resorted to a diversification of their activities to compensate for the loss of purchasing power resulting from lower cocoa prices.

Based on the economic performance and the short- and long-term needs of families, these activities were an addition to, or a partial substitution of, the work period earlier devoted to cocoa cultivation: the afternoon. The morning remained reserved for coconut harvesting activities.

15.2.9.1 Initial Processing of Cocoa

The initial processing of cocoa consists of converting fresh beans to dry beans using fermentation and drying. The sale of dried beans increases income by 50 % compared to fresh beans. Initial processing requires little equipment and can easily be done by the farmer, making it a good way for him to offset falling prices. The Federal Agriculture Marketing Authority started actively disseminating and popularizing fermentation techniques in 1980. Although very few producers dried beans before selling them in the early 1980s, a majority now sells dried beans.

15.2.9.2 Pasture-Fattening of Livestock

Livestock activities could be increased through the establishment of grazing areas between the village and the sea. The government has allocated such areas to poor families since 1970 within the framework of the poverty eradication policy. Coconut plantations came up here as coconut was the only species that could tolerate this brackish soil. These new lands led to an increase in the demand for casual labour for harvesting coconuts. Although these lands were not fit for intercropping, they could be used as extensive grazing areas. The drop in cocoa prices reduced the wages for labour for cocoa cultivation to a level comparable to those paid for livestock rearing. Consequently, type I and type II farmers did not stand to lose much by according a lower priority to cocoa, which they continued to harvest in order to sustain this other activity.

The economic parameters of livestock breeding are diverse and involved all types of farmers: livestock is a capital and represents the piggy bank that is broken for one-time big expenditures such as buying a motorcycle or a house, or paying for a child's university admission. Livestock rearing can generate wealth rapidly. If a livestock breeder is in a position to postpone the regular income from livestock, then by selling three heads of cattle, he can buy back five. The herd then increases geometrically in relation to the number of livestock rotations. The rotation period ranges from 12 to 18 months.

This aspect explains the attraction livestock rearing has for the youth (type III). It is possible to start this activity without capital, by first rearing cows as a sharecropper. Coconut plantation pastures are available for use to livestock breeders in exchange for the maintenance of the plots, which is, in any case, essential for livestock farming. For farmers—mainly type-I older farmers—who give out their cattle through a sharecropping model, livestock remains, despite the inherent risks involved, a high-returns investment and one that does not conflict with Islamic injunctions on financial interest.

15.2.9.3 Banana Cultivation

Banana cultivation is characterized by low labour requirements and a fortnightly income. Bananas are currently associated with cocoa where, for example, they serve to replace fallen cocoa trees.

The decline in cocoa prices resulted in lower land rents for coconut plantations. In the future, banana plantations may even come up in old abandoned cocoa plots. This is already the case with some farmers who have large cocoa plantations, but no longer have sufficient labour to replant cocoa and maintain new plots. They take advantage of the slightest demand for land and lease it out for banana plantations. Although the per-hectare income from banana cultivation is lower than that of cocoa, it fetches them better returns per labour hour since the labour requirement is low.

15.2.9.4 Fishing

Shrimp fishing is done at night in village drains, and the catch is either home consumed or sold. It can offset the effects of a decline in cocoa prices on the standard of living, even though the total labour time is increased. After their busy work day, type-II farmers, who also represent the most disadvantaged lot, sometimes spend a large part of the night catching shrimp.

In contrast, clam fishing is a seasonal activity that coincides with the peaking of cocoa production, thus directly impacting the harvest of cocoa pods. It is mostly type-III farmers who are involved in clam fishing.

In addition to new coconut plantations that gradually enter into production, all these activities are extensification factors for cocoa cultivation. They compete with cocoa in terms of the labour time and the intermediary inputs devoted to them. Cocoa, which is a productive asset, may increasingly become a sleeping asset.

The access farmers have to diversification activities varies; it depends on their resources and geographical location within the village. Thus, some have no alternative but to increase the number of hours spent as casual labour for harvesting coconuts, and to ask their wives to help out with the cocoa cultivation. Farmers make the best they can of the flexibility of a non-specialised production system to deal with climatic and economic risks.

15.2.9.5 Agricultural Exodus and Cocoa Cultivation in Parit Haji Abdul Salam

The end of the 1960s and the 1970s witnessed a shift in occupation at Parit Haji Abdul Salam of most type-II and type-III labourers. They moved from being casual workers for coconut harvesting to being salaried employees in the tertiary and secondary sectors. The few remaining workers diversified their activities, by constructing their houses, for example. This diversification was the result of the local mode of payment (half of the harvest), which helped coconut labourers manage the effects of price fluctuations. There was no longer enough labour available to meet the harvesting requirements of all the coconut plantations, and the workers opportunely chose to work in the largest plantations.

Thus, the older type-I farmers had no alternative but to undertake a part of the harvest and process the nuts by themselves. They ultimately became dependent on their children to fulfil most of their needs. A few type-II farmers continued growing cocoa after 1974 with government aid for revival of farms. They replicated the phenomenon observed at Sungai Batang.

15.2.9.6 Various Opportunities of Diversifying from Coconut Cultivation

The three types of agricultural diversification we have examined are: coconut intercropped with cocoa, with coffee, and the substitution of coconut with oil palm. These three cropping systems all begin producing from the fourth year after planting. Within the framework of a cocoa project, cocoa saplings, inputs and processing equipment were offered to farmers at a 50 % subsidized rate starting in 1985. In addition, a collective extension, processing and marketing programme was also implemented.

The economic characteristics of unsubsidized cocoa and coffee cultivation are similar: high per-hectare requirement of labour. According to the farmers, a full-time labourer can handle a plot of 1.5–2 ha where intensive methods, including chemical weeding, are practiced. This work is, however, physically possible for everyone, including women, children and the elderly. The per-hour labour wage is low compared to that obtained in oil palm cultivation.

The gross margin per hectare of cocoa and coffee crops grown under coconut trees is also significantly lower than that for oil palm. This stems from a partial or non-harvest of coconuts and a drop in cocoa prices. Although coffee harvesting is a more delicate operation than that of cocoa, it can tolerate soils that are more acidic and hydromorphic. Coffee is more resistant to defoliating caterpillars, a known pest in the region, and tends to offset the advantages of cocoa cultivated with the help of subsidized chemical inputs.

The most remunerative crop per hectare, in terms of money and, even more, labour time, is oil palm. Even though the investment required for setting up an oil palm plantation is very high, it is possible to start recovering it from the fourth year if the farmer himself provides the labour. Although palm trees can be maintained easily, harvesting can, however, only be done by younger men.

15.2.10 Behaviours of Different Types of Farmers

A minority of land owners consists of current or former salaried employees who are more than 50 years old. Most of them inherited a small coconut plantation. Having a salary, pension or savings when they quit their job, or sometimes even help from their children, their plantations are not essential to their livelihoods. Despite their low profitability, cocoa and coffee are the only crops that these farmers can, or will be able to, physically maintain and harvest by themselves, once they retire.

In 1985, subsidies and relatively high cocoa prices impelled these farmers to extend the cultivated areas under this crop onto the most marginal soils. This explains why more than half (5/8) of the rural coconut plantations are officially intercropped with cocoa. Some of these cocoa trees, however, will never enter into production.

It is actually on these marginal soils that some farmers have converted part of their coconut plantations, before or after planting cocoa, into oil palm plantations. In this way, they have truly diversified their plantations. Such conversions translate strategies of mitigation of risks arising from wide price fluctuations of agricultural products and the scarcity of casual labour to harvest palm bunches.

This last point explains why the farmers who were the quickest to implement this partial conversion were actually those who had teenage or adult sons still living at home. A few farmers without a young labour force at hand have already converted their unproductive cocoa plantations to coffee. The other farmers own only a small fraction of the cultivated area.

Young salaried workers who own land also plant oil palm. Although such a worker's permanent job only allows him to grow a crop that demands a minimum of labour, it gives him access to credit required to plant oil palm, which is difficult to obtain for other crops. Some even take the money lent by their employer to buy their parents' plot. These latter then either invest the money or use it directly to sustain themselves in their old age.

The few casual workers who opted to remain in agriculture were often the most disadvantaged ones, both in terms of access to land and access to credit. Their main objective, even as they get older, is to maximize income from their labour, even if means having to put in painful hours of labour.

When the cocoa project was launched, they entered into 8 year contracts which allowed them to plant and cultivate cocoa plantations belonging to non-resident or busy land owners who wanted to create a retirement nest egg. The ownership of the cocoa trees reverted to the owner at the end of the contractual period.

Following a decline in cocoa prices, these farmers preferred to be sharecroppers on oil palm plantations. They so preferred to cultivate this crop on their own lands that they even took on loans at usurious rates.

Cropping systems adopted by most farmers are intensive in chemical inputs and labour. This is due to several factors:

- cocoa has replaced coconut as retirement capital;
- no heritage capital status forbids the regulation of shading for cocoa, usually undertaken through the poisoning of the extra coconut trees;
- many farmers do not depend on cocoa for their survival. They do not mind a small profit or even none.

The decline in cocoa prices only affects those farmers whose incompressible needs have to be covered by cocoa income. The result is a reduction in spending on inputs. This saving is often achieved by substituting chemical weeding by manual weeding in plantations with a sufficient availability of family labour for the area concerned.

Under current conditions, given the young age of cocoa plantations (3–5 years) and an elderly labour force obliged to cultivate cocoa, the production of cocoa can only increase when the plantations become mature. Cocoa cultivation, intensive for the most part, has been established and maintained with the help of external funds.

15.3 Conclusion

Malaysia's cocoa boom-and-bust cycle was one of the fastest in history. At its peak, cocoa production exceeded 200,000 tonnes, catapulting Malaysia to the 4th or 5th position of global cocoa producers. Today, however, Malaysia's contribution to world cocoa production is marginal, with an annual output of 25,000 tonnes in the 2000s. The question, however, is whether the cycle of cocoa diversification was truly without value for the country and for family farming?

This short history of plantations led to significant investments in the processing industry (butter and powder) at national level. This is an asset for Malaysia as well as for south-east Asia, and an excellent outlet for the Indonesian cocoa sector that has developed in the meantime.

In 2000, family farms in Malaysia produced almost all of the 25,000 tonnes of the country's cocoa production. Large plantations had, in fact, already switched to oil palm or fruit trees. Cocoa cultivation played an important role at a particular time in the country's history: it created jobs for young people and helped preserve a social structure, at least for a while. It also helped reduce rural poverty in a country undergoing rapid urbanization. Just as in the case of Côte d'Ivoire and Ghana (Chaps. 2 and 8), the Malaysian experience introduced the important concept of retirement capital: cocoa (or some other perennial crops, such as rubber) can represent a diversification of activities and incomes for salaried people in urban areas who are planning for their retirement in their villages.

The basic hypothesis of the book is reflected in the determinants of this diversification cycle, in conjunction with prices: the essential role of ecological change; rents of the natural environment and their exhaustion; life cycles of trees and of families.

Thus, cocoa was introduced on the west coast when coconut orchards were in full decline. Plantations were owned by aging farmers who wanted to convince their sons to stay with them. The coconut cycle belonged to the parents. The diversification through a new cocoa cycle, however, pertained to the new generation. This new diversification encompassed cocoa, banana and even fishing, and occurred when cocoa trees were aging and prices of the crop falling. The newest developments in the 2000s seemed to favour oil palm cultivation. This is, once again, a new diversification and a new cycle, which is helping draw value from degraded lands and underpinning the aspirations of a new generation.

References

Chayanov AV (1966) The theory of peasant economy. In: Thorner D, Kerblay B, Smith REF (eds) Irwin, Homewood, IL

Dupraz P (1989) Étude économique des comportements des petits planteurs de cacao sous cocotier en Malaisie péninsulaire à partir de l'hypothèse de cycles de vie. Diplom Thesis in Agronomy. Ensam de Montpellier—DSA, Cirad

Modigliani F (1966) The life cycle hypothesis of saving, the demand for wealth and the supply of capital. Soc Res 33(2):160–217

Modigliani F (1980) The life cycle hypothesis of saving. Andrew Abel, Mit Press, Cambridge, 527 p

Ruf F (1987) Éléments pour une théorie sur l'agriculture des régions tropicales humides. I—De la forêt, rente différentielle au cacaoyer, capital travail. Agronomie tropicale 42(3):218–232

Chapter 16
Diversification and Perennial-Crop Cycles in Aceh, Indonesia

Florie Paul, François Ruf and Yoddang

Aceh province, spread over an area of 55,392 km^2, is located on the northern-most point of the island of Sumatra and therefore at the top of the Indonesian archipelago. The history and politics of Aceh reflect this geographical marginality. The central government and the Suharto regime appropriated the revenue from large gas reserves discovered in 1971 at Arun. This prompted the people of Aceh to revolt in 1976 and found the Free Aceh Movement (GAM[1]) to push for political independence and reclaim their right to gas revenues. Several years of violent clashes between GAM and the Indonesian army followed. This situation often made it extremely difficult for the villagers to cultivate their farms. GAM and the government finally signed a peace accord in August 2005.

Aceh province is also known for the catastrophic tsunami of December 2004. While it resulted in a massive loss of human life and damage to property in the province, it also served as a catalyst to the resolution of the conflict (Hyndman and Waizenegger 2010). Not only have the villagers and villages in Aceh seen trying times in the recent past, they have also suffered through a long period of isolation from the rest of Indonesia. Nevertheless, cocoa production has increased steadily in recent years. In spite of all the geographical and historical circumstances, visitors travelling on roads in the north of the province come across a typically Indonesian landscape on the plains of the Pidie and Pidie Jaya districts: vast and perfect rice

[1]GAM: *Gerakan Aceh Merdeka* or Free Aceh Movement, which fought for the independence of Aceh for almost 30 years.

F. Paul (✉)
13 Résidence de La Chatellenie, 62840 Fleubaix, France
e-mail: paulflorie@gmail.com

F. Ruf
UMR Innovation, Cirad-ES, TA C-85/15 (Bât. 15, Bur. 223), 73 Rue Jean-François Breton, 34 398 Montpellier Cedex 5, France
e-mail: francois.ruf@cirad.fr

Yoddang
Cirad, Jakarta, Indonesia
e-mail: cirad@idola.net.id

F. Ruf and G. Schroth (eds.), *Economics and Ecology of Diversification*,
DOI 10.1007/978-94-017-7294-5_16

protected areas (Leuser and Bukit Barisan national parks). Aceh Tenggara is a district that has attracted immigrants for a long time now. Indigenous people here are called Alas,[2] and immigrants are people of various ethnic origins (Batak,[3] Gayo,[4] Jawa,[5] etc.).

The district draws migrants because of its location (bordering North Sumatra province), well-known for soil fertility and land availability (low population pressure).

A qualitative approach has been adopted for the study, based on qualitative surveys with 90 farmers (30 in each of the three districts).

16.2 Results and Discussion

16.2.1 Diversification to Cocoa

Statistics confirm an increase in cocoa production in Indonesia from 17,000 tonnes in 2007 to 26,000 tonnes in 2009. The production of rubber and oil palm are also on the increase. Coffee production, which was thought to have stabilized between 100,000 and 110,000 tonnes over the past decade has, however, declined to less than 40,000 tonnes in 2009. It should be noted that these statistics are subject to several errors. For example, they do not take into account the transfer and marketing of production from one province to another. Nonetheless, these trends reveal a reality of a strong growth of cocoa cultivation, a small growth in rubber, a large increase in areas under oil palm cultivation and a decline in coffee. Granted that these three districts studied are not representative of all the crops but even if one crop such as the oil palm has originated from large private plantations, we will be able to discern these trends and attempt to explain them on the basis of diversification decisions taken by farmers.

Cocoa cultivation was introduced to Pidie and Pidie Jaya by a private plantation in the mid-1970 s and was adopted by smallholders in the 1980s. It was adopted in Aceh Tenggara a decade later, in the 1990s. Surveys conducted in 2008–2010 confirmed that cocoa was adopted in all the three districts, and thus indicate a dynamic of diversification towards cocoa in recent years (Figs. 16.2 and 16.3).

---

[2]Alas: original people from Aceh Tenggara (Alas ethnic group).

[3]Batak: person of the Batak ethnic group, a native of North Sumatra province.

[4]Gayo: person from Gayo Lues district, north of Aceh Tenggara in Aceh province.

[5]Jawa: person from the island of Java.

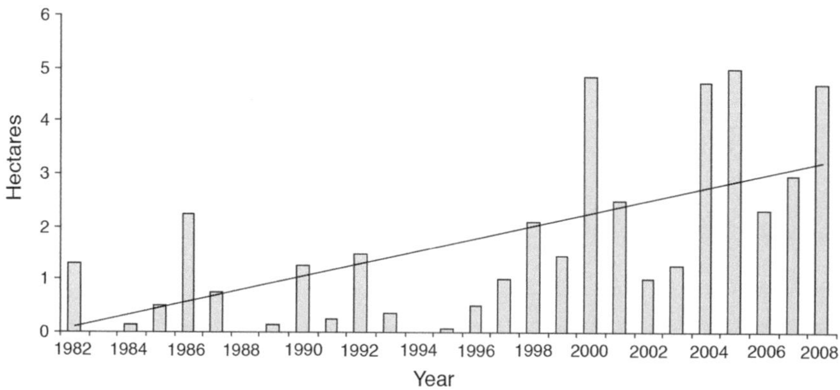

Fig. 16.2 Number of cocoa plantations established by villagers in Pidie district between 1982 and 2008. (*Source* Survey by authors 2008)

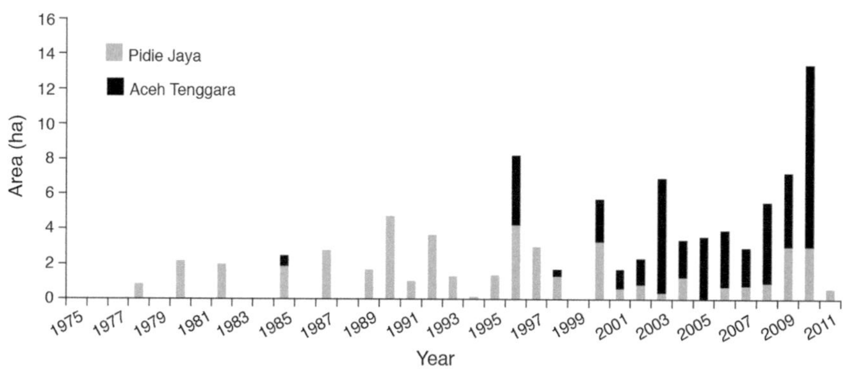

Fig. 16.3 Number of cocoa plantations established by villagers in Pidie Jaya and Aceh Tenggara districts between 1978 and 2010. (*Source* Survey by authors 2011)

16.2.1.1 Pidie and Pidie Jaya

The agricultural system present in these two districts was diversified at the farm and plot levels even before cocoa was introduced here. In the context of an armed conflict, families continued to cultivate rice and food crops (maize, soybean, groundnut, chilly and vegetables), which satisfied their food needs and also earned them a little income. Furthermore, more market-oriented perennial crops were intercropped in these plots: Areca palm (areca or betel), coconut, clove, nutmeg and various fruit trees.

Simultaneous Diversification at the Level of the Plot, Farm, Region

Starting in the 1980s, cocoa began to be introduced into existing cultivation systems: under Areca palms, coconut and fruit trees (durian, rambutan, etc.). Cocoa

Table 16.1 Crops preceding cocoa in Pidie Jaya

Crops preceding cocoa	1980–1985	1986–1990	1991–1995	1996–2000	2001–2005	2006–2010	Total
Perennial crops/insertion into an agroforestry system (%)	40	9	27	40	38	20	26
Forest (%)	60	83	55	53	38	60	59
Food crop plots (%)	–	–	18	7	12	7	9
Fallow with *Imperata* (%)	–	8	–	–	12	13	6
Total (%)	100	100	100	100	100	100	100

Source Survey by authors 2011

trees thus became integrated into a diverse agroforestry system at the plot level (Table 16.1).

At the same time and subsequently, cocoa trees were also planted after clearing of forest lands, away from roads, on the hills—this is the prevalent planting method. While young, cocoa trees can be intercropped with banana trees (the former gets shade and the farmer gets a marketable product) and other food crops such as cassava, vegetables and groundnuts. After 2 or 3 years, when the cocoa is grown, the system evolves towards plantations with a simple association with bananas, then, finally, to a monospecific one. In this case, diversification takes place no longer at the plot scale, but at the level of the farm or the landscape.

Farmers can also revert to a diversification at the plot level by reproducing agroforestry systems, by intercropping Areca palms with cocoa in the same year. Some agricultural experts on cocoa believe that an association between Areca palm and cocoa is not good for managing soil fertility. In fact, the Areca palm consumes a substantial amount of soil minerals without any significant recharge as compared to other species: few leaves fall to the ground, and those that do take time to degrade. However, the association seems effective in terms of light management as the two trees occupy different canopy layers: Areca palms grow rapidly to more than 20 m and provide a moderate shade for cocoa trees, which generally do not grow taller than 3 m. While benefitting from this complementarity, the farmers' strategy is to diversify market risks.

On the other hand, coffee is gradually being phased out and is being replaced by cocoa. Coffee, in fact, requires more work and occupies the same agroforestry canopy level as cocoa. Moreover, it is attacked by an insect. Likewise, cloves are also being phased out due to disease (Box 16.1).

Diversification into cocoa can thus take various forms and different types of cropping systems. It also reflects the influence of several determinants.

Box 16.1. Smallholder diversification itineraries
In the 1970s, Muhamad N. was primarily a clove and coffee farmer in the village of Ulle Gunung, Pidie district. Both were primary crops in the region, until they were attacked by pests and disease. Clove trees which provided the shade for coffee were the first to be affected, followed by the coffee plants themselves.

It was thanks to the planting material provided for free by extension services in 1990 that Muhamad N. grew cocoa, replacing clove and much of the coffee. The plot, however, remained an agroforestry system with Areca palms and fruit trees.

However, according to Muhamad N., tens of hectares of cocoa were being established by clearing the forest behind the mountains. Access remained difficult and visitors were not always welcome (Survey by authors 2008).

Role of Public Policy

The majority of those who adopted it early readily affirm that cocoa was introduced through a project of the extension services in 1982. In fact, that was the year when a national project was launched in several provinces to promote cocoa cultivation by handing out subsidies (cocoa planting material and grants for tools, sometimes even in cash). This project, generally speaking, had little impact, except in specific areas where smallholders had started adopting cocoa a few years earlier, mainly in Sulawesi (Ruf and Yoddang 2004). In the case of Aceh, with the start of hostilities between GAM and the Indonesian military in 1976, the central government and extension services viewed the promotion of cocoa cultivation as a means of consolidating their influence over the villagers.

Nevertheless, social pressure in the village was palpably obvious behind these pro-government interactions. During interviews, many farmers eventually recognized other factors in adopting cocoa, mainly the availability of information and planting material, which are major enablers of diversification.

Plantation Companies

The transfer of planting material from an industrial plantation to the village by villagers hired as labourers, with or without the knowledge of the plantation company, is a widespread factor in the adoption of a new crop, and thus of diversification. From the time coffee came to Sumatra, during the colonial period, to the arrival of cocoa in Central Sulawesi in the 1990s, we have often observed this type of unplanned diversification. It is induced by rural labourers who 'divert' some of the planting material for planting in their villages (Ruf and Yoddang 2000). Indeed, the private company PT Gotong Royong was one of the first to plant cocoa in Aceh in the 1970s (in the Ulim sub-district, Pidie). A section of the villagers working in the plantation were the first to try out cocoa cultivation in their own farms.

Individual Initiatives, Migrants, NGOs and the GAM Rebellion

The private plantation of PT Gotong Royong is probably one of the few large cocoa plantations in Aceh, even though the neighbouring province of North Sumatra had several. According to the surveys, the first person to have planted cocoa was a native of Medan, the capital of North Sumatra province. This Javanese farmer had observed cocoa cultivation and its profitability in the Medan area. When he moved to Pidie Jaya to work in the construction sector, he planted cocoa trees and encouraged his neighbours to do the same.

Furthermore, four villages in Pidie (Kambo Nica, Kambo Peapi, Senadi and Blang Unci) received planting material and assistance through a project funded by Japanese aid in 1998.

Finally, at least in Pidie, the villagers from Alue Glumpong mentioned receiving plants distributed by GAM members in the 1990s. This unexpected role of the rebellion, at least by some of its members, illustrated the significance of agricultural extension for officials involved in a territory's governance. It was a matter of gaining influence over the rural population. This is one of the aspects of the interaction between public policies and diversification.

Life Cycles of Perennial Crops and Ecological Change

Planting material and information on cocoa were progressively used in a context of environmental change. At the time when cocoa was introduced in Aceh, clove cultivation was afflicted by disease, leading to a decline in production. Farmers were looking for a solution. There was also the impact of insect damage on *Arabica* coffee, at least in Pidie (Box 16.1).

Interaction Between the Market and Policies

Several mechanisms pertaining to prices and income combined with ecological change and plant physiology. They contributed to diversification towards cocoa, even a partial replacement of clove and coffee by it. This was a process observed in several parts of Indonesia, especially Sulawesi (Ruf 1995, 2000). These mechanisms were also observed in Aceh:

- in the 1990s, the procurement price of cloves paid to producers across Indonesian provinces collapsed as a result of a stabilization fund imposed by the son of President Suharto. Revenue from clove soon became lower than that from cocoa (Fig. 16.4);

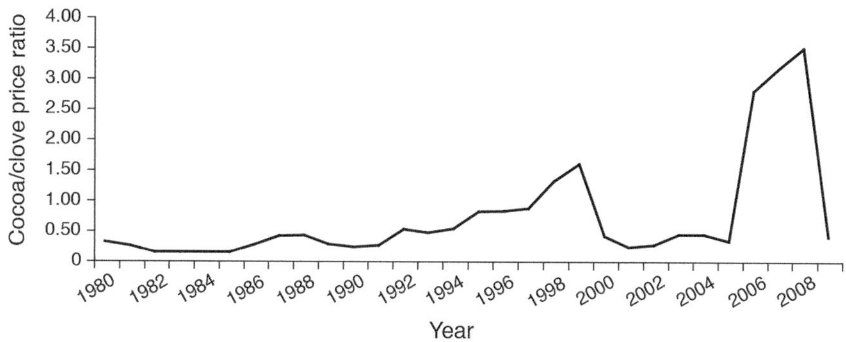

Fig. 16.4 Cocoa/clove price ratio in Indonesia, 1990–2009. (*Source* FAO stat)

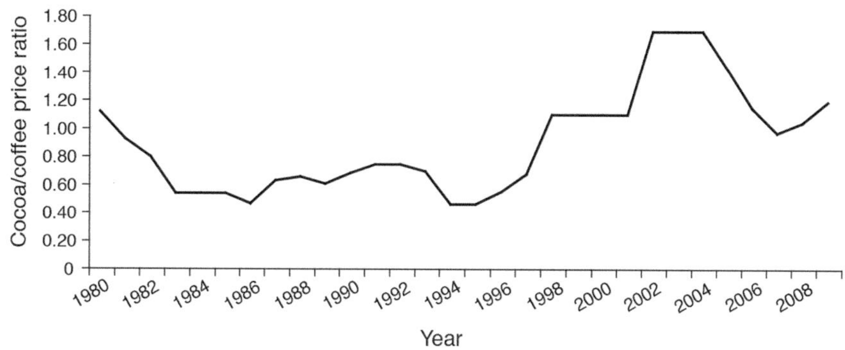

Fig. 16.5 Cocoa/coffee price ratio in Indonesia, 1990–2009. (*Source* FAO stat)

- except for a brief period between 1994 and 1996, the ratio between cocoa and coffee prices increased (Fig. 16.5);
- regularity of income. Clove and coffee could be harvested only once a year, at least in Aceh. This resulted in difficulties in managing income. Cocoa brings in a certain income monthly due to the conditions of regular rainfall prevalent in Aceh;
- finally, clove harvesting, which is done on ladders, is dangerous. Coffee harvesting work is considered long and tedious. In contrast, harvesting work for cocoa, when it is not affected by pests and disease, is considered very easy.

Effects of Imitation

Following the economic success of the early adopters of cocoa, manifested primarily by the construction of new houses, the other villagers followed suit. Since cocoa comes into production only after 3 or 4 years, this typical process of imitation took 5–10 years. Thus, the great imitation wave occurred between 1987 and 1990 since the pioneers had planted cocoa in the early 1980s.

With the conflict intensifying in Pidie Jaya after 1990, planting activities became erratic. They often slowed down when the intensity of the conflict rendered access to plots difficult and picked up when things were quieter. Cocoa plantations accelerated in 2000 in spite of a flare-up of the conflict between 2002 and 2003 in some districts (Figs. 16.2 and 16.3). The revival of cocoa production was brought about mainly by relocating the cocoa production areas towards the hills.

Process of Relocation of Production Areas

By 2011, many forest plots were already cleared and burned in the interior regions (for example, near Jijem village, on the hills west of Pidie Jaya district). They were ready for or had already been planted with cocoa, such as the agricultural frontiers. New cocoa plantations were set up almost without shading.

Farmers or their son set up *pondoks* on their plots. These are big cabins which serve as second homes when they work for several days on their plots. This was not limited to farmers from Jijem village alone; farmers from other villages up to 15 km distant from Jijem and up to the plains also participated.

An Aceh native, even if he was from a different village, could easily acquire land here, often through a simple authorization from the village chief to clear forest land. Migrants other than Aceh natives could not acquire land in this manner; they therefore had to purchase it.

Many farmers were willing to move away from their homes to obtain land near Jijem and plant cocoa. The benefits of this extraction of value from cocoa added to the reasons for moving away from the plains: the soil of a forest patch that has just been cleared is considered to be very rich. In addition, these frontier areas did not yet have problems of disease or pests (except of monkeys, on forest edges).

In addition, clearing a forest plot accords a kind of a right to the property, provided that that forest land was never cleared before. This means that it is continuously necessary to move deeper into the hills.

In this process of relocation of the cocoa production areas, we come across the twin phenomenon of seeking a forest rent and a kind of land rent (Ruf 1987, 1995). Many of the farmers in Jijem who appropriated new forest plots in the hills were already cocoa growers. These farmers had often abandoned their old plantations on the plains.

Population Growth and Generation Change

This process of relocation of cocoa production areas was given further impetus by population growth and generation change. Most of the young villagers who were interested in cocoa attempted to establish plantations in the hills (Box 16.2).

Box 16.2. Examples of the combination of the 'young farmer—young plantations' cycles

Thirty-year-old M.A., currently a technician with the Peka cocoa rehabilitation project, is a farmer in Pidie Jaya. This project could help him with his old 1 ha agroforestry plantation in the village which he inherited from his father. However, he has just cleared a forest patch in the hills to establish a 1 ha plot with young cocoa.

N.M., a 28 year old farmer in Jijem, in the hills of Pidie Jaya, negotiated to acquire a 3 ha plot of forest with the village chief in 2009. He benefitted from the fertility of the soil following a slash-and-burn process to get a good crop of rice and soybean, even as the young cocoa trees were growing well. His main tool now is the brush cutter. His primary objective is to fund his wife's studies at the University of Banda Aceh (Survey by authors 2011).

16.2.1.2 Aceh Tenggara

The process of diversification towards cocoa is similar to those observed in Pidie and Pidie Jaya. Before cocoa cultivation was adopted in Aceh Tenggara, the main crops were rice, maize and perennial crops like candlenut, coffee (*Arabica* and *Robusta*) and clove. These three perennial crops are often intercropped on the same plot. Like in Pidie, farmers discontinued clove once it was affected by disease; they first replaced it with candlenut and coffee, before introducing cocoa. However, some older farmers mentioned that cocoa came to Aceh Tenggara much before it was brought to Pidie Jaya: they remembered it from their childhood, towards the end of the colonial period. Cocoa trees were brought by the Dutch and were, at that time, merely seen as ornamental tree which also produced a bean covered with candy-like sweet mucilage that children liked.

On the plains, near homes, cocoa was also grown as an agroforestry tree, intercropped with fruit trees that were either older than cocoa (durian, coconut, jackfruit, rambutan, banana) or of the same age (Areca palm).

In 1990, an employee of the agricultural extension service brought home, on his own initiative, some cocoa plants for trial. He planted some on his plots and gave away some to other farmers in his village. In 1991, the extension service took up the movement, and implemented a project to promote cocoa. As part of this project, it distributed planting material, disseminated information on how to grow the crop and, sometimes, even gave out a few tools and subsidies. Despite this, only a few villagers chose to play the role of pioneers in planting cocoa; most of the farmers knew nothing about the crop, and no cocoa market existed locally. To compound this, world cocoa prices were at their lowest at the time.

From 1994, the imitation effect started to work. Prices and incomes had a marked effect starting in 1997, even as the Asian currency crisis (*krismon* in

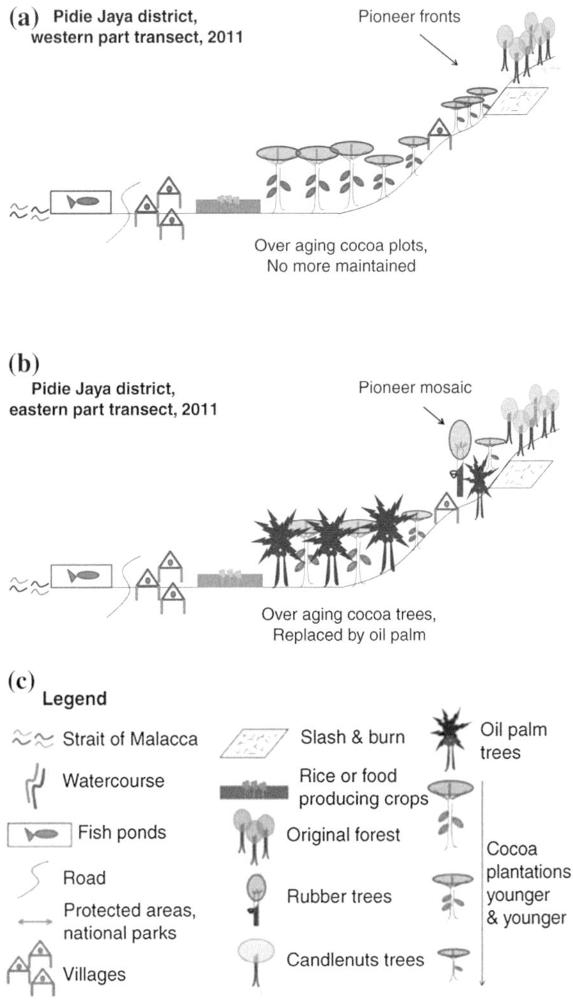

Fig. 16.6 Landscape transects west and east of Pidie Jaya. (*Source* Paul 2011)

Indonesian) greatly favoured export-oriented agriculture, that of cocoa in particular. In the 1990s, the situation in Aceh Tenggara was less conflictual than in Pidie and Pidie Jaya, and villagers here could easily go about their fields and plant cocoa, replacing not only clove and coffee (Fig. 16.4), but also the rubber trees in the district. More recently, cocoa trees, planted in almost exclusive monocultural plots, were being planted at the expense of forest and food crops, especially rice (Table 16.2 and Fig. 16.7).

Despite a decline in interest due to the difficulty in controlling cocoa pests and diseases, farmers continued to invest in cocoa in 2011, on valley floors, instead of rice or on mountain slopes. Moreover, these slopes were most often located in

Table 16.2 Crops preceding cocoa in Aceh Tenggara

Crops preceding cocoa	1980–1995	1996–2000	2006–2010	Total
Perennial crops/insertion into an agroforestry system (%)	100	86	59	62
Forest (%)	–	–	15	13
Food crop plots (%)	–	14	21	19
Fallow with *Imperata* (%)	–	–	6	5
Total	100	100	100	100

(*Sources* Survey by authors 2011)

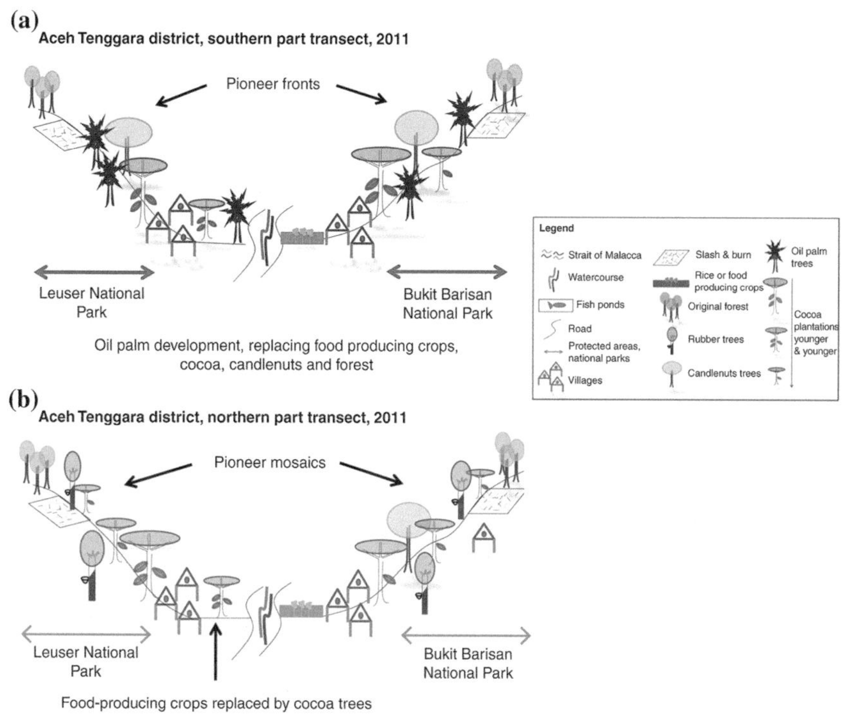

Fig. 16.7 Landscape transects south and north of Aceh Tenggara. (*Source* Paul 2011)

protected forests and national parks (Table 16.2). However, it still seemed easy enough for indigenous people and even the sons or grandsons of migrants to acquire land in the mountains. Here too, it sufficed for villagers to get the permission of the village chief to clear forest areas. Sometimes, even this authorization was not necessary.

The partial conversion of paddy fields or *sawah*[6] to cocoa, which constitutes a kind of diversification as the conversion is partial, was due to several reasons according to the farmers:

- irrigation became difficult;
- income from rice is lower than that from cocoa;
- there was no more uncultivated land available near some villages for planting cocoa;
- uncultivated land that was privately owned by villagers (those who had cleared it or their inheritors) could still be found near the village. If a farmer wished to acquire such a plot, he had to buy it.

To overcome the obstacle of start-up land rent, farmers had to move ever further into the mountains. The work there was all the more difficult as they had to travel to the plot, usually on mountainous terrain. Some farmers, therefore, chose to convert their *sawah*, which is often close to the village, to cocoa.

Nevertheless, the largest share of cocoa production came from the hilly areas after forests had been cleared. Cocoa buyers confirmed these observations. The process of relocation of production areas remained dominant.

16.2.2 Diversification from Cocoa to Other Perennial Crops

16.2.2.1 Pidie and Pidie Jaya

Opportunities for Diversification

Villages on the plains near the coast no longer seemed to be interested in their, often aging, cocoa trees. Several plantations seemed to be in a state of neglect. Farmers attributed this to the growing number of rodents, pests and diseases: squirrels, pod borer and black pod (*Phytophthora*, a fungus responsible for cocoa pod rot). Cocoa yields decreased with age of the tree, even as they required ever-increasing maintenance.

The effort put in maintaining an agroforestry system can be justified and funded by the income from several trees—for example, Areca palms and coconuts here—in addition to cocoa. However, even in these systems that could theoretically have balancing factors, the cocoa grown on the plains in Aceh is not immune to the ‘end of the cycle’ principle and the spiral of recession.

The universal strategy of relocating production areas towards the forest is still the predominant solution adopted in Pidie Jaya. Re-diversification to another perennial crop, by replacing cocoa, is rarely seen in the western part of this district

---

[6] *Sawah:* lowlands on which Indonesians grow rice in the rainy season (or all year round if the *sawah* is irrigated) and other short-cycle crops in the dry season.

because access to forests in the hills is relatively easier here (Fig. 16.6, western transect).

By contrast, a process of re-diversification to oil palm could start in the district's eastern part. In fact, cocoa farmers of the sub-districts more to the east have opted for oil palm. The opportunity was provided by a private plantation, the same one that introduced cocoa to Pidie and Pidie Jaya in the 1970s. The private plantation is currently replacing its old cocoa trees with oil palms. When this survey was conducted in 2011, about 70 ha were already planted with oil palm.

16.2.2.2 Explanation by Farmers: Available Capital

Farmers expressed the determinants of diversification very clearly. Cocoa cultivation can be adopted with little capital as long as forest land is available and accessible. Indeed, establishing a cocoa plantation is not an expensive proposition if farmers can acquire forest land by clearing it, and then get cocoa planting material from extension services or, alternatively, from their first plantations. Besides, fertilization of cocoa plantations is not crucial, especially if they benefit from the forest rent.

On the other hand, for oil palm, hybrid plants and fertilization seem almost indispensable to farmers. This requires capital for investment and operating costs. The diversification process can be taken up, especially if a private company provides support. We find exactly the same processes in Western and Central Africa (Chaps. 2–10).

16.2.2.3 Aceh Tenggara

When this survey was being conducted, in mid-2011, interest in cocoa was waning due to frequent insect attacks and price fluctuations. The total production in the district had been falling for 2 or 3 years (according to cocoa buyers).

Even though some villagers continued to plant cocoa, they also turned to other crops, mainly rubber. Several farmers who have been 'discouraged by cocoa' chose to either expand their farms by planting rubber, or by replacing cocoa by rubber. In such a case, the process of replanting and diversification is progressive and is partly based on an agroforestry technique: the principle of relay cultivation. They intercrop rubber trees with cocoa trees, and cut down the latter when rubber trees enter into production.

Among the benefits, farmers cited the absence of disease and easy maintenance work, even if the plantations are far from home. In addition, the price was very favourable at the time of the survey: 15,000 IDR/kg[7] as against 18,000 IDR/kg only for cocoa and with more problems. The only drawback with rubber plantations is

[7]IDR: Indonesian rupiah, 1 € was equivalent to about 12,000 IDR in 2011.

the harvesting of rubber: it is a time-consuming process which requires a certain expertise and is partially dependent on the rains.

Regardless of the status of a protected forest, farmers continued to clear forest land and plant cash crops in 2011. Three types of crops were in competition in Aceh Tenggara: cocoa, rubber and oil palm. The farmers themselves suggested the following determinants:

- selling price of products;
- cost of establishing each crop;
- the difficulty of the work to maintain the plantation;
- level of attacks by pests and diseases;
- number of harvests per year which determines income regularity;
- knowledge that farmers have about the crop.

The survey did not observe any competition between cocoa and oil palm in the northern part of the district. The further we went into the hills, the younger were the cocoa plants and more they tended to be grown as a 'monoculture', with very little shade. Plantations thus created a 'pioneer mosaic' landscape with rubber plots and forests. Competition between cocoa and oil palm existed in the southern part of the district, near North Sumatra province, both in the valley and on the slopes (Fig. 16.7). Nevertheless, despite saying they will remove cocoa in the near future due to the pressure of insects, rodents and diseases, oil palm growers did solve the issue of competition by intercropping the two crops on 40 % of their plantation area. Diversification into oil palm took place mainly at the expense of candlenut, which was also affected by the two major determinants of diversification for perennial crops: declining prices of the product and the aging of the trees which reduces yields irreversibly and thus affects income.

16.3 Conclusion

Just like other regions of Indonesia, Aceh exhibits multiple diversification processes. These include diversification through agroforestry systems at the plot level —such plots in Indonesia are very old,—at the village level and at the landscape level. The latter are based more on monospecific plots and on the opening of new agricultural frontiers. Cocoa cultivation finds place in the two major systems, both as a new crop for diversification that is expanding rapidly, and a crop in decline—neglected or partially replaced by new perennial crops.

As a new diversification crop, cocoa inserted itself into an agroforestry system by partly replacing clove and coffee. It adapted best under the tall Areca palms. Since it is capable of establishing itself rapidly, it was used to open up new agricultural frontiers and for expanding areas under cultivation. To a lesser degree, cocoa can also replace paddies that are little irrigated, or not irrigated at all. A plot-level conversion can then take place. However, because of the marginal nature of this form of diversification, we can safely affirm that farm- or village-level diversification is the most common.

Like other perennial crop, cocoa too is prone to a recession phase, at which time it is either neglected in the agroforestry system or is gradually replaced by rubber and oil palm. Here too, there can be partial abandonment or conversion to another crop at the plot level, thus leaving little room for diversification. However, due to the simultaneous expansion of cocoa cultivation in adjacent forest areas, we witness a diversification at the level of farms, landscapes and 'village terroirs'.

Despite differences in geographical, social and political situations between districts in the north and the south-east, we found a combination of major determinants of diversification. There is the quasi-structural ecological change, in the sense that after several years of expansion and growth, all perennial crops are weakened eventually by age and the arrival or adaptation of pests and diseases. This situation naturally encourages the adoption of a new crop. This is indeed one of the principles of the 'unfolding of a cycle'. We notice how diversification reflects the manner in which this works around villages: first clove, then coffee, and finally cocoa, with a new cycle emerging subsequently, that of oil palm and rubber.

We also saw a key determinant, 'price', which led farmers to expect good future income with a new diversification crop when the 'old' crop suffered from low prices for several years. We observed the attraction to an income that came in with greater regularity. In Aceh, this factor was the basis of the attraction, between 1980 and 2000, of cocoa over coffee or clove which are harvested but once a year. Even now, in the 2010s, the consistent regularity of palm bunch and rubber harvests attracts both Indonesian cocoa farmers and their African counterparts.

We also saw the key role of public policies and private plantations as disseminators of information and planting material, which are naturally essential to diversification.

Public policy, chronologically the first factor in Aceh, was affected by continued armed conflict for more than 30 years. This paradox is partly explained by attempts made by the then government to garner the goodwill of rural people, or at least turn them away from the rebels. A second 'public policy' paradox pertains to the antagonistic effect of the conflict, with the army ordering farmers not to go to their fields, which delayed cocoa diversification until the late 2000s.

Finally, as in Africa and Sulawesi, Aceh enters this quasi-universal transition phase in the late 2000s. The dynamics swing between:

- the decline of cocoa. This is both the cause and consequence of the start of diversification to oil palm and rubber. The agroforestry strategy gave no prior indication of this;
- the expansion of cocoa in the form of quasi-monoculture agricultural frontiers.

The issue of replanting, rehabilitation and intensification of existing old plantations resurfaced once again. It is an age-old problem that extension services and projects have been trying to solve for decades but without much success. This will be so as long as farmers have the opportunity to open up new agricultural frontiers. Several projects have been launched in Aceh following the tsunami, the most ambitious of them being *Peka*. In 2011, relying greatly on the experience and success of the Mars company in Sulawesi, this project successfully drew the attention of farmers in Aceh to the concept of rehabilitation by grafting. However, as in the case of many rehabilitation projects, we can only hope for change rather than seek a demonstration, because past experience shows that the diversification process often takes precedence over those of rehabilitation and replanting. The subject is relevant today and will remain so for foreseeable future.

Meanwhile, the strategies of the expansion of agricultural frontiers and diversification into other perennial crops seem to be gaining in strength. This could ultimately prove unfortunate. However, considering the state of technology and policies, we can also conclude that this dual strategy of short-term specialization and opportunistic diversification is very effective, one of the best possible strategies for family farming in Aceh.

References

Adi Prawoto A, Jacquet M, Paulin D, Undang Fadjar (2008) Cocoa commodity development technological improvement of post-harvest and recovery of the cocoa sector in the Pidie district. LPRI and Cirad, Jakarta et Montpellier, 57 pp + annexes

Hyndman J, Waizenegger A (2010) Two solitudes: post-tsunami and post-conflict Aceh. Disasters 34(3):787–808

Jacquet M, Paulin D (2006) Development of cocoa in Pidie district. preparation of masterplan and training in the sub districts of pilot project. Cirad, Montpellier, 55 pp

Jacquet M, Paulin D (2007) Développement du secteur cacao dans le district de Pidie. Aide-mémoire, 4th mission, 7 pp

Paul F (2011) Évaluation des conditions d'adoption de pratiques culturales, dont le greffage, sur le cacaoyer et caractérisation des dynamiques concernant le cacao à Aceh, Indonésie. Engineer Thesis, Lille, Institut supérieur d'agriculture, Lille, 153 pp

Ruf F (1987) Éléments pour une théorie sur l'agriculture des régions tropicales humides. I - De la forêt, rente différentielle au cacaoyer, capital travail. Agronomie tropicale 42(3):218–232

Ruf F (1995) Booms et crises du cacao : les vestiges de l'or brun. Paris: Éditions Karthala. Collection Économie et développement, 455 pp

Ruf F (2000). Déterminants sociaux et économiques de la replantation. Dossier : Afrique, plantation et développement. OCL 7(2):189–196

Ruf F, Yoddang (2004) The Sulawesi case. Deforestation, pre-cocoa and cocoa migrations. In: Babin D (ed) From tropical deforestation to forest cover dynamics and forest development. Unesco, Paris, pp 277–295

MIX
Papier aus verantwortungsvollen Quellen
Paper from responsible sources
FSC® C105338

If you have any concerns about our products,
you can contact us on
ProductSafety@springernature.com

In case Publisher is established outside the EU,
the EU authorized representative is:
Springer Nature Customer Service Center GmbH
Europaplatz 3, 69115 Heidelberg, Germany

Printed by Libri Plureos GmbH
in Hamburg, Germany